A U R O R A

DOVER MODERN MATH ORIGINALS

Dover Publications is pleased to announce the publication of the first volumes in our new Aurora Series of original books in mathematics. In this series we plan to make available exciting new and original works in the same kind of well-produced and affordable editions for which Dover has always been known.

Aurora titles currently in the process of publication are:

Optimization in Function Spaces by Amol Sasane. (978-0-486-78945-3)

The Theory and Practice of Conformal Geometry by Steven G. Krantz. (978-0-486-79344-3)

Numbers: Histories, Mysteries, Theories by Albrecht Beutelspacher. (978-0-486-80348-7)

Elementary Point-Set Topology: A Transition to Advanced Mathematics by André L. Yandl and Adam Bowers. (978-0-486-80349-4)

Additional volumes will be announced periodically.

OPTIMIZATION IN FUNCTION SPACES

AMOL SASANE

DEPARTMENT OF MATHEMATICS
LONDON SCHOOL OF ECONOMICS

DOVER PUBLICATIONS, INC.
MINEOLA, NEW YORK

Bibliographical Note

Optimization in Function Spaces is a new work, first published by Dover
Publications, Inc., in 2016, as part of the "Aurora: Dover Modern Math
Originals" series.

International Standard Book Number

ISBN-13: 978-0-486-78945-3
ISBN-10: 0-486-78945-4

Manufactured in the United States by RR Donnelley
78945401   2016
www.doverpublications.com

Contents

Introduction

The subject matter of this book is

> Optimization in Function Spaces.

It is thus natural to begin by explaining what we mean by this.

If we ignore the "in function spaces" part, we see that the subject matter is a part of optimization. In optimization, we know that the basic object of study is a real-valued function

$$f : S \to \mathbb{R},$$

defined on a set S, and the central problem is that of maximizing or minimizing f:

> Find x_* such that $f(x) \geq f(x_*)$ for all $x \in S$.

This is the case of minimizing f. In maximization problems, the inequality above is reversed. There is no real difference between maximization and minimization problems: indeed, if we learn to solve minimization problems, we also know how to solve maximization problems, because we can just look at $-f$ instead of f.

Exercise 0.1. Let $f : S \to \mathbb{R}$ be a given function on a set S, and define $-f : S \to \mathbb{R}$ by

$$(-f)(x) = -f(x), \quad x \in S.$$

Show that $x_* \in S$ is a maximizer for f if and only if x_* is a minimizer for $-f$.

Why bother with optimization problems? In applications, we are often faced with choices or options, that is, different ways of doing the same thing. Imagine, for example, traveling to a certain place by rail, bus or air. Given the choice, it makes sense that we then choose the "best possible" option. For example, one might wish to seek the cheapest means of transport. The set S in our optimization problem is thus the set of possible choices. To each choice $x \in S$, we assign a number $f(x) \in \mathbb{R}$ (measuring how "good" that choice is), and this gives the "cost" function $f : S \to \mathbb{R}$ to be minimized. In an optimization problem of minimizing $f : S \to \mathbb{R}$, the set S is called the *feasible set*, and the function f is called the *objective function*.

Thus, in optimization, one studies the problem

$$\begin{cases} \text{minimize} & f(x) \\ \text{subject to} & x \in S. \end{cases}$$

Depending on the nature of S and f, the broad subject of optimization is divided into subdisciplines such as

> combinatorial/discrete optimization,
>
> finite dimensional/static optimization,
>
> stochastic optimization,

and so on. In this book, we will study dynamic optimization in continuous-time. Roughly, this is the subject where the set S is the collection of functions on a real interval, for example,

$$S = C[a,b] \text{ (all continuous functions } \mathbf{x} : [a,b] \to \mathbb{R}).$$

Let us consider a simple example to see that such problems do arise in real life situations.

Example 0.2. Imagine a copper mining company that is mining in a mountain, which has an estimated amount of Q tonnes of copper, over a period of T years. Suppose that $\mathbf{x}(t)$ denotes the total amount of copper removed up to time $t \in [0,T]$. Since the operation is over a large time period, we may assume that this $\mathbf{x}$ is a function living on the "continuous-time" interval $[0,T]$. The company has the freedom to choose its mining operation: $\mathbf{x}$ can be any nondecreasing function on $[0,T]$ such that $\mathbf{x}(0) = 0$ (no copper removed initially) and $\mathbf{x}(T) = Q$ (all copper removed at the end of the mining operations).

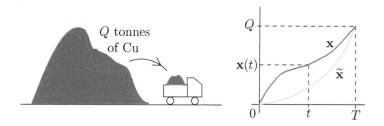

Figure 1. Several possible mining operations: $\mathbf{x}, \widetilde{\mathbf{x}}, \cdots$.

The cost of extracting copper per unit tonne at time t is given by

$$\mathbf{c}(t) = a\mathbf{x}(t) + b\mathbf{x}'(t), \quad t \in [0,T].$$

Here a, b are given positive constants. The expression is reasonable, since the term $a \cdot \mathbf{x}(t)$ accounts for the fact that when more and more copper is taken out, it becomes more and more difficult to find the left over copper, while the term $b \cdot \mathbf{x}'(t)$

accounts for the fact that if the rate of removal of copper is high, then the costs increase (for example, due to machine replacement costs). We don't need to follow the exact reasoning behind this formula; this is just a model that the optimizer has been given.

If the company decides on a particular mining operation, say $\mathbf{x} : [0, T] \to \mathbb{R}$, then the overall cost $f(\mathbf{x}) \in \mathbb{R}$ over the whole mining period $[0, T]$ is given by

$$f(\mathbf{x}) = \int_0^T \Big(a\mathbf{x}(t) + b\mathbf{x}'(t) \Big) \mathbf{x}'(t) dt.$$

Indeed, $\mathbf{x}'(t)dt$ is the incremental amount of copper removed at time t, and if we multiply this by $\mathbf{c}(t)$, we get the incremental cost at time t. The total cost should be the sum of all these incremental costs over the interval $[0, T]$, and so we obtain the integral expression for $f(\mathbf{x})$ given above.

Hence, the mining company is faced with the following natural problem: Which mining operation $\mathbf{x}$ incurs the least cost? In other words,

$$\begin{cases} \text{minimize} & f(\mathbf{x}) \\ \text{subject to} & \mathbf{x} \in S, \end{cases}$$

where S denotes the set of all (continuously differentiable) functions $\mathbf{x} : [0, T] \to \mathbb{R}$ such that $\mathbf{x}(0) = 0$ and $\mathbf{x}(T) = Q$. $\diamond$

Exercise 0.3. In Example 0.2, calculate $f(\mathbf{x}_1), f(\mathbf{x}_2)$ where:

$$\mathbf{x}_1(t) = Q\frac{t}{T} \quad \text{and} \quad \mathbf{x}_2(t) = Q\Big(\frac{t}{T}\Big)^2, \quad t \in [0, T].$$

Which is smaller among $f(\mathbf{x}_1)$ and $f(\mathbf{x}_2)$? Which mining operation among $\mathbf{x}_1$ and $\mathbf{x}_2$ will be preferred?

In the subject of optimization in function spaces, we will learn to solve optimization problems where the feasible set is a set of functions on an interval $[a, b]$. This interval is thought of as the "continuous-time" variable, but one needn't always have this interpretation with time. We hope that the above example convinces the reader that optimization problems in function spaces do arise in practice, and we will see many more examples from diverse application areas such as economics, engineering, physics, and life sciences.

How do we solve optimization problems in function spaces? Suppose that instead of an optimization problem in a function space, we consider a much simpler problem:

$$\begin{cases} \text{minimize} & x^2 - 2x + 5 \\ \text{subject to} & x \in \mathbb{R}. \end{cases}$$

Then we know how to solve this. Indeed, from ordinary calculus, we know the following two facts.

Fact 1. If $x_* \in \mathbb{R}$ is a minimizer of f, then $f'(x_*) = 0$.

Fact 2. If $f''(x) \geq 0$ for all $x \in \mathbb{R}$ and $f'(x_*) = 0$, then x_* is a minimizer of f.

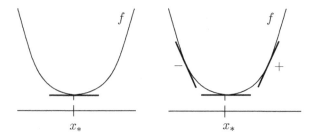

Figure 2. Fact 1 says that at a minimizer x_*, the tangent to the graph of f must be horizontal. Fact 2 says that if f' is increasing ($f'' \geq 0$), and if $f'(x_*) = 0$, then for points x to the left of x_*, f' must be nonpositive, and so f must be decreasing there, and similarly f must be increasing to the right of x_*. This has the consequence that x_* is a minimizer of f.

Fact 1 allows us to narrow our choice of possible minimizers, since we can calculate

$$f'(x) = 2x - 2,$$

and note that $2x - 2 = 0$ if and only if $x = 1$. So if at all there is a minimizer, then it must be $x_* = 1$. On the other hand,

$$f''(x) = 2 > 0 \text{ for all } x \in \mathbb{R},$$

and so Fact 2 confirms that $x_* = 1$ is a minimizer. Thus using these two facts, we have completely solved the optimization problem: we know that $x_* = 1$ is the only minimizer of f.

Optimization problems in function spaces do not fit into the usual framework of ordinary calculus, where the feasible set is a subset of $\mathbb{R}$ or more generally a *finite* dimensional real vector space like $\mathbb{R}^n$. Indeed, vector spaces of functions such as $C[a, b]$ are *infinite* dimensional vector spaces. Nevertheless, it is natural to try solving our optimization problem in function spaces in the same manner. But we immediately run into a problem. We don't yet know what the derivative $f'(\mathbf{x}_*)$ at $\mathbf{x}_*$ is for a real-valued function f whose domain is itself a space of functions like $C[a, b]$! So we first develop calculus in this setting. Our approach will be to proceed abstractly, where we will first learn about calculus in vector spaces (which are possibly infinite dimensional, like $C[a, b]$, or the set $C^1[a, b]$ of all continuously differentiable functions on $[a, b]$), and we will also learn the analogues of Facts 1 and 2 in this more general setting. We will then be in a position to apply these general results to more specialized problems, and see what they give there. The initial abstract approach we follow allows us to solve many problems in one go, that is, seemingly different problems are considered as one basic problem in a unified setting by the removal of inessential details. In the second part of the course, we will also learn about more complicated optimization problems in function spaces, namely "optimal control" problems, in which one has an objective function given by an integral, where the integrand involves functions that satisfy a differential equation constraint. The broad outline of the content in this book is as follows:

(1) Calculus in Normed Spaces.

(2) The Euler-Lagrange Equation.

(3) Preliminaries on Ordinary Differential Equations (ODEs).

(4) The Pontryagin Minimum Principle.

(5) The Hamilton-Jacobi-Bellman Equation.

The book is elementary, assuming no prerequisites beyond knowledge of linear algebra and ordinary calculus (with ϵ-δ arguments), and so it should be accessible to a wide spectrum of students. Preliminary versions of the book were used for a course with the same title as that of the book, for the students in the third year of the BSc in Mathematics and/with Economics programs at the London School of Economics. The book contains many exercises, which form an integral part of the text. Detailed solutions appear at the end of the book. Also in the book, a project is outlined, which involves some "theoretical" exercises, together with a "practical" component of writing a program in MATLAB to implement the resulting numerical algorithm.

This book relies heavily on some of the sources mentioned in the bibliography. I am thankful to Erik G.F. Thomas, Sara Maad Sasane, and Adam Ostaszewski for several useful discussions.

Amol Sasane
London, 2013.

Part 1

Calculus in Normed Spaces;
Euler-Lagrange Equation

Chapter 1

Calculus in Normed Spaces

As we had discussed in the previous chapter, we wish to differentiate functions living on vector spaces (such as $C[a,b]$) whose elements are functions, and taking values in $\mathbb{R}$. In order to talk about the derivative, we need a notion of closeness between points of a vector space so that the derivative can be defined. It turns out that vector spaces such as $C[a,b]$ can be equipped with a "norm," and this provides a "distance" between two vectors. Having done this, we have the familiar setting of calculus, and we can talk about the derivative of a function living on a normed space. We then also have analogues of the two facts from ordinary calculus relevant to optimization that were mentioned in the previous chapter, namely the vanishing of the derivative for minimizers, and the sufficiency of this condition for minimization when the function has an increasing derivative. Thus the outline of this chapter is as follows:

(1) We will learn the notion of a "normed space," that is a vector space equipped with a "norm," enabling one to measure distances between vectors in the vector space. This makes it possible to talk about concepts from calculus, and in particular the notion of differentiability of functions between normed spaces.

(2) We will define what we mean by the derivative of a function $f : X \to Y$, where X, Y are normed spaces, at a point $x_0 \in X$. In other words, we will explain what we mean by "$f'(x_0)$."

(3) We will prove the following two fundamental results in optimization for a real-valued function f on a normed space X:

Fact 1. If $x_* \in X$ is a minimizer of f, then $f'(x_*) = 0$.

Fact 2. If f is convex and $f'(x_*) = 0$, then x_* is a minimizer of f.

1.1. Normed Spaces

We would like to develop calculus in the setting of vector spaces of functions such as $C[a, b]$. Underlying all the fundamental notions in ordinary calculus is the notion of closeness between points. For example, recall that a sequence $(a_n)_{n \in \mathbb{N}}$ is said to *converge with limit* $L \in \mathbb{R}$ if, for every $\epsilon > 0$, there exists an $N \in \mathbb{N}$ such that whenever $n > N$, $|a_n - L| < \epsilon$. In other words, the sequence in $\mathbb{R}$ converges to L if no matter what *distance* $\epsilon > 0$ is given, one can guarantee that all the terms of the sequence beyond a certain index N are at a distance of at most ϵ away from L (this is the inequality $|a_n - L| < \epsilon$). So we notice that in this definition of "convergence of a sequence" indeed the notion of distance played a crucial role. After all, we want to say that the terms of the sequence get "close" to the limit, and to measure closeness, we use the distance between points of $\mathbb{R}$. A similar thing happens with continuity and differentiability. For example, recall that a function $f : \mathbb{R} \to \mathbb{R}$ is said to be *continuous at* $x_0 \in \mathbb{R}$ if, for every $\epsilon > 0$, there exists a $\delta > 0$ such that whenever $|x - x_0| < \delta$, there holds that $|f(x) - f(x_0)| < \epsilon$. Roughly, given any distance ϵ, I can find a distance δ such that whenever I choose an x not farther than a distance δ from x_0, I am guaranteed that $f(x)$ is not farther than a distance of ϵ from $f(x_0)$. Again, notice the key role played by the distance in this definition. We observe that the distance between two real numbers $x, y \in \mathbb{R}$ is given by $|x - y|$, where $|\cdot| : \mathbb{R} \to \mathbb{R}$ is the absolute value function: for $x \in \mathbb{R}$,

$$|x| = \begin{cases} x & \text{if } x \geq 0, \\ -x & \text{if } x < 0. \end{cases}$$

So in order to generalize the notions from ordinary calculus (where we work with real numbers and where the absolute value is used to measure distances) to the situation of vector spaces like $C[a, b]$, we need a notion of distance between elements of the vector space. We will see that this can be done with the introduction of a "norm" in a vector space, which is a real-valued function $\|\cdot\|$ defined on the vector space, analogous to the role played by the absolute value in $\mathbb{R}$. Once we have a norm on a vector space X (in other words a "normed space"), then the distance between $x, y \in X$ will be taken as $\|x - y\|$. But let us first recall the definition of a vector space.

1.1.1. Vector spaces. Recall that, roughly speaking, a vector space is a set of elements, called "vectors," in which any two vectors can be "added," resulting in a new vector, and any vector can be multiplied by an element from $\mathbb{R}$, so as to give a new vector. The precise definition is given below.

Definition 1.1 (Vector space).

A *vector space* over $\mathbb{R}$ is a set V together with two functions, $+ : V \times V \to V$, called *vector addition*, and $\cdot : \mathbb{R} \times V \to V$, called *scalar multiplication*, that satisfy the following:

(V1) For all $x_1, x_2, x_3 \in V$, $x_1 + (x_2 + x_3) = (x_1 + x_2) + x_3$.

(V2) There exists an element, denoted by $\mathbf{0}$ (called the *zero vector*), such that for all $x \in V$, $x + \mathbf{0} = x = \mathbf{0} + x$.

(V3) For every $x \in V$, there exists an element in V, denoted by $-x$ ($\in V$), such that $x + (-x) = (-x) + x = \mathbf{0}$.

(V4) For all x_1, x_2 in V, $x_1 + x_2 = x_2 + x_1$.

(V5) For all $x \in V$, $1 \cdot x = x$.

(V6) For all $x \in V$ and all $\alpha, \beta \in \mathbb{R}$, $\alpha \cdot (\beta \cdot x) = (\alpha\beta) \cdot x$.

(V7) For all $x \in V$ and all $\alpha, \beta \in \mathbb{R}$, $(\alpha + \beta) \cdot x = \alpha \cdot x + \beta \cdot x$.

(V8) For all $x_1, x_2 \in V$ and all $\alpha \in \mathbb{R}$, $\alpha \cdot (x_1 + x_2) = \alpha \cdot x_1 + \alpha \cdot x_2$.

Here are a few examples of vector spaces.

Example 1.2.

(1) $\mathbb{R}$ is a vector space over $\mathbb{R}$, with vector addition being the usual addition of real numbers, and scalar multiplication being the usual multiplication of real numbers. It follows from the usual rules of arithmetic that the vector space axioms (V1)-(V8) are satisfied.

(2) $\mathbb{R}^n$ is a vector space over $\mathbb{R}$, with addition and scalar multiplication defined as follows:

$$\text{if } \begin{bmatrix} x_1 \\ \vdots \\ x_n \end{bmatrix}, \begin{bmatrix} y_1 \\ \vdots \\ y_n \end{bmatrix} \in \mathbb{R}^n, \text{ then } \begin{bmatrix} x_1 \\ \vdots \\ x_n \end{bmatrix} + \begin{bmatrix} y_1 \\ \vdots \\ y_n \end{bmatrix} = \begin{bmatrix} x_1 + y_1 \\ \vdots \\ x_n + y_n \end{bmatrix};$$

$$\text{if } \alpha \in \mathbb{R} \text{ and } \begin{bmatrix} x_1 \\ \vdots \\ x_n \end{bmatrix} \in \mathbb{R}^n, \text{ then } \alpha \cdot \begin{bmatrix} x_1 \\ \vdots \\ x_n \end{bmatrix} = \begin{bmatrix} \alpha x_1 \\ \vdots \\ \alpha x_n \end{bmatrix}.$$

It can then be checked that the vector space axioms (V1)-(V8) are satisfied.

(3) It can be checked that the set $\mathbb{R}^{m \times n}$ of all matrices having real entries, with m rows and n columns, and with the usual matrix addition and multiplication of matrices by scalars is a vector space. The zero element in $\mathbb{R}^{m \times n}$ is the zero matrix with all entries equal to 0.

(4) Let $a, b \in \mathbb{R}$ and $a < b$. Define

$$C[a, b] = \{\mathbf{x} : [a, b] \to \mathbb{R} \mid \mathbf{x} \text{ is continuous on } [a, b]\}.$$

So far we have a set, and now we define

$$\text{addition } + : C[a, b] \times C[a, b] \to \mathbb{R} \text{ and}$$
$$\text{scalar multiplication } \cdot : \mathbb{R} \times C[a, b] \to C[a, b]$$

as follows. If $\mathbf{x}_1, \mathbf{x}_2 \in C[a, b]$, then $\mathbf{x}_1 + \mathbf{x}_2 \in C[a, b]$ is the new function defined by

$$(\mathbf{x}_1 + \mathbf{x}_2)(t) = \mathbf{x}_1(t) + \mathbf{x}_2(t), \quad t \in [a, b]. \tag{1.1}$$

If $\alpha \in \mathbb{R}$ and $\mathbf{x} \in C[a, b]$, then $\alpha \cdot \mathbf{x} \in C[a, b]$ is the function given by

$$(\alpha \cdot \mathbf{x})(t) = \alpha \mathbf{x}(t), \quad t \in [a, b]. \tag{1.2}$$

It can be checked that the vector space axioms (V1)-(V8) are satisfied. Thus each vector in $C[a, b]$ is a function from $[a, b]$ to $\mathbb{R}$. In particular, the zero vector is the zero function $\mathbf{0} \in C[a, b]$: $\mathbf{0}(t) = 0$, $t \in [a, b]$.

(5) Let $C^1[a, b]$ denote the space of continuously differentiable functions on $[a, b]$:

$$C^1[a, b] = \{\mathbf{x} : [a, b] \to \mathbb{R} \mid \mathbf{x} \text{ is continuously differentiable on } [a, b]\}.$$

(Recall that a function $\mathbf{x} : [a, b] \to \mathbb{R}$ is *continuously differentiable* if for all $t \in [a, b]$, the derivative of x at t, namely $\mathbf{x}'(t)$, exists, and the map $t \mapsto \mathbf{x}'(t) : [a, b] \to \mathbb{R}$ is a continuous function.) We note that

$$C^1[a, b] \subset C[a, b],$$

because whenever a function $\mathbf{x} : [a, b] \to \mathbb{R}$ is differentiable at a point $t \in [a, b]$, $\mathbf{x}$ is continuous at t. In fact, $C^1[a, b]$ is a *subspace* of $C[a, b]$ because it is closed under addition and scalar multiplication, and is nonempty:

(S1) For all $\mathbf{x}_1, \mathbf{x}_2 \in C^1[a, b]$, $\mathbf{x}_1 + \mathbf{x}_2 \in C^1[a, b]$.

(S2) For all $\alpha \in \mathbb{R}$, $\mathbf{x} \in C^1[a, b]$, $\alpha \cdot \mathbf{x} \in C^1[a, b]$.

(S3) $\mathbf{0} \in C^1[a, b]$.

Thus $C^1[a, b]$ is a vector space with the induced operations from $C[a, b]$, namely the same pointwise operations as defined in (1.1) and (1.2). $\Diamond$

Exercise 1.3. Let $S := \{\mathbf{x} \in C^1[a, b] : \mathbf{x}(a) = y_a \text{ and } \mathbf{x}(b) = y_b\}$, where $y_a, y_b \in \mathbb{R}$. Prove that S is a subspace of $C^1[a, b]$ if and only if $y_a = y_b = 0$. (So we see that S is a vector space with pointwise operations if and only if $y_a = y_b = 0$.)

Exercise 1.4. Show that $C[0, 1]$ is not a finite dimensional[1] vector space.

Hint: One can prove this by contradiction. Let $C[0, 1]$ be a finite dimensional vector space with dimension d. First show that the set $B = \{t, t^2, \ldots, t^d\}$ is linearly independent. Then B is a basis for $C[0, 1]$, and so the constant function 1 should be a linear combination of the functions from B. Derive a contradiction.

(So we see that $C[a, b]$, which is isomorphic[2] to $C[0, 1]$, is not finite dimensional either. Typically vector spaces of functions such as $C[a, b]$, $C^1[a, b]$, etc. are not finite dimensional. Hence, morally our optimization problems in function spaces are not problems which can be converted to some equivalent optimization problem in $\mathbb{R}^n$ with a choice of basis.)

---

[1] The student who wishes to recall the definitions of linear dependence/independence of a set of vectors, and their span, of the basis and the dimension of a vector space is referred to [**Ar**, pages 87-94].

[2] For example, via the isomorphism $\mathbf{x} \mapsto \mathbf{x}\left(\frac{\cdot - a}{b - a}\right) : C[a, b] \to C[0, 1]$.

1.1.2. Normed spaces. In order to do calculus (that is, speak about limiting processes, convergence, continuity, and differentiability) in vector spaces, we need a notion of "distance" or "closeness" between the vectors of the vector space. This is provided by the notion of a norm, which allows one to do this, as we shall see below. A "normed space" is just a vector space with a norm.

Definition 1.5 (Norm). Let X be a vector space over $\mathbb{R}$. A *norm* on X is a function $\|\cdot\| : X \to \mathbb{R}$ such that:

   N1. (*Positive definiteness*) For all $x \in X$, $\|x\| \geq 0$. If $x \in X$ and $\|x\| = 0$, then $x = \mathbf{0}$.

   N2. For all $\alpha \in \mathbb{R}$ and for all $x \in X$, $\|\alpha x\| = |\alpha| \|x\|$.

   N3. (*Triangle inequality*) For all $x, y \in X$, $\|x + y\| \leq \|x\| + \|y\|$.

A *normed space* is a vector space X equipped with a norm.

Distance in a normed space. Just like in $\mathbb{R}$, with the absolute value and where the distance between $x, y \in \mathbb{R}$ is $|x - y|$, now if we have a normed space $(X, \|\cdot\|)$, then for $x, y \in X$, then the number $\|x - y\|$ is taken as the distance between $x, y \in X$. Thus $\|x\| = \|x - \mathbf{0}\|$ is the distance of x from the zero vector $\mathbf{0}$ in X.

Remark 1.6. The reader familiar with "metric spaces" may notice that in a normed space $(X, \|\cdot\|)$, if we define $d : X \times X \to \mathbb{R}$ by $d(x, y) = \|x - y\|$ for $x, y \in X$, then (X, d) is a metric space with the metric/distance function d.

We now give a few examples of normed spaces.

Example 1.7. $\mathbb{R}$ is a vector space over $\mathbb{R}$, and if we define $\|\cdot\| : \mathbb{R} \to \mathbb{R}$ by

$$\|x\| = |x|, \quad x \in \mathbb{R},$$

then it becomes a normed space. ◇

Example 1.8. $\mathbb{R}^n$ is a vector space over $\mathbb{R}$, and let us define the *Euclidean norm* $\|\cdot\|_2$ by

$$\|x\|_2 = \sqrt{x_1^2 + \cdots + x_n^2}, \quad x = (x_1, \cdots, x_n) \in \mathbb{R}^n.$$

Then $\mathbb{R}^n$ is a normed space (see Exercise 1.15 on page 17; the verification of (N3) requires the Cauchy-Schwarz inequality).

This is not the only norm that can be defined on $\mathbb{R}^n$. For example,

$$\|x\|_1 = |x_1| + \cdots + |x_n|, \text{ and } \|x\|_\infty = \max\{|x_1|, \ldots, |x_n|\}, \quad x \in \mathbb{R}^n,$$

are also examples of norms (see Exercise 1.15 on page 17).

Note that $(\mathbb{R}^n, \|\cdot\|_2)$, $(\mathbb{R}^n, \|\cdot\|_1)$ and $(\mathbb{R}^n, \|\cdot\|_\infty)$ are all *different* normed spaces. This illustrates the important fact that from a given vector space, we can obtain various normed spaces by choosing different norms. What norm is considered depends on the particular application at hand. We illustrate this in the next paragraph.

Imagine a city (like New York) in which there are streets and avenues with blocks in between, forming a square grid as shown in Figure 1. Then if we take a taxi/cab to go from point A to point B in the city, it is clear that it isn't the Euclidean norm in $\mathbb{R}^2$ which is relevant, but rather the $\|\cdot\|_1$-norm in $\mathbb{R}^2$. (It is for this reason that the $\|\cdot\|_1$-norm is sometimes called the *taxicab norm*.)

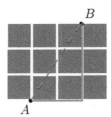

Figure 1. The taxicab norm versus the Euclidean norm.

So what norm one uses depends on the situation at hand, and is something that the modeler decides. It is not something that falls out of the sky!   ◇

Example 1.9. $C[a,b]$ with the supremum norm $\|\cdot\|_\infty$. Let $a, b \in \mathbb{R}$ and $a < b$. Consider the vector space $C[a,b]$ comprising functions that are continuous on $[a,b]$, with addition and scalar multiplication defined earlier by (1.1)-(1.2). Define

$$\|\mathbf{x}\|_\infty = \sup_{t \in [a,b]} |\mathbf{x}(t)| = \max_{t \in [a,b]} |\mathbf{x}(t)|, \quad \mathbf{x} \in C[a,b]. \tag{1.3}$$

Then $\|\cdot\|_\infty$ is a norm on $C[a,b]$, and is referred to as the "supremum norm." The second equality above, guaranteeing that the supremum is attained, that is, that there is a $c \in [a,b]$ such that

$$|\mathbf{x}(c)| = \sup_{t \in [a,b]} |\mathbf{x}(t)|,$$

follows from the Extreme Value Theorem for continuous functions.

Exercise 1.10. In $C[0,1]$ equipped with the $\|\cdot\|_\infty$-norm, calculate the norms of t, $-t$, t^n and $\sin(2\pi nt)$, where $n \in \mathbb{N}$.

Let us check that $\|\cdot\|_\infty$ on $C[a,b]$ does satisfy (N1), (N2), (N3).

(N1) For $\mathbf{x} \in C[a,b]$, clearly

$$\|\mathbf{x}\|_\infty = \max_{t \in [a,b]} |\mathbf{x}(t)| \geq 0,$$

since for each $t \in [a,b]$, $|\mathbf{x}(t)| \geq 0$. Also, if $\mathbf{x} \in C[a,b]$ is such that $\|\mathbf{x}\|_\infty = 0$, then for each $t \in [a,b]$,

$$0 \leq |\mathbf{x}(t)| \leq \max_{t \in [a,b]} |\mathbf{x}(t)| = \|\mathbf{x}\|_\infty = 0.$$

So for each $t \in [a, b]$, $|\mathbf{x}(t)| = 0$, and so $\mathbf{x}(t) = 0$. In other words, $\mathbf{x} = \mathbf{0}$, the zero function in $C[a, b]$.

(N2) If $\alpha \in \mathbb{R}$ and $\mathbf{x} \in C[a, b]$, then

$$|(\alpha \cdot \mathbf{x})(t)| = |\alpha \mathbf{x}(t)| = |\alpha| |\mathbf{x}(t)|, \quad t \in [a, b],$$

and so $\|\alpha \cdot \mathbf{x}\|_\infty = \max_{t \in [a,b]} |\alpha| |\mathbf{x}(t)| = |\alpha| \max_{t \in [a,b]} |\mathbf{x}(t)| = |\alpha| \|\mathbf{x}\|_\infty$.

(N3) Let $\mathbf{x}_1, \mathbf{x}_2 \in C[a, b]$. If $t \in [a, b]$, then

$$\begin{aligned}
|(\mathbf{x}_1 + \mathbf{x}_2)(t)| &= |\mathbf{x}_1(t) + \mathbf{x}_2(t)| \le |\mathbf{x}_1(t)| + |\mathbf{x}_2(t)| \\
&\le \max_{s \in [a,b]} |\mathbf{x}_1(s)| + \max_{s \in [a,b]} |\mathbf{x}_2(s)| = \|\mathbf{x}_1\|_\infty + \|\mathbf{x}_2\|_\infty.
\end{aligned}$$

As this holds for *all* $t \in [a, b]$, we obtain also that

$$\max_{t \in [a,b]} |(\mathbf{x}_1 + \mathbf{x}_2)(t)| \le \|\mathbf{x}_1\|_\infty + \|\mathbf{x}_2\|_\infty,$$

that is, $\|\mathbf{x}_1 + \mathbf{x}_2\|_\infty \le \|\mathbf{x}_1\|_\infty + \|\mathbf{x}_2\|_\infty$.

So $C[a, b]$ with the supremum norm $\|\cdot\|_\infty$ is a normed space. Thus we can use $\|\mathbf{x}_1 - \mathbf{x}_2\|_\infty$ as the distance between $\mathbf{x}_1, \mathbf{x}_2 \in C[a, b]$.

Geometric meaning of the distance in $C[a, b]$ equipped with the supremum norm. We ask the question: what does it mean geometrically when we say that $\mathbf{x}$ is close to $\mathbf{x}_0$? In other words, what does the set of points $\mathbf{x}$ that are close to (say within a distance of ϵ from) $\mathbf{x}_0$ look like?

In $(\mathbb{R}, |\cdot|)$, we know that the set of points x whose distance to x_0 is less than ϵ is an interval:

$$|x - x_0| < \epsilon \iff [x - x_0 < \epsilon \text{ and } -(x - x_0) < \epsilon] \iff x_0 - \epsilon < x < x_0 + \epsilon,$$

and so $\{x \in \mathbb{R} : |x - x_0| < \epsilon\} = (x_0 - \epsilon, x_0 + \epsilon)$. See Figure 2.

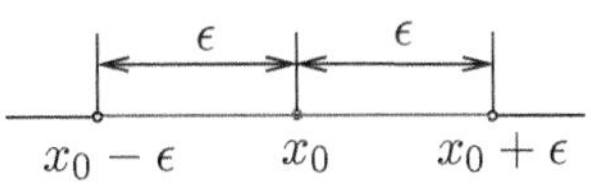

Figure 2. The set of points in $(\mathbb{R}, |\cdot|)$ that are within a distance of ϵ to x_0 is the interval $(x_0 - \epsilon, x_0 + \epsilon)$.

Now we ask: can we visualize the set $\{\mathbf{x} \in C[a, b] : \|\mathbf{x} - \mathbf{x}_0\|_\infty < \epsilon\}$? We have that

$$\begin{aligned}
\|\mathbf{x} - \mathbf{x}_0\|_\infty < \epsilon \quad &\iff \quad \max_{t \in [a,b]} |\mathbf{x}(t) - \mathbf{x}_0(t)| < \epsilon \\
&\iff \quad \text{for all } t \in [a, b], \ |\mathbf{x}(t) - \mathbf{x}_0(t)| < \epsilon \\
&\iff \quad \text{for all } t \in [a, b], \ \mathbf{x}(t) \in (\mathbf{x}_0(t) - \epsilon, \mathbf{x}_0(t) + \epsilon).
\end{aligned}$$

We can imagine translating the graph of $\mathbf{x}_0$ upward by a distance of ϵ, and downward through a distance of ϵ, so as to obtain the shaded strip depicted in Figure 3. Then the graph of $\mathbf{x}$ lies in this shaded strip. This is because at each t, $\mathbf{x}_0(t)-\epsilon < \mathbf{x}(t) < \mathbf{x}_0(t)+\epsilon$: so for example at the particular t indicated in Figure 3, $\mathbf{x}(t)$ has to lie on the line segment AB. Since this has to happen at each $t \in [a, b]$, we see that the graph of $\mathbf{x}$ lies in the shaded strip.

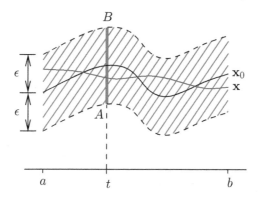

Figure 3. The set of all continuous functions $\mathbf{x}$ whose graph lies between the two dotted lines is the 'ball' $B(\mathbf{x}_0, \epsilon) = \{\mathbf{x} \in C[a, b] : \|\mathbf{x} - \mathbf{x}_0\|_\infty < \epsilon\}$.

Here are examples of some other frequently used norms in $C[a, b]$:

$$\|\mathbf{x}\|_1 \quad := \quad \int_a^b |\mathbf{x}(t)| dt, \tag{1.4}$$

$$\|\mathbf{x}\|_2 \quad := \quad \sqrt{\int_a^b |\mathbf{x}(t)|^2 dt}, \tag{1.5}$$

for $\mathbf{x} \in C[a, b]$. The $\| \cdot \|_1$-norm can be thought of as a continuous analogue of the taxicab norm, while the $\| \cdot \|_2$ norm is the continuous analogue of the Euclidean norm. However, in this course, unless otherwise stated, we will always equip $C[a, b]$ with the supremum norm. ◇

Example 1.11. $C^1[a, b]$. Recall our optimal mining problem from Example 0.2, where the function to be minimized was defined on a subset of the subspace $C^1[a, b]$. So we see that the space $C^1[a, b]$ also arises naturally in applications. What norm do we use in $C^1[a, b]$? In general, if X is a normed space and Y is a subspace of the vector space X, then we can make Y into a normed space by using the restriction of the norm in X to Y. This is called the induced norm in Y, and in Exercise 1.14, we will see that this does give a norm on Y. So surely $C^1[a, b]$, being a subspace of $C[a, b]$ (which is a normed space with the supremum norm), is also a normed space with the supremum norm $\| \cdot \|_\infty$. However, it turns out that in applications, this is not a good choice, essentially because the differentiation map

$$\frac{d}{dt} : C^1[a,b] \to C[a,b], \quad \mathbf{x} \mapsto \mathbf{x}' \quad (\mathbf{x} \in C^1[a,b])$$

is not continuous, which we will see later on. There is a different norm on $C^1[a,b]$ given below, which we shall use:

$$\|\mathbf{x}\|_{\infty,1} = \|\mathbf{x}\|_\infty + \|\mathbf{x}'\|_\infty = \max_{t\in[a,b]} |\mathbf{x}(t)| + \max_{t\in[a,b]} |\mathbf{x}'(t)|, \quad \mathbf{x} \in C^1[a,b].$$

Roughly, two functions in $(C^1[a,b], \|\cdot\|_{\infty,1})$ are regarded as close together if both the functions themselves *and* their first derivatives are close together. Indeed, $\|\mathbf{x}_1 - \mathbf{x}_2\|_{\infty,1} < \epsilon$ implies that

$$|\mathbf{x}_1(t) - \mathbf{x}_2(t)| < \epsilon \quad \text{and} \quad |\mathbf{x}_1'(t) - \mathbf{x}_2'(t)| < \epsilon \quad \text{for all } t \in [a,b], \tag{1.6}$$

and conversely, (1.6) implies that $\|\mathbf{x}_1 - \mathbf{x}_2\|_{\infty,1} < 2\epsilon$. We will see that the differentiation mapping from $C^1[a,b]$ to $C[a,b]$ is continuous if $C^1[a,b]$ is equipped with the $\|\cdot\|_{\infty,1}$-norm and $C[a,b]$ is equipped with the $\|\cdot\|_\infty$-norm. In $C^1[0,1]$, for example

$$\|t\|_{1,\infty} = \|t\|_\infty + \|1\|_\infty = 1 + 1 = 2,$$
$$\|t^n\|_{1,\infty} = \|t^n\|_\infty + \|nt^{n-1}\|_\infty = 1 + n, \quad n \in \mathbb{N}. \qquad \diamond$$

Exercise 1.12. Let $(X, \|\cdot\|)$ be a normed space. Prove that for all $x, y \in X$,

$$\big|\|x\| - \|y\|\big| \le \|x - y\|.$$

Exercise 1.13. If $x \in \mathbb{R}$, then let $\|x\| = |x|^2$. Is $\|\cdot\|$ a norm on $\mathbb{R}$?

Exercise 1.14. Let X be a normed space with norm $\|\cdot\|_X$, and Y be a subspace of X. Prove that Y is also a normed space with the norm $\|\cdot\|_Y$ defined simply as the restriction of the norm $\|\cdot\|_X$ to Y. This norm on Y is called the *induced norm*.

Exercise 1.15. The *Cauchy-Schwarz Inequality* says that if $x_1, \ldots, x_n$ and $y_1, \ldots, y_n$ are any real numbers, then

$$(x_1 y_1 + \cdots + x_n y_n)^2 \le (x_1^2 + \cdots + x_n^2)(y_1^2 + \cdots + y_n^2).$$

Let $n \in \mathbb{N}$. Define for $x = (x_1, \cdots, x_n) \in \mathbb{R}^n$

$$\|x\|_p = \sqrt[p]{|x_1|^p + \cdots + |x_n|^p} \quad \text{if } p = 1 \text{ or } 2, \text{ and}$$
$$\|x\|_\infty = \max\{|x_1|, \ldots, |x_n|\}. \tag{1.7}$$

Show that the function $x \mapsto \|x\|_p$ is a norm on $\mathbb{R}^n$. (Use Cauchy-Schwarz inequality for establishing the triangle inequality in the $p = 2$ case.)

Now suppose that $n = 2$. Depict the following sets pictorially:

$$B_2(\mathbf{0}, 1) := \{x \in \mathbb{R}^2 : \|x\|_2 < 1\},$$
$$B_1(\mathbf{0}, 1) := \{x \in \mathbb{R}^2 : \|x\|_1 < 1\},$$
$$B_\infty(\mathbf{0}, 1) := \{x \in \mathbb{R}^2 : \|x\|_\infty < 1\}.$$

Exercise 1.16. A subset C of a vector space is said to be *convex* if for all $x, y \in C$, and all $\alpha \in (0, 1)$, $(1 - \alpha)x + \alpha y \in C$; see Figure 4.

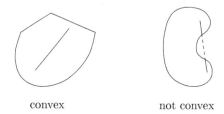

convex   not convex

Figure 4. Examples of convex and nonconvex sets in $\mathbb{R}^2$.

(1) Show that the unit ball $B(\mathbf{0}, 1) = \{x \in X : \|x\| < 1\}$ is convex in any normed space $(X, \| \cdot \|)$.

(2) Depict the set $\{(x_1, x_2) \in \mathbb{R}^2 : \sqrt{|x_1|} + \sqrt{|x_2|} = 1\}$ in the plane.

(3) Prove that $\|x\|_{1/2} := (\sqrt{|x_1|} + \sqrt{|x_2|})^2$, $x = (x_1, x_2) \in \mathbb{R}^2$, does not define a norm on $\mathbb{R}^2$.

Exercise 1.17. Show that (1.4) defines a norm on $C[a, b]$.

Exercise 1.18. Let $C^n[a, b]$ denote the space of n times continuously differentiable functions on $[a, b]$:

$$C^n[a, b] = \{\mathbf{x} : [a, b] \to \mathbb{R} \text{ such that } \mathbf{x}', \mathbf{x}'', \cdots, \mathbf{x}^{(n)} \in C[a, b]\},$$

equipped with the norm

$$\|\mathbf{x}\|_{n,\infty} = \|\mathbf{x}\|_\infty + \|\mathbf{x}'\|_\infty + \cdots + \|\mathbf{x}^{(n)}\|_\infty, \quad \mathbf{x} \in C^n[a, b]. \tag{1.8}$$

Show that (1.8) defines a norm on $C^n[a, b]$.

1.2. Continuity, Linear Transformations, and Continuous Linear Transformations

We would like to differentiate functions $f : X \to Y$, where X, Y are normed spaces. We will see later on that if $x_0 \in X$, then the derivative $f'(x_0)$ is a *continuous* linear transformation from X to Y. So we will first study continuity of maps between normed spaces.

1.2.1. Continuity. In everyday life, by "continuity", we mean "no breaks at any point." If a break occurs, it occurs at some point. We will first define what we mean by a function being continuous *at a point*, and then we will say that a function is continuous if it is continuous at each point in its domain.

Let us revisit the definition of continuity of a function $f : \mathbb{R} \to \mathbb{R}$ at a point $x_0 \in \mathbb{R}$ in ordinary calculus. If a function has a break at a point, say x_0, then even if points x are close to x_0, the points $f(x)$ do not get close to $f(x_0)$. See Figure 5.

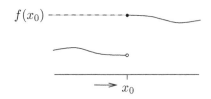

Figure 5. A function with a break at x_0. If x lies to the left of x_0, then $f(x)$ is not close to $f(x_0)$, no matter how close x comes to x_0.

This motivates the familiar definition of continuity in ordinary calculus, which guarantees that if a function is continuous at a point x_0, then we can make $f(x)$ as close as we like to $f(x_0)$, by choosing x sufficiently close to x_0. See Figure 6.

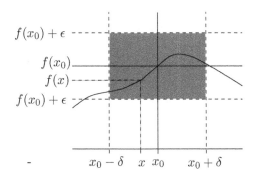

Figure 6. The definition of the continuity of a function at point x_0. If the function is continuous at x_0, then given any $\epsilon > 0$ (which determines a strip around the line $y = f(x_0)$ of width 2ϵ), there exists a $\delta > 0$ (which determines an interval of width 2δ around the point x_0) such that whenever x lies in this width (so that x satisfies $|x - x_0| < \delta$) and then $f(x)$ satisfies $|f(x) - f(x_0)| < \epsilon$.

Definition 1.19. A map $f : \mathbb{R} \to \mathbb{R}$ is *continuous at x_0* if for every $\epsilon > 0$, there exists a $\delta > 0$ such that for all $x \in \mathbb{R}$ satisfying $|x - x_0| < \delta$, $|f(x) - f(x_0)| < \epsilon$.

A function $f : \mathbb{R} \to \mathbb{R}$ is *continuous* if for every $x_0 \in \mathbb{R}$, f is continuous at x_0.

We observe that in the definition of continuity in ordinary calculus, if x, y are real numbers, then $|x - y|$ is a measure of the distance between them, and that the absolute value $|\cdot|$ is a norm in the finite (1) dimensional normed space $\mathbb{R}$. So it is natural to define continuity in arbitrary normed spaces by simply replacing the absolute values by the corresponding norms, since the norm provides the notion of distance between vectors.

Definition 1.20. Let X and Y be normed spaces over $\mathbb{R}$, and $x_0 \in X$. A map $f : X \to Y$ is said to be *continuous at x_0* if for every $\epsilon > 0$, there exists a $\delta > 0$ such that whenever $x \in X$ satisfies $\|x - x_0\| < \delta$, there holds that $\|f(x) - f(x_0)\| < \epsilon$. The map $f : X \to Y$ is called *continuous* if for all $x_0 \in X$, f is continuous at x_0.

Example 1.21. Define $D : C^1[0, 1] \to C[0, 1]$ by

$$(D\mathbf{x})(t) = \mathbf{x}'(t), \quad t \in [0, 1], \quad \mathbf{x} \in C^1[0, 1].$$

(So D is "differentiation pointwise.") We claim that D is continuous when $C[0, 1]$ is equipped with the norm $\|\mathbf{x}\|_\infty$ and $C^1[0, 1]$ is equipped with the norm $\|\cdot\|_{1,\infty}$. Let $\mathbf{x}_0 \in C^1[0, 1]$. Let $\epsilon > 0$. Set $\delta = \epsilon$. Then $\delta > 0$ and for all $\mathbf{x} \in C^1[0, 1]$ satisfying $\|\mathbf{x} - \mathbf{x}_0\|_\infty < \delta$, we have

$$\begin{aligned} \|D\mathbf{x} - D\mathbf{x}_0\|_\infty &= \|\mathbf{x}' - \mathbf{x}_0'\|_\infty \\ &\leq \|\mathbf{x}' - \mathbf{x}_0'\|_\infty + \|\mathbf{x} - \mathbf{x}_0\|_\infty \\ &= \|\mathbf{x} - \mathbf{x}_0\|_{1,\infty} < \delta = \epsilon. \end{aligned}$$

Hence D is continuous at $\mathbf{x}_0 \in C^1[0, 1]$. As the choice of $\mathbf{x}_0 \in C^1[0, 1]$ was arbitrary, it follows that D is continuous on $C^1[0, 1]$. $\Diamond$

Exercise 1.22. Let $(X, \|\cdot\|)$ be a normed space. Show that the norm $\|\cdot\| : X \to \mathbb{R}$ is a continuous map.

Exercise 1.23. This exercise concerns the norm on $C^1[a, b]$ we have chosen to use. Since we want to be able to use ordinary analytic operations such as passage to the limit, then, given a function $f : C^1[0, 1] \to \mathbb{R}$, it is reasonable to choose a norm such that f is continuous. As our f, let us take the arc length function given by

$$f(\mathbf{x}) = \int_0^1 \sqrt{1 + (\mathbf{x}'(t))^2} dt, \quad \mathbf{x} \in C^1[a, b].$$

We show in the following sequence of exercises that f is not continuous if we equip $C^1[0, 1]$ with the supremum norm $\|\cdot\|_\infty$ induced from $C[0, 1]$.

(1) Calculate $f(\mathbf{0})$. (The arc length of the constant curve taking value 0 everywhere on $[0, 1]$ is obviously 1, and check that the above formula delivers this.)

(2) Now consider $\mathbf{x}_n := \dfrac{\cos(2\pi nt)}{\sqrt{n}}$, $t \in [0, 1]$, $n \in \mathbb{N}$. Using the fact that

$$\sin(2\pi nt) \geq \frac{1}{\sqrt{2}} \text{ for } t \in \left[\frac{1}{8n}, \frac{3}{8n}\right],$$

and the periodicity of $\sin(2\pi nt)$ (the graph of $\sin(2\pi nt)$ on $[0, 1/n]$ is repeated n times in $[0, 1]$), conclude that

$$f(\mathbf{x}_n) \geq n \cdot \int_{\frac{1}{8n}}^{\frac{3}{8n}} \sqrt{1 + 4\pi^2 n(\sin(2\pi nt))^2} dt \geq \frac{\sqrt{1 + 2\pi^2 n}}{4}.$$

(3) Show that f is not continuous at $\mathbf{0}$. (Prove this by contradiction. Note that by taking larger and larger n, $\|\mathbf{x}_n - \mathbf{0}\|_\infty$ can be made as small as we please, but $f(\mathbf{x}_n)$ doesn't stay close to $f(\mathbf{0})$.)

Show that the arc length function f is continuous if we equip $C^1[0, 1]$ with the norm $\|\cdot\|_{1,\infty}$. It may be useful to note that by using the triangle inequality in $(\mathbb{R}^2, \|\cdot\|_2)$, we have

$$\begin{aligned} \left| \|(1, a)\|_2 - \|(1, b)\|_2 \right| &= |\sqrt{1^2 + a^2} - \sqrt{1^2 + b^2}| \\ &\leq \|(1, a) - (1, b)\|_2 \\ &= \sqrt{(1 - 1)^2 + (a - b)^2} = |a - b|, \quad a, b \in \mathbb{R}. \end{aligned}$$

1.2.2. Linear transformations. We will see later that the derivative $f'(x_0)$ of $f : X \to Y$ (where X, Y are normed spaces), at a point $x_0 \in X$, is a continuous *linear transformation* from X to Y. Let us recall what a linear transformation between vector spaces is.

Definition 1.24 (Linear transformation). Let X and Y be real vector spaces. A map $T : X \to Y$ is called a *linear transformation* if it satisfies the following:

(L1) For all $x_1, x_2 \in X$, $T(x_1 + x_2) = T(x_1) + T(x_2)$.

(L2) For all $x \in X$ and all $\alpha \in \mathbb{R}$, $T(\alpha \cdot x) = \alpha \cdot T(x)$.

Here are a few examples.

Example 1.25. $D : C^1[0, 1] \to C[0, 1]$ given by $D\mathbf{x} = \mathbf{x}'$, $\mathbf{x} \in C^1[0, 1]$, is a linear transformation:

(L1) For all $\mathbf{x}_1, \mathbf{x}_2 \in C^1[0, 1]$,

$$D(\mathbf{x}_1 + \mathbf{x}_2) = (\mathbf{x}_1 + \mathbf{x}_2)' = \mathbf{x}_1' + \mathbf{x}_2' = D(\mathbf{x}_1) + D(\mathbf{x}_2).$$

(L2) For all $\mathbf{x} \in C^1[0, 1]$ and all $\alpha \in \mathbb{R}$,

$$D(\alpha \cdot \mathbf{x}) = (\alpha \cdot \mathbf{x})' = \alpha \cdot \mathbf{x}' = \alpha \cdot (D\mathbf{x}).$$

$\diamond$

Example 1.26. Let $S := \{\mathbf{h} \in C^1[a, b] : \mathbf{h}(a) = \mathbf{h}(b) = 0\}$. Then we have seen in Exercise 1.3 that S is a subspace of $C^1[a, b]$. Let $\mathbf{A}, \mathbf{B} \in C[a, b]$ be two fixed functions. Consider the map $L : S \to \mathbb{R}$ given by

$$L(\mathbf{h}) = \int_a^b \Big(\mathbf{A}(t)\mathbf{h}(t) + \mathbf{B}(t)\mathbf{h}'(t) \Big) dt, \quad \mathbf{h} \in S.$$

Let us check that L is a linear transformation. We have:

(L1) For all $\mathbf{h}_1, \mathbf{h}_2 \in S$,

$$L(\mathbf{h}_1 + \mathbf{h}_2)$$
$$= \int_a^b \Big(\mathbf{A}(t)(\mathbf{h}_1 + \mathbf{h}_2)(t) + \mathbf{B}(t)(\mathbf{h}_1 + \mathbf{h}_2)'(t) \Big) dt$$
$$= \int_a^b \Big(\mathbf{A}(t)(\mathbf{h}_1(t) + \mathbf{h}_2(t)) + \mathbf{B}(t)(\mathbf{h}_1'(t) + \mathbf{h}_2'(t)) \Big) dt$$
$$= \int_a^b \Big(\mathbf{A}(t)\mathbf{h}_1(t) + \mathbf{B}(t)\mathbf{h}_1'(t) + \mathbf{A}(t)\mathbf{h}_2(t) + \mathbf{B}(t)\mathbf{h}_2'(t) \Big) dt$$
$$= \int_a^b \Big(\mathbf{A}(t)\mathbf{h}_1(t) + \mathbf{B}(t)\mathbf{h}_1'(t) \Big) dt + \int_a^b \Big(\mathbf{A}(t)\mathbf{h}_2(t)) + \mathbf{B}(t)\mathbf{h}_2'(t) \Big) dt$$
$$= L(\mathbf{h}_1) + L(\mathbf{h}_2).$$

(L2) For all $\mathbf{h} \in C^1[0,1]$ and all $\alpha \in \mathbb{R}$,

$$
\begin{aligned}
L(\alpha \cdot \mathbf{h}) &= \int_a^b \Big(\mathbf{A}(t)(\alpha \cdot \mathbf{h})(t) + \mathbf{B}(t)(\alpha \cdot \mathbf{h})'(t) \Big) dt \\
&= \int_a^b \Big(\mathbf{A}(t)\alpha \mathbf{h}(t) + \mathbf{B}(t)\alpha \mathbf{h}'(t) \Big) dt \\
&= \alpha \int_a^b \Big(\mathbf{A}(t)\mathbf{h}(t) + \mathbf{B}(t)\mathbf{h}'(t) \Big) dt = \alpha L(\mathbf{h}).
\end{aligned}
$$

Thus L is a linear transformation.   ◇

Example 1.27. $f : C[0,1] \to \mathbb{R}$ defined by

$$
f(\mathbf{x}) = \int_0^1 (\mathbf{x}(t))^2 dt, \quad \mathbf{x} \in C[0,1],
$$

is *not* a linear transformation. For example, if $\mathbf{1}$ is the constant function taking value 1 everywhere, then we have

$$
f(\mathbf{1}) = \int_0^1 (\mathbf{1}(t))^2 dt = \int_0^1 1 dt = 1,
$$

and so $2 \cdot I(1) = 2$, but this is not equal to $f(2 \cdot \mathbf{1}) = 4 \neq 2 = 2 \cdot f(\mathbf{1})$, violating (L2). Hence f is not a linear transformation.   ◇

Exercise 1.28. Is the map $f : C[0,1] \to \mathbb{R}$ given below a linear transformation?

$$
f(\mathbf{x}) := \int_0^1 \mathbf{x}(t^2) dt, \quad \mathbf{x} \in C[0,1].
$$

1.2.3. Continuous linear transformations. Since the derivative at a point of a function between normed spaces is going to be a linear transformation that is continuous, the following result will be useful, which gives a characterization of continuity for linear transformations.

Theorem 1.29. *Let X and Y be normed spaces over $\mathbb{R}$, and let $T : X \to Y$ be a linear transformation. Then the following are equivalent:*

(1) *T is continuous.*

(2) *T is continuous at 0.*

(3) *There exists an $M > 0$ such that for all $x \in X$, $\|Tx\| \leq M\|x\|$.*

The useful part is (3)⇒(1), since by just showing the existence of the bound M, we can conclude the continuity of the given linear transformation. So we don't have to go through the rigmarole of verifying the ϵ-δ definition. Note also that it seems miraculous that continuity at just one point (at $\mathbf{0}$) delivers continuity everywhere on X! This miracle happens because the map T is not any old map, but rather a *linear transformation*.

Proof. We will show the three implications (1)$\Rightarrow$(2), (2)$\Rightarrow$(3), and (3)$\Rightarrow$(1), which are enough to get all the three equivalences (and six implications) given in the statement of the theorem.

(1)$\Rightarrow$(2). This is just the definition of continuity on X. Indeed, T has to be continuous at each point in X, and in particular at $\mathbf{0} \in X$.

(2)$\Rightarrow$(3). Take $\epsilon := 1 > 0$. Then there exists a $\delta > 0$ such that $\|x - \mathbf{0}\| = \|x\| < \delta$ implies $\|Tx - T\mathbf{0}\| = \|Tx - \mathbf{0}\| = \|Tx\| < 1$. Let us check that this yields:

$$\|Tx\| \leq \frac{2}{\delta}\|x\| \quad \text{for all } x \in X. \tag{1.9}$$

First consider $x = \mathbf{0}$. Then

$$\begin{aligned}
\text{the left-hand side} &= \|Tx\| = \|T\mathbf{0}\| = \|\mathbf{0}\| = 0, \text{ while} \\
\text{the right-hand side} &= \frac{2}{\delta}\|x\| = \frac{2}{\delta}\|\mathbf{0}\| = 0.
\end{aligned}$$

And so the claim in (1.9) holds because we have in fact an equality.

On the other hand, now suppose that $x \neq \mathbf{0}$. Set $y := \dfrac{\delta}{2\|x\|} \cdot x$. Then

$$\|y\| = \left\|\frac{\delta}{2\|x\|} \cdot x\right\| = \frac{\delta}{2\|x\|}\|x\| = \frac{\delta}{2} < \delta,$$

and so $\|Ty\| < 1$, that is

$$\left\|T\left(\frac{\delta}{2\|x\|} \cdot x\right)\right\| = \left\|\frac{\delta}{2\|x\|} \cdot Tx\right\| = \frac{\delta}{2\|x\|}\|Tx\| < 1.$$

Upon rearranging, we obtain (1.9).

So the claim in (3) holds with $M = \dfrac{2}{\delta}$.

(3)$\Rightarrow$(1). Let $M > 0$ be such that for all $x \in X$, $\|Tx\| \leq M\|x\|$. Let $x_0 \in X$, and $\epsilon > 0$. Set

$$\delta := \frac{\epsilon}{M} > 0.$$

Then whenever $\|x - x_0\| < \delta$, we have

$$\|Tx - Tx_0\| = \|T(x - x_0)\| \leq M\|x - x_0\| < M \cdot \delta = M \cdot \frac{\epsilon}{M} = \epsilon.$$

So T is continuous at x_0. But as $x_0 \in X$ was arbitrary, T is continuous. $\qquad\square$

Here are a few examples.

Example 1.30. Consider Example 1.26 again. Let

$$S := \{\mathbf{h} \in C^1[a, b] : \mathbf{h}(a) = \mathbf{h}(b) = 0\}.$$

Then we have seen in Exercise 1.3 that S is a subspace of $C^1[a, b]$. Let $\mathbf{A}, \mathbf{B} \in C[a, b]$ be two fixed functions. Consider the map $L : S \to \mathbb{R}$ given by

$$L(\mathbf{h}) = \int_a^b \Big(\mathbf{A}(t)\mathbf{h}(t) + \mathbf{B}(t)\mathbf{h}'(t)\Big)dt, \quad \mathbf{h} \in S.$$

We had seen in Example 1.26 that L is a linear transformation. Now we ask: is L continuous? We will use Theorem 1.29 to check that L is continuous. Indeed, we have for $\mathbf{h} \in S$ that

$$
\begin{aligned}
|L(\mathbf{h})| &= \left| \int_a^b \Big(\mathbf{A}(t)\mathbf{h}(t) + \mathbf{B}(t)\mathbf{h}'(t)\Big)dt \right| \\
&\leq \int_a^b |\mathbf{A}(t)\mathbf{h}(t) + \mathbf{B}(t)\mathbf{h}'(t)|dt \\
&\leq \int_a^b (|\mathbf{A}(t)||\mathbf{h}(t)| + |\mathbf{B}(t)||\mathbf{h}'(t)|)dt \\
&\leq \int_a^b (|\mathbf{A}(t)|\|\mathbf{h}\|_\infty + |\mathbf{B}(t)|\|\mathbf{h}'\|_\infty)dt \\
&\leq \int_a^b (|\mathbf{A}(t)|\|\mathbf{h}\|_{1,\infty} + |\mathbf{B}(t)|\|\mathbf{h}\|_{1,\infty})dt \\
&\leq \left(\int_a^b (|\mathbf{A}(t)| + |\mathbf{B}(t)|)dt \right) \|\mathbf{h}\|_{1,\infty} = M\|\mathbf{h}\|_{1,\infty},
\end{aligned}
$$

with $M := \int_a^b (|\mathbf{A}(t)| + |\mathbf{B}(t)|)dt$. In the above, we have used

$$
\begin{aligned}
\|\mathbf{h}\|_{1,\infty} &= \|\mathbf{h}\|_\infty + \|\mathbf{h}'\|_\infty \geq \|\mathbf{h}\|_\infty \text{ and} \\
\|\mathbf{h}\|_{1,\infty} &= \|\mathbf{h}\|_\infty + \|\mathbf{h}'\|_\infty \geq \|\mathbf{h}'\|_\infty.
\end{aligned}
$$

$\diamond$

Example 1.31. Consider Example 1.21 again. We had seen that the mapping $D : C^1[0, 1] \to C[0, 1]$ defined by

$$(D\mathbf{x})(t) = \mathbf{x}'(t), \quad t \in [0, 1], \quad \mathbf{x} \in C^1[0, 1],$$

is continuous if $C^1[0, 1]$ is equipped with the $\|\cdot\|_{1,\infty}$-norm, and $C[0, 1]$ is equipped with the $\|\cdot\|_\infty$-norm.

We now show that D is *not* continuous if we equip $C^1[0, 1]$ with the induced $\|\cdot\|_\infty$ norm. Suppose on the contrary that D is continuous. As D is a linear transformation, it follows from Theorem 1.29 that there exists an $M > 0$ such that for all $\mathbf{x} \in C^1[0, 1]$,

$$\|D\mathbf{x}\|_\infty = \|\mathbf{x}'\|_\infty \leq M\|\mathbf{x}\|_\infty.$$

But if we take $\mathbf{x} = t^n$ $(n \in \mathbb{N})$, then we have

$$\|\mathbf{x}\|_\infty = \|t^n\|_\infty = 1, \text{ and}$$
$$\|\mathbf{x}'\|_\infty = \|nt^{n-1}\|_\infty = n\|t^{n-1}\|_\infty = n,$$

and so $\|D\mathbf{x}\|_\infty = \|\mathbf{x}'\|_\infty = n \leq M\|\mathbf{x}\|_\infty = M \cdot 1$, that is, $n \leq M$ for all $n \in \mathbb{N}$, which is clearly not true. So D is not continuous. $\diamond$

Exercise 1.32. Show that the linear transformation $f : C[0,1] \to \mathbb{R}$ given below is continuous when $C[0,1]$ is equipped with the $\|\cdot\|_\infty$-norm:

$$f(\mathbf{x}) := \int_0^1 \mathbf{x}(t^2)dt, \quad \mathbf{x} \in C[0,1].$$

1.3. Differentiation

In this section, we will study differentiation: we will define the derivative of a map $f : X \to Y$ at a point $x_0 \in X$. Roughly speaking, the derivative $f'(x_0)$ of f at a point x_0 will be a continuous linear transformation $f'(x_0) : X \to Y$ that provides an approximation of f in the vicinity of x_0.

Let us first revisit the situation in ordinary calculus, where we have $f : \mathbb{R} \to \mathbb{R}$, and let us rewrite the definition of the derivative of f at $x_0 \in \mathbb{R}$ in a manner that lends itself to generalization to the case of maps between normed spaces.

Recall that for a function $f : \mathbb{R} \to \mathbb{R}$, the derivative at a point x_0 is the approximation of f around x_0 by a straight line. See Figure 7.

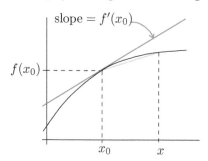

Figure 7. The derivative of f at x_0.

Let $f : \mathbb{R} \to \mathbb{R}$ and let $x_0 \in \mathbb{R}$. f is differentiable at x_0 with derivative $f'(x_0) \in \mathbb{R}$ if

$$\lim_{x \to x_0} \frac{f(x) - f(x_0)}{x - x_0} = f'(x_0),$$

that is, for every $\epsilon > 0$, there exists a $\delta > 0$ such that whenever $x \in \mathbb{R}$ satisfies $0 < |x - x_0| < \delta$, there holds that

$$\left| \frac{f(x) - f(x_0)}{x - x_0} - f'(x_0) \right| < \epsilon,$$

that is,

$$\frac{|f(x) - f(x_0) - f'(x_0)(x - x_0)|}{|x - x_0|} < \epsilon.$$

If we now imagine f instead to be a map from a normed space X to another normed space Y, then bearing in mind that the norm is a generalization of the absolute value in $\mathbb{R}$, we may try mimicking the above definition and replace the denominator in the inequality above by $\|x - x_0\|$, and the numerator absolute value can be replaced by the norm in Y, since we see that $f(x) - f(x_0)$ lives in Y. But what object should be there in the box below?:

$$\frac{\|f(x) - f(x_0) - \boxed{f'(x_0)}(x - x_0)\|_Y}{\|x - x_0\|_X} < \epsilon.$$

Since $f(x), f(x_0)$ live in Y, we expect the term $f'(x_0)(x - x_0)$ to be also in Y. As $x - x_0$ is in X, $f'(x_0)$ should take this into Y. So we see that it is natural that we should not expect $f'(x_0)$ to be a number (as was the case when $X = Y = \mathbb{R}$), but rather it we expect it should be a certain mapping from X to Y. We will in fact want it to be a continuous linear transformation from X to Y. Why? We will see this later, but a short answer is that with this definition, we can prove analogous theorems from ordinary calculus, and we can use these theorems in applications to solve real-life problems. After this rough motivation, let us now see the precise definition.

Definition 1.33. Let X, Y be normed spaces, $f : X \to Y$ be a map, and $x_0 \in X$. f is said to be *differentiable at x_0* if there exists a continuous linear transformation $L : X \to Y$ such that for every $\epsilon > 0$, there exists a $\delta > 0$ such that whenever $x \in X$ satisfies $0 < \|x - x_0\| < \delta$, we have

$$\frac{\|f(x) - f(x_0) - L(x - x_0)\|}{\|x - x_0\|} < \epsilon.$$

If f is differentiable at x_0, then it can be shown that there can be at most one continuous linear transformation L such that the above statement holds. We will prove this below in Theorem 1.38. This unique continuous linear transformation L is denoted by $f'(x_0)$, and is called the *derivative of f at x_0*. If f is differentiable at every point $x \in X$, then f is simply said to be *differentiable*.

Before we see simple illustrative examples on the calculation of the derivative, let us check that this is a genuine extension of the notion of differentiability from ordinary calculus. Over there the concept of derivative was very simple, and $f'(x_0)$ was just a number. But now we will see that over there too, it was actually a continuous linear transformation, but it just so happens that any continuous linear transformation from $\mathbb{R}$ to $\mathbb{R}$ is just given by multiplication by a fixed number. We explain this below.

Coincidence of our new definition with the old definition when we have $X = Y = \mathbb{R}$, $f : \mathbb{R} \to \mathbb{R}$, $x_0 \in \mathbb{R}$.

(1) Differentiable in the old sense $\Rightarrow$ differentiable in the new sense.

Suppose that
$$\lim_{x \to x_0} \frac{f(x) - f(x_0)}{x - x_0}$$
exists and equals the number $f'_{\text{old}}(x_0) \in \mathbb{R}$. Define $L : \mathbb{R} \to \mathbb{R}$ by $L(v) = f'_{\text{old}}(x_0) \cdot v$, $v \in \mathbb{R}$. Then L is a linear transformation because

(L1) For every $v_1, v_2 \in \mathbb{R}$,
$$L(v_1 + v_2) = f'_{\text{old}}(x_0) \cdot (v_1 + v_2) = f'_{\text{old}}(x_0) \cdot v_1 + f'_{\text{old}}(x_0) \cdot v_2 = L(v_1) + L(v_2).$$

(L2) For every $\alpha \in \mathbb{R}$ and every $v \in V$,
$$L(\alpha \cdot v) = f'_{\text{old}}(x_0) \cdot (\alpha \cdot v) = \alpha \cdot (f'_{\text{old}}(x_0) \cdot v) = \alpha L(v).$$

L is continuous because
$$|L(v)| = |f'_{\text{old}}(x_0) \cdot v| = \underbrace{|f'_{\text{old}}(x_0)|}_{=:M} |v| \text{ for all } v \in \mathbb{R}.$$

We know that
$$\lim_{x \to x_0} \frac{f(x) - f(x_0)}{x - x_0} = f'_{\text{old}}(x_0),$$
that is, for every $\epsilon > 0$, there exists a $\delta > 0$ such that whenever $x \in \mathbb{R}$ satisfies $0 < |x - x_0| < \delta$, we have
$$\left| \frac{f(x) - f(x_0)}{x - x_0} - f'_{\text{old}}(x_0) \right| = \frac{|f(x) - f(x_0) - f'_{\text{old}}(x_0) \cdot (x - x_0)|}{|x - x_0|} < \epsilon,$$
that is,
$$\frac{|f(x) - f(x_0) - L(x - x_0)|}{|x - x_0|} < \epsilon.$$
So f is differentiable in the new sense too, and $f'_{\text{new}}(x_0) = L$, that is, we have $f'_{\text{new}}(x_0)(v) = f'_{\text{old}}(x_0) \cdot v$, $v \in \mathbb{R}$.

(2) Differentiable in the new sense $\Rightarrow$ differentiable in the old sense.

Suppose there is a continuous linear transformation $f'_{\text{new}}(x_0) : \mathbb{R} \to \mathbb{R}$ such that for every $\epsilon > 0$, there exists a $\delta > 0$ such that whenever $x \in \mathbb{R}$ satisfies $0 < |x - x_0| < \delta$, we have
$$\frac{|f(x) - f(x_0) - f'_{\text{new}}(x_0)(x - x_0)|}{|x - x_0|} < \epsilon,$$
Define $f'_{\text{old}}(x_0) := f'_{\text{new}}(x_0)(1) \in \mathbb{R}$. Then if $x \in \mathbb{R}$, we have
$$\begin{aligned} f'_{\text{new}}(x_0)(x - x_0) &= f'_{\text{new}}(x_0)((x - x_0) \cdot 1) = (x - x_0) \cdot f'_{\text{new}}(x_0)(1) \\ &= f'_{\text{old}}(x_0) \cdot (x - x_0). \end{aligned}$$

So there exists a number, namely $f'_{\text{old}}(x_0)$, such that for every $\epsilon > 0$, there exists a $\delta > 0$ such that whenever $x \in \mathbb{R}$ satisfies $0 < |x - x_0| < \delta$, we have

$$\left| \frac{f(x) - f(x_0)}{x - x_0} - f'_{\text{old}}(x_0) \right| = \frac{|f(x) - f(x_0) - f'_{\text{old}}(x_0) \cdot (x - x_0)|}{|x - x_0|}$$

$$= \frac{|f(x) - f(x_0) - f'_{\text{new}}(x_0)(x - x_0)|}{|x - x_0|} < \epsilon.$$

Consequently, f is differentiable at x_0 in the old sense, and $f'_{\text{old}}(x_0) = f'_{\text{new}}(x_0)(1)$.

The derivative as a local linear approximation. We know that in ordinary calculus, for a function $f : \mathbb{R} \to \mathbb{R}$ that is differentiable at $x_0 \in \mathbb{R}$, the number $f'(x_0)$ has the interpretation of being the slope of the tangent line to the graph of the function at the point $(x_0, f(x_0))$, and the tangent line itself serves as a local linear approximation to the graph of the function. (Imagine zooming into the point $(x_0, f'(x_0))$ using lenses of greater and greater magnification: then there is little difference between the graph of the function and the tangent line.) We now show that also in the more general setup when f is a map from a normed space X to a normed space Y that is differentiable at a point $x_0 \in X$, $f'(x_0)$ can be interpreted as giving a local linear approximation to the mapping f near the point x_0, and we explain this below. Let $\epsilon > 0$. Then we know that for all x close enough to x_0 and distinct from x_0, we have

$$\frac{\|f(x) - f(x_0) - f'(x_0)(x - x_0)\|}{\|x - x_0\|} < \epsilon,$$

that is, $\|f(x) - f(x_0) - f'(x_0)(x - x_0)\| < \epsilon\|x - x_0\|$. So for all x close enough to x_0,

$$\|f(x) - f(x_0) - f'(x_0)(x - x_0)\| \approx 0,$$

that is, $f(x) - f(x_0) - f'(x_0)(x - x_0) \approx \mathbf{0} \in X$, and upon rearranging,

$$f(x) - f(x_0) \approx f'(x_0)(x - x_0).$$

The above says that near x_0, $f(x) - f(x_0)$ looks like the action of the *linear transformation* $f'(x_0)$ acting on $x - x_0$. We will keep this important message in mind because it will help us calculate the derivative in concrete examples. Given an f, for which we need to find $f'(x_0)$, our starting point will always be to start with calculating $f(x) - f(x_0)$ and trying to guess what linear transformation L would give that $f(x) - f(x_0) \approx L(x - x_0)$ for x near x_0. So we would start by writing

$$f(x) - f(x_0) = L(x - x_0) + \text{error},$$

and then showing that the error term is mild enough so that the derivative definition can be verified. We will soon see this in action below, but first let us make an important remark.

Remark 1.34. In our definition of the derivative, why do we insist the derivative $f'(x_0)$ of $f : X \to Y$ at $x_0 \in X$ should be a *continuous* linear transformation—that is, why not settle just for it being a linear transformation (without demanding continuity)?

The answer to this question is tied to wanting

$$\boxed{\text{Differentiability at } x_0 \quad \Rightarrow \quad \text{continuity at } x_0.}$$

We know this holds with the usual derivative concept in ordinary calculus when $f : \mathbb{R} \to \mathbb{R}$. If we want this property to hold also in our more general setting of normed spaces, then just having $f'(x_0)$ as a linear transformation won't do, but in addition we also need the continuity. For the purposes of optimization, it turns out that even if one doesn't have differentiability at a point implying continuity at the point, one can prove useful optimization theorems using the weaker notion of the derivative. The weaker notion is called the Gateux derivative, while our stronger notion is the Fréchet derivative. As we will exclusively work with just the Fréchet derivative in these notes, we will just refer to our "Fréchet derivative" as "derivative." The reader interested in delving into further details on this may wish to refer to the book by Luenberger [**L**].

Example 1.35. Let X, Y be normed spaces, and let $T : X \to Y$ be a continuous linear transformation. We ask: Is T differentiable at $x_0 \in X$? If so, what is $T'(x_0)$?

Let us do some rough work first. We would like to fill the question mark in the box below with a continuous linear transformation so that

$$T(x) - T(x_0) \approx \boxed{?}(x - x_0),$$

for x close to x_0. But owing to the linearity of T, we know that for all $x \in X$,

$$T(x) - T(x_0) = T(x - x_0),$$

and (the left-hand side) T is already a continuous linear transformation. So we make a *guess* that $T'(x_0) = T$! Let us check this now.

Let $\epsilon > 0$. Choose any $\delta > 0$, for example, $\delta = 399$. Then whenever $x \in X$ satisfies $0 < \|x - x_0\| < \delta = 399$, we have

$$\frac{\|T(x) - T(x_0) - T(x - x_0)\|}{\|x - x_0\|} = \frac{\|\mathbf{0}\|}{\|x - x_0\|} = \frac{0}{\|x - x_0\|} = 0 < \epsilon.$$

Hence $T'(x_0) = T$. Note that as the choice of x_0 was arbitrary, we have in fact obtained that for all $x \in X$, $T'(x) = T$! This is analogous to the observation in ordinary calculus that a linear function $x \mapsto m \cdot x$ has the same slope at all points, namely the number m. $\Diamond$

Example 1.36. Consider $f : C[a, b] \to \mathbb{R}$ defined by

$$f(\mathbf{x}) = \int_a^b (\mathbf{x}(t))^2 dt, \quad \mathbf{x} \in C[a, b].$$

Let $\mathbf{x}_0 \in C[a, b]$. What is $f'(\mathbf{x}_0)$?

As before, we begin with some rough work to make a guess for $f'(\mathbf{x}_0)$ and we seek a continuous linear transformation L so that for $\mathbf{x} \in C[a,b]$ near $\mathbf{x}_0$, we have

$$f(\mathbf{x}) - f(\mathbf{x}_0) \approx L(\mathbf{x} - \mathbf{x}_0).$$

We have

$$
\begin{aligned}
f(\mathbf{x}) - f(\mathbf{x}_0) &= \int_a^b \left((\mathbf{x}(t))^2 - (\mathbf{x}_0(t))^2 \right) dt \\
&= \int_a^b (\mathbf{x}(t) + \mathbf{x}_0(t))(\mathbf{x}(t) - \mathbf{x}_0(t)) dt \\
&\approx \int_a^b (\mathbf{x}_0(t) + \mathbf{x}_0(t))(\mathbf{x}(t) - \mathbf{x}_0(t)) dt \\
&= \int_a^b 2\mathbf{x}_0(t)(\mathbf{x}(t) - \mathbf{x}_0(t)) dt = L(\mathbf{x} - \mathbf{x}_0),
\end{aligned}
$$

where $L : C[a,b] \to \mathbb{R}$ is given by

$$L(\mathbf{h}) = \int_a^b 2\mathbf{x}_0(t)\mathbf{h}(t) dt, \quad \mathbf{h} \in C[a,b].$$

This L is a continuous linear transformation, since this is a special case (when $\mathbf{A} := 2 \cdot \mathbf{x}_0$ and $\mathbf{B} = \mathbf{0}$) of the Examples 1.26 and 1.30. Let us now check that the "ϵ-δ definition of differentiability" holds with this L. For $x \in C[a,b]$, we have

$$
\begin{aligned}
f(\mathbf{x}) -&f(\mathbf{x}_0) - L(\mathbf{x} - \mathbf{x}_0) \\
&= \int_a^b \left((\mathbf{x}(t))^2 - (\mathbf{x}_0(t))^2 \right) dt - \int_a^b 2\mathbf{x}_0(t)(\mathbf{x}(t) - \mathbf{x}_0(t)) dt \\
&= \int_a^b (\mathbf{x}(t) + \mathbf{x}_0(t) - 2\mathbf{x}_0(t))(\mathbf{x}(t) - \mathbf{x}_0(t)) dt \\
&= \int_a^b (\mathbf{x}(t) - \mathbf{x}_0(t))^2 dt,
\end{aligned}
$$

and so

$$
\begin{aligned}
|f(\mathbf{x}) - f(\mathbf{x}_0) - L(\mathbf{x} - \mathbf{x}_0)| &= \left| \int_a^b (\mathbf{x}(t) - \mathbf{x}_0(t))^2 dt \right| \leq \int_a^b |(\mathbf{x} - \mathbf{x}_0)(t)|^2 dt \\
&\leq \int_a^b \|\mathbf{x} - \mathbf{x}_0\|_\infty^2 dt = (b-a)\|\mathbf{x} - \mathbf{x}_0\|_\infty^2.
\end{aligned}
$$

So if $0 < \|\mathbf{x} - \mathbf{x}_0\|_\infty$, we have

$$\frac{|f(\mathbf{x}) - f(\mathbf{x}_0) - L(\mathbf{x} - \mathbf{x}_0)|}{\|\mathbf{x} - \mathbf{x}_0\|_\infty} \leq (b-a)\|\mathbf{x} - \mathbf{x}_0\|_\infty.$$

Let $\epsilon > 0$. Set $\delta := \epsilon/(b-a)$. Then $\delta > 0$ and if $0 < \|\mathbf{x} - \mathbf{x}_0\|_\infty < \delta$,

$$\frac{|f(\mathbf{x}) - f(\mathbf{x}_0) - L(\mathbf{x} - \mathbf{x}_0)|}{\|\mathbf{x} - \mathbf{x}_0\|_\infty} \leq (b-a)\|\mathbf{x} - \mathbf{x}_0\|_\infty < (b-a)\delta = \epsilon.$$

So $f'(\mathbf{x}_0) = L$. In other words, $f'(\mathbf{x}_0)$ is the continuous linear transformation from $C[a, b]$ to $\mathbb{R}$ given by

$$(f'(\mathbf{x}_0))(\mathbf{h}) = \int_a^b 2\mathbf{x}_0(t)\mathbf{h}(t)dt, \quad \mathbf{h} \in C[a, b].$$

So as opposed to the ordinary calculus case, one must stop thinking of the derivative as being a mere number, but instead, in the context of maps between normed spaces, the derivative at a point is itself a map, in fact a continuous linear transformation. So the answer to the question

"What is $f'(\mathbf{x}_0)$?"

should always begin with the phrase

"$f'(\mathbf{x}_0)$ is the continuous linear transformation from X to Y given by $\cdots$."

To emphasize this, let us see some particular cases of our calculation of $f'(\mathbf{x}_0)$ above, for specific choices of $\mathbf{x}_0$.

In particular, we have that the derivative of f at the zero function $\mathbf{0}$, namely $f'(\mathbf{0})$, is the zero linear transformation $\mathbf{0} : C[a, b] \to \mathbb{R}$ that sends every $\mathbf{h} \in C[a, b]$ to the number 0:

$$\mathbf{0}(\mathbf{h}) = 0, \quad \mathbf{h} \in C[a, b].$$

(With a slight abuse of notation, we denote this map also by $\mathbf{0}$, just like the zero element in $C[a, b]$, but from the context it should be clear what is meant.)

As another example, we have that the derivative of f at the constant function $\mathbf{1}$, namely $f'(\mathbf{1})$, is the continuous linear transformation

$$\mathbf{h} \mapsto 2\int_a^b \mathbf{h}(t)dt : C[a, b] \to \mathbb{R}. \qquad\qquad \Diamond$$

Exercise 1.37. Consider $f : C[0, 1] \to \mathbb{R}$ given by $f(\mathbf{x}) := \int_0^1 \mathbf{x}(t^2)dt$, $\mathbf{x} \in C[0, 1]$. Let $\mathbf{x}_0 \in C[0, 1]$. What is $f'(\mathbf{x}_0)$? What is $f'(\mathbf{0})$?

We now prove that something we had mentioned earlier, but which we haven't proved yet: if f is differentiable at x_0, then its derivative is unique.

Theorem 1.38. *Let X, Y be normed spaces. If $f : X \to Y$ is differentiable at $x_0 \in X$, then there is a unique continuous linear transformation L such that for every $\epsilon > 0$, there is a $\delta > 0$ such that whenever $x \in X$ satisfies $0 < \|x - x_0\| < \delta$, there holds*

$$\frac{\|f(x) - f(x_0) - L(x - x_0)\|}{\|x - x_0\|} < \epsilon.$$

Proof. Suppose that $L_1, L_2 : X \to Y$ are two continuous linear transformations such that for every $\epsilon > 0$, there is a $\delta > 0$ such that whenever $x \in X$ satisfies $0 < \|x - x_0\| < \delta$, there holds

$$\frac{\|f(x) - f(x_0) - L_1(x - x_0)\|}{\|x - x_0\|} < \epsilon, \tag{1.10}$$

$$\frac{\|f(x) - f(x_0) - L_2(x - x_0)\|}{\|x - x_0\|} < \epsilon. \tag{1.11}$$

Suppose that $L_1(h_0) \neq L_2(h_0)$ for some $h_0 \in X$. Clearly $h_0 \neq \mathbf{0}$ (for otherwise we would have $L_1(\mathbf{0}) = \mathbf{0} = L_2(\mathbf{0})$!). Take $\epsilon = 1/n$ for some $n \in \mathbb{N}$. Then there is a $\delta_n > 0$ such that whenever $x \in X$ satisfies $0 < \|x - x_0\| < \delta_n$, the inequalities (1.10), (1.11) hold. Set

$$x := x_0 + \frac{\delta_n}{2\|h_0\|} h_0 \in X.$$

Then $x \neq x_0$, and

$$\|x - x_0\| = \frac{\delta_n}{2\|h_0\|} \|h_0\| = \frac{\delta_n}{2} < \delta_n,$$

so that (1.10), (1.11) hold. But then by the triangle inequality it follows that

$$\frac{\|L_1(x - x_0) - L_2(x - x_0)\|}{\|x - x_0\|} = \frac{\|L_1(h_0) - L_2(h_0)\|}{\|h_0\|} < 2\epsilon = \frac{2}{n}.$$

Upon rearranging, we obtain $\|L_1(h_0) - L_2(h_0)\| \leq \frac{2}{n}\|h_0\|$.

As the choice of $n \in \mathbb{N}$ was arbitrary, it follows that $\|L_1(h_0) - L_2(h_0)\| = 0$, and so $L_1(h_0) = L_2(h_0)$, a contradiction. This completes the proof. $\qquad\square$

Exercise 1.39. Let X, Y be normed spaces. Prove that if $f : X \to Y$ is differentiable at x_0, then f is continuous at x_0.

Exercise 1.40. Consider $f : C^1[0,1] \to \mathbb{R}$ defined by $f(\mathbf{x}) = (\mathbf{x}'(1))^2$, $\mathbf{x} \in C^1[0,1]$. Is f differentiable? If so, compute $f'(\mathbf{x}_0)$ at $\mathbf{x}_0 \in C^1[0,1]$.

Exercise 1.41 (Chain rule). Given distinct x_1, x_2 in a normed space X, define the straight line $\gamma : \mathbb{R} \to X$ passing through x_1, x_2 by

$$\gamma(t) = (1 - t)x_1 + tx_2.$$

Prove that if $f : X \to \mathbb{R}$ is differentiable at $\gamma(t_0)$, for some $t_0 \in \mathbb{R}$, then $f \circ \gamma : \mathbb{R} \to \mathbb{R}$ is differentiable at t_0 and

$$\frac{d}{dt}(f \circ \gamma)(t_0) = f'(\gamma(t_0))(x_2 - x_1).$$

Deduce that if $g : X \to \mathbb{R}$ is differentiable and $g'(x) = \mathbf{0}$ at every $x \in X$, then g is constant.

1.4. Fundamental Theorems of Optimization

From ordinary calculus, we know the following two facts that enable one to solve optimization problems for $f : \mathbb{R} \to \mathbb{R}$.

Fact 1. If $x_* \in \mathbb{R}$ is a minimizer of f, then $f'(x_*) = 0$.

Fact 2. If $f''(x) \geq 0$ for all $x \in \mathbb{R}$ and $f'(x_*) = 0$, then x_* is a minimizer of f.

The first fact gives a necessary condition for minimization (and allows one to narrow the possibilities for minimizers — together with the knowledge of the existence of a minimizer, this is a very useful result since it then tells us that the minimizer x_* *has to be* one which satisfies $f'(x_*) = 0$). On the other hand, the second fact gives a sufficient condition for minimization.

Analogously, we will prove the following two results in this section, but now for a real-valued function $f : X \to \mathbb{R}$ on a *normed space* X.

Fact 1. If $x_* \in X$ is a minimizer of f, then $f'(x_*) = \mathbf{0}$.

Fact 2. If f is convex and $f'(x_*) = 0$, then x_* is a minimizer of f.

Before proceeding, let us fix some terminology that we will use throughout the book, for a function $f : S \to \mathbb{R}$, and for an $x_* \in S$, which is given in the following table.

Phrase	Meaning
"f has a minimum at x_*" "x_* is a minimizer of f"	For all $x \in S$, $f(x) \geq f(x_*)$.
"f has a maximum at x_*" "x_* is a maximizer of f"	For all $x \in S$, $f(x_*) \geq f(x)$.

Also, we mention again that there is no loss of generality in assuming that we have a *minimization* problem, as opposed to a *maximization* one. (See Exercise 0.1.)

1.4.1. Optimization: necessity of vanishing derivative.

Theorem 1.42. *Let X be a normed space, and let $f : X \to \mathbb{R}$ be a function that is differentiable at $x_* \in X$. If f has a minimum at x_*, then $f'(x_*) = \mathbf{0}$.*

Let us first clarify what $\mathbf{0}$ above means: $\mathbf{0} : X \to \mathbb{R}$ is the continuous linear transformation that sends everything in X to the number $0 \in \mathbb{R}$:

$$\mathbf{0}(h) = 0, \quad h \in X.$$

It is good to keep in mind a picture for this, as shown in Figure 8.

So to say that "$f'(x_*) = \mathbf{0}$" is the same as saying that for all $h \in X$, $(f'(x_*))(h) = 0$.

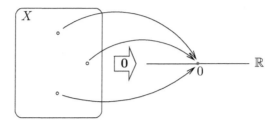

Figure 8. The zero linear transformation **0**.

Proof. Suppose that $f'(x_*) \neq \mathbf{0}$. Then there exists a vector $h_0 \in X$ such that $(f'(x_*))(h_0) \neq 0$. Clearly this h_0 must be a nonzero vector (because the linear transformation $f'(x_0)$ takes the zero vector in X to the zero vector in $\mathbb{R}$, which is 0). Let $\epsilon > 0$. Then there exists a $\delta > 0$ such that whenever $x \in X$ satisfies $0 < \|x - x_*\| < \delta$, we have

$$\frac{|f(x) - f(x_*) - (f'(x_*))(x - x_*)|}{\|x - x_*\|} < \epsilon.$$

Thus whenever $0 < \|x - x_*\| < \delta$, we have

$$\frac{-(f'(x_*))(x - x_*)}{\|x - x_*\|} \leq \overbrace{\frac{f(x) - f(x_*) - (f'(x_*))(x - x_*)}{\|x - x_*\|}}^{\geq 0}$$
$$\leq \frac{|f(x) - f(x_*) - (f'(x_*))(x - x_*)|}{\|x - x_*\|} < \epsilon.$$

Hence whenever $0 < \|x - x_*\| < \delta$, $-\dfrac{(f'(x_*))(x - x_*)}{\|x - x_*\|} < \epsilon$.

Now we will construct a special x using the h_0 from before. Take

$$x := x_* - \underbrace{\left(\frac{\delta}{2} \cdot \frac{f'(x_*)(h_0)}{|f'(x_*)(h_0)|} \cdot \frac{1}{\|h_0\|} \right)}_{\text{a scalar}} \cdot h_0 \in X.$$

Then $x \neq x_*$ and $\|x - x_*\| = \dfrac{\delta}{2} \cdot \dfrac{|f'(x_*)(h_0)|}{|f'(x_*)(h_0)|} \cdot \dfrac{1}{\|h_0\|} \|h_0\| = \dfrac{\delta}{2} < \delta.$

Using the linearity of $f'(x_0)$, we obtain

$$-\frac{(f'(x_*))(x - x_*)}{\|x - x_*\|} = \frac{\dfrac{\delta}{2} \cdot \dfrac{(f'(x_*)(h_0))^2}{|f'(x_*)(h_0)|} \cdot \dfrac{1}{\|h_0\|}}{\dfrac{\delta}{2}} < \epsilon.$$

Thus, $|f'(x_*)(h_0)| < \epsilon\|h_0\|$. But $\epsilon > 0$ was arbitrary. Hence $|f'(x_*)(h_0)| = 0$, and so we obtain $f'(x_*)(h_0) = 0$, a contradiction. This completes the proof. $\square$

We remark that this condition is a *necessary* condition for x_* to be a minimizer, but it is not sufficient. This is analogous to the situation to optimization in $\mathbb{R}$: if we look at $f : \mathbb{R} \to \mathbb{R}$ given by $f(x) = x^3$, $x \in \mathbb{R}$, then with $x_* := 0$, we have that $f'(x_*) = 3x_*^2 = 3 \cdot 0^2 = 0$, but clearly $x_* = 0$ is not a minimizer of f. See Figure 9.

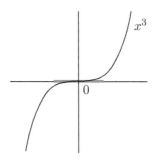

Figure 9. $f'(0) = 0$, but 0 is not a minimizer for the map f.

Example 1.43. Consider $f : C[0,1] \to \mathbb{R}$ given by $f(\mathbf{x}) = (\mathbf{x}(1))^3$, $\mathbf{x} \in C[0,1]$. Then $f'(\mathbf{0}) = \mathbf{0}$. (Here the $\mathbf{0}$ on the left-hand side is the zero function in $C[0,1]$, while the $\mathbf{0}$ on the right-hand side is the zero linear transformation $\mathbf{0} : C[0,1] \to \mathbb{R}$.) Indeed, given $\epsilon > 0$, we may set $\delta := \min\{\epsilon, 1\}$, and then we have that whenever $\mathbf{x} \in C[0,1]$ satisfies $0 < \|\mathbf{x} - \mathbf{0}\|_\infty < \delta$, there holds that

$$\frac{|f(\mathbf{x}) - f(\mathbf{0}) - \mathbf{0}(\mathbf{x} - \mathbf{0})|}{\|\mathbf{x} - \mathbf{0}\|_\infty} = \frac{|\mathbf{x}(1)^3|}{\|\mathbf{x} - \mathbf{0}\|_\infty} \leq \frac{\|\mathbf{x}\|_\infty^3}{\|\mathbf{x}\|_\infty} = \|\mathbf{x}\|_\infty^2 < \delta \cdot \delta \leq 1 \cdot \epsilon.$$

But $\mathbf{0}$ is not a minimizer for f. For example, with $\mathbf{x} := -\alpha \cdot \mathbf{1} \in C[0,1]$, where $\alpha > 0$, we have $f(\mathbf{x}) = (-\alpha)^3 = -\alpha^3 < 0 = f(\mathbf{0})$, showing that $\mathbf{0}$ is not[3] a minimizer.   ◊

Exercise 1.44. Let $f : C[a,b] \to \mathbb{R}$ be defined by

$$f(\mathbf{x}) = \int_0^1 (\mathbf{x}(t))^2 dt, \quad \mathbf{x} \in C[a,b],$$

which was considered in Example 1.36, where we showed that if $\mathbf{x}_0 \in C[a,b]$, then $f'(\mathbf{x}_0)$ is given by

$$(f'(\mathbf{x}_0))h = 2 \int_a^b \mathbf{x}_0(t)\mathbf{h}(t)dt, \quad \mathbf{h} \in C[a,b].$$

Find all $\mathbf{x}_0 \in C[a,b]$ for which $f'(\mathbf{x}_0) = \mathbf{0}$. If we know that $\mathbf{x}_* \in C[a,b]$ is a minimizer for f, what can we say about $\mathbf{x}_*$?

---

[3]In fact, not even a "local" minimizer because $\|\mathbf{x} - \mathbf{0}\|_\infty = \alpha$ can be chosen as small as we please.

1.4.2. Optimization: sufficiency in the convex case. In this section, we will show that if $f : X \to \mathbb{R}$ is a convex function, then a vanishing derivative at some point is enough to conclude that the function has a minimum at that point. Thus the second derivative condition

$$\text{``}f''(x) \geq 0 \text{ for all } x \in \mathbb{R}\text{''}$$

in the case of ordinary calculus when $X = \mathbb{R}$ is now replaced by the condition that

$$\text{``}f \text{ is convex''}$$

when X is a general normed space. We will see below that in the special case when $X = \mathbb{R}$ (and when f is twice continuously differentiable), convexity is precisely characterized by the second derivative condition above. We begin by giving the definition of a convex function.

Definition 1.45 (Convex set; convex function). Let X be a normed space.

(1) A subset $C \subset X$ is said to be a *convex set* if for every $x_1, x_2 \in C$, and all $\alpha \in (0, 1)$, $(1 - \alpha) \cdot x_1 + \alpha \cdot x_2 \in C$.

(2) Let C be a convex subset of X. A map $f : C \to \mathbb{R}$ is said to be a *convex function* if for every $x_1, x_2 \in C$, and all $\alpha \in (0, 1)$,

$$f((1 - \alpha) \cdot x_1 + \alpha \cdot x_2) \leq (1 - \alpha)f(x_1) + \alpha f(x_2). \tag{1.12}$$

The geometric interpretation of the inequality (1.12), when $X = \mathbb{R}$, is shown in Figure 10: the graph of a convex function lies above all possible chords.

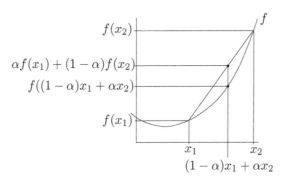

Figure 10. The convexity inequality.

Exercise 1.46. Let $a < b$, y_a, y_b be fixed real numbers, and define

$$S := \{ \mathbf{x} \in C^1[a, b] : \mathbf{x}(a) = y_a \text{ and } \mathbf{x}(b) = y_b \}.$$

Show that S is a convex set.

Exercise 1.47. If X is a normed space, then prove that the norm $x \mapsto \|x\| : X \to \mathbb{R}$ is a convex function.

Exercise 1.48. Let X be a normed space, C be a convex subset of X, and let $f : X \to \mathbb{R}$ be a function. Define the *epigraph* of f by

$$U(f) = \bigcup_{x \in C} \{(x, y) : y \geq f(x)\} \subset X \times \mathbb{R}.$$

This is the "region above the graph of f." Show that f is convex if and only if $U(f)$ is a convex subset of $X \times \mathbb{R}$.

Exercise 1.49. Suppose that $f : X \to \mathbb{R}$ is a convex function on a normed space X. If $n \in \mathbb{N}$, $x_1, \cdots, x_n \in X$, then

$$f\left(\frac{1}{n} \cdot (x_1 + \cdots + x_n)\right) \leq \frac{f(x_1) + \cdots + f(x_n)}{n}.$$

1.4.3. How to check convexity. Checking convexity will turn out to be useful in what follows. We will first learn about checking convexity of functions defined on $\mathbb{R}$, then in $\mathbb{R}^n$, and we will see examples to see how these results can be used to check convexity of commonly occurring classes of functions living on normed spaces like $C[a, b]$.

Convexity of functions living in $\mathbb{R}$. If one were to use the definition alone, then the verification can be cumbersome. Consider for example the function $f : \mathbb{R} \to \mathbb{R}$ given by $f(x) = x^2$, $x \in \mathbb{R}$. To verify that this function is convex, we note that for $x_1, x_2 \in \mathbb{R}$ and $\alpha \in (0, 1)$,

$$
\begin{aligned}
f((1-\alpha)x_1 + \alpha x_2) &= ((1-\alpha)x_1 + \alpha x_2)^2 \\
&= (1-\alpha)^2 x_1^2 + 2\alpha(1-\alpha)x_1 x_2 + \alpha^2 x_2^2 \\
&= (1-\alpha)x_1^2 + \alpha x_2^2 + ((1-\alpha)^2 - (1-\alpha))x_1^2 + (\alpha^2 - \alpha)x_2^2 \\
&\quad + 2\alpha(1-\alpha)x_1 x_2 \\
&= (1-\alpha)x_1^2 + \alpha x_2^2 - \alpha(1-\alpha)(x_1^2 + x_2^2 - 2x_1 x_2) \\
&= (1-\alpha)x_1^2 + \alpha x_2^2 - \alpha(1-\alpha)(x_1 - x_2)^2 \\
&\leq (1-\alpha)x_1^2 + \alpha x_2^2 - 0 = (1-\alpha)f(x_1) + \alpha f(x_2).
\end{aligned}
$$

On the other hand, we will now prove the following result.

Theorem 1.50. *Let $f : \mathbb{R} \to \mathbb{R}$ be twice continuously differentiable. Then f is convex if and only if for all $x \in \mathbb{R}$, $f''(x) \geq 0$.*

The convexity of $x \mapsto x^2$ is now immediate, as $\dfrac{d^2}{dx^2} x^2 = 2 > 0$, $x \in \mathbb{R}$.

Example 1.51. We have

$$\frac{d^2}{dx^2} e^x = e^x > 0, \quad x \in \mathbb{R},$$

and so $x \mapsto e^x$ is convex. Consequently, for all $x_1, x_2 \in \mathbb{R}$ and all $\alpha \in (0, 1)$, $e^{(1-\alpha)x_1 + \alpha x_2} \leq (1-\alpha)e^{x_1} + \alpha e^{x_2}$. $\Diamond$

Exercise 1.52. Consider $f : \mathbb{R} \to \mathbb{R}$ given by $f(x) = \sqrt{1 + x^2}$, $x \in \mathbb{R}$. Show that f is convex.

Proof of Theorem 1.50. ("Only if" part) Let $x < u < y$. Set

$$\alpha = \frac{u - x}{y - x}.$$

Then $\alpha \in (0, 1)$, and

$$1 - \alpha = \frac{y - u}{y - x}.$$

From the convexity of f,

$$\frac{y - u}{y - x} f(x) + \frac{u - x}{y - x} f(y) \geq f\left(\frac{y - u}{y - x} x + \frac{u - x}{y - x} y\right) = f(u),$$

that is,

$$(y - x)f(u) \leq (u - x)f(y) + (y - u)f(x). \tag{1.13}$$

From (1.13), we obtain $(y - x)f(u) \leq (u - x)f(y) + (y - x + x - u)f(x)$, that is,

$$(y - x)f(u) - (y - x)f(x) \leq (u - x)f(y) - (u - x)f(x),$$

and so

$$\frac{f(u) - f(x)}{u - x} \leq \frac{f(y) - f(x)}{y - x}. \tag{1.14}$$

From (1.13), we also have $(y - x)f(u) \leq (u - y + y - x)f(y) + (y - u)f(x)$, that is,

$$(y - x)f(u) - (y - x)f(y) \leq (u - y)f(y) - (u - y)f(x),$$

and so

$$\frac{f(y) - f(x)}{y - x} \leq \frac{f(y) - f(u)}{y - u}. \tag{1.15}$$

Combining (1.14) and (1.15), we obtain

$$\frac{f(u) - f(x)}{u - x} \leq \frac{f(y) - f(x)}{y - x} \leq \frac{f(y) - f(u)}{y - u}.$$

Passing the limit as $u \searrow x$ and $u \nearrow y$, we obtain

$$f'(x) \leq \frac{f(y) - f(x)}{y - x} \leq f'(y).$$

Hence f' is increasing, and so

$$f''(x) = \lim_{y \searrow x} \frac{f'(y) - f'(x)}{y - x} \geq 0.$$

Consequently, for all $x \in \mathbb{R}$, $f''(x) \geq 0$.

("If part") Since $f''(x) \geq 0$ for all $x \in \mathbb{R}$, it follows that f' is increasing. Indeed by the Fundamental Theorem of Calculus, for $x < y$,

$$f'(y) - f'(x) = \int_x^y \underbrace{f''(\xi)}_{\geq 0} \, d\xi \geq 0.$$

Now let $x, y \in \mathbb{R}$ be such that $x < y$, and let $\alpha \in (0, 1)$. Define $u := (1 - \alpha)x + \alpha y$. Then $x < u < y$. By the Mean Value Theorem,

$$\frac{f(u) - f(x)}{u - x} = f'(v) \text{ for some } v \in (x, u).$$

Similarly,

$$\frac{f(y) - f(u)}{y - u} = f'(w) \text{ for some } w \in (u, y).$$

As $w > v$, we have $f'(w) \geq f'(v)$, and so

$$\frac{f(y) - f(u)}{(1 - \alpha)(y - x)} = \frac{f(y) - f(u)}{y - u} \geq \frac{f(u) - f(x)}{u - x} = \frac{f(u) - f(x)}{\alpha(y - x)}.$$

Rearranging, we obtain

$$(1 - \alpha)f(x) + \alpha f(y) \geq (1 - \alpha)f(u) + \alpha f(u) = f(u) = f((1 - \alpha)x + \alpha y).$$

Thus f is convex.   □

Example 1.53. Consider the function $f : C[a, b] \to \mathbb{R}$ given by

$$f(\mathbf{x}) = \int_a^b (\mathbf{x}(t))^2 dt, \quad \mathbf{x} \in C[a, b].$$

We ask: Is f convex? We will show below that f is convex, using the convexity of the map $\xi \mapsto \xi^2 : \mathbb{R} \to \mathbb{R}$. Let $\mathbf{x}_1, \mathbf{x}_2 \in C[a, b]$ and $\alpha \in (0, 1)$. Then for all $a, b \in \mathbb{R}$, $((1 - \alpha)a + \alpha b)^2 \leq (1 - \alpha)a^2 + \alpha b^2$. Hence for each $t \in [a, b]$, with $a := \mathbf{x}_1(t)$, $b := \mathbf{x}_2(t)$, we obtain $((1 - \alpha)\mathbf{x}_1(t) + \alpha \mathbf{x}_2(t))^2 \leq (1 - \alpha)(\mathbf{x}_1(t))^2 + \alpha(\mathbf{x}_2(t))^2$. Thus

$$\begin{aligned}
f((1 - \alpha)\mathbf{x}_1 + \alpha \mathbf{x}_2) &= \int_a^b \Big((1 - \alpha)\mathbf{x}_1(t) + \alpha \mathbf{x}_2(t)\Big)^2 dt \\
&\leq \int_a^b \Big((1 - \alpha)(\mathbf{x}_1(t))^2 + \alpha(\mathbf{x}_2(t))^2\Big) dt \\
&= (1 - \alpha)\int_a^b (\mathbf{x}_1(t))^2 dt + \alpha \int_a^b (\mathbf{x}_2(t))^2 dt \\
&= (1 - \alpha)f(\mathbf{x}_1) + \alpha f(\mathbf{x}_2).
\end{aligned}$$

Consequently, f is convex.   ◊

Exercise 1.54. Let $f : C^1[0, 1] \to \mathbb{R}$ be given by

$$f(\mathbf{x}) = \int_0^1 \sqrt{1 + (\mathbf{x}'(t))^2} dt, \quad \mathbf{x} \in C^1[0, 1].$$

Prove that f is convex.

Convexity of functions living in $\mathbb{R}^n$. Here is a generalization of the second derivative test for convexity. First we give some terminology: we call a subset U of $\mathbb{R}^n$ *open* if for every $x_0 \in U$ there exists an $r > 0$ such that

$$B(x_0, r) := \{x \in \mathbb{R}^n : \|x - x_0\|_2 < r\} \subset U.$$

Theorem 1.55. *Let*

(1) *C be an open set,*

(2) *$f : C \to \mathbb{R}$ be a twice continuously differentiable function such that for all $x \in C$,*

$$H_f(x) := \begin{bmatrix} \dfrac{\partial^2 f}{\partial x_1^2}(x) & \cdots & \dfrac{\partial^2 f}{\partial x_1 \partial x_n}(x) \\ \vdots & \ddots & \vdots \\ \dfrac{\partial^2 f}{\partial x_n \partial x_1}(x) & \cdots & \dfrac{\partial^2 f}{\partial x_n^2}(x) \end{bmatrix}$$

is positive semi-definite[4],

Then f is convex.

$H_f(x)$ is called the *Hessian of f at x*.

Proof. Let $x, y \in C$ and $d := y - x$. Let φ be defined by $\varphi(t) := f(x + td)$, $t \in [0, 1]$. Then we have

$$\varphi'(t) = \nabla f(x + td) \cdot d, \text{ and } \varphi''(t) = d^\top H_f(x + td)d.$$

By Taylor's Formula, $\varphi(1) = \varphi(0) + \varphi'(0) + \dfrac{\varphi''(\theta)}{2}$, for some $\theta \in (0, 1)$. Thus

$$f(y) = f(x) + \nabla f(x) \cdot d + \frac{1}{2} \underbrace{d^\top H_f(x + \theta d)d}_{\geq 0},$$

and so $f(y) \geq f(x) + \nabla f \cdot (y - x)$. Let $u, v \in C$, and $t \in (0, 1)$. Fix $x = (1 - t)u + tv$. And first let us take $y = u$ to obtain

$$f(u) \geq f(x) + t\nabla f(x) \cdot (u - v). \tag{1.16}$$

Next take $y = v$ to obtain

$$f(v) \geq f(x) - (1 - t)\nabla f(x) \cdot (u - v). \tag{1.17}$$

By multiplying (1.16) by $1 - t$, (1.17) by t, and adding the resulting inequalities, we get that

$$(1 - t)f(u) + tf(v) \geq (1 - t)f(x) + tf(x) + 0 = f(x) = f((1 - t)u + tv).$$

So f is convex. $\qquad\square$

---

[4]That is, $v^\top H_f(x)v \geq 0$ for all $v \in \mathbb{R}^n$

We will consider an example below, but first we make a linear algebraic digression on how one checks positive semi-definiteness of symmetric matrices. If $M \in \mathbb{R}^{n \times n}$ is a *symmetric matrix* (that is, M is equal to its own *transpose* $M^\top$), then it is easy to see that M is positive semi-definite if and only if for any $k \in \mathbb{N}$ and any elementary matrices $E_1, \cdots, E_k$ (corresponding to elementary row transformations; see for example [**Ar**, pages 9-18]), $E_k \cdots E_2 E_1 M E_1^\top E_2^\top \cdots E_k^\top$ is positive semi-definite. Thus if there are a bunch of elementary matrices $E_1, \cdots, E_k$ such that $E_k \cdots E_2 E_1 M E_1^\top E_2^\top \cdots E_k^\top$ is a *diagonal* matrix, then it is easy to check the positive semi-definiteness of M. Recall that a matrix E is called an *elementary matrix* if it is of one of the following three types:

(1) For $i \neq j$, E is a matrix with α ($\in \mathbb{R}$) in the ith row and jth column (the (i, j)th entry), all diagonal entries equal to 1, and all remaining entries equal to 0, that is,

$$
E = \begin{bmatrix} 1 & & & & \\ & \ddots & & & \\ & & \alpha & \ddots & \\ & & & & 1 \end{bmatrix}.
$$

Then EM is the matrix obtained by performing the row operation on M of adding $\alpha \cdot (\text{row } j)$ to $(\text{row } i)$ of M. Similarly, $ME^\top$ corresponds to performing the column operation on M of adding $\alpha \cdot (\text{column } j)$ to $(\text{column } i)$ of M.

(2) For $i \neq j$, E has (i, j) and (j, i) entries equal to 1, the diagonal entries (i, i) and (j, j) equal to 0, all other diagonal entries equal to 1, and all remaining entries equal to 0, that is,

$$
E = \begin{bmatrix} 1 & & & & & & & & & \\ & \ddots & & & & & & & & \\ & & 1 & & & & & & & \\ & & & 0 & & 1 & & & & \\ & & & & 1 & & & & & \\ & & & & & \ddots & & & & \\ & & & & & & 1 & & & \\ & & & 1 & & & & 0 & & \\ & & & & & & & & 1 & \\ & & & & & & & & \ddots & \\ & & & & & & & & & 1 \end{bmatrix}.
$$

Then EM is the matrix obtained by performing the row operation on M of interchanging $(\text{row } j)$ with $(\text{row } i)$ of M. Similarly, $ME^\top$ corresponds to performing the column operation on M of interchanging $(\text{column } j)$ with $(\text{column } i)$ of M.

(3) E is a diagonal matrix with a nonzero scalar c as the (i,i)th diagonal entry, and all other diagonal entries equal to 1, that is,

$$E = \begin{bmatrix} 1 & & & & & & \\ & \ddots & & & & & \\ & & 1 & & & & \\ & & & c & & & \\ & & & & 1 & & \\ & & & & & \ddots & \\ & & & & & & 1 \end{bmatrix}.$$

Then EM is the matrix obtained by performing the row operation on M of multiplying (row i) of M by c. Similarly, $ME^{\top}$ corresponds to performing the column operation on M of multiplying (column i) of M by c.

(We remark that this method of checking positive semi-definiteness M, relying on elementary matrices to bring M to a diagonal form, is superior to the traditional way of diagonalizing M using eigenvalues and eigenvectors. Indeed, finding eigenvalues by hand may not be possible for big matrices having a characteristic polynomial of a high degree!)

Example 1.56. Let $F : \mathbb{R}^2 \to \mathbb{R}$ be given by $F(\xi, \eta) = \xi^2 + \xi\eta + \eta^2$, $(\xi, \eta) \in \mathbb{R}^2$. Then

$$\frac{\partial F}{\partial \xi} = 2\xi + \eta \text{ and } \frac{\partial F}{\partial \eta} = \xi + 2\eta.$$

Thus

$$H_F(\xi, \eta) = \begin{bmatrix} 2 & 1 \\ 1 & 2 \end{bmatrix}.$$

For a suitable elementary matrix E,

$$EH_F(\xi, \eta) = \begin{bmatrix} 2 & 1 \\ 0 & 3/2 \end{bmatrix},$$

and

$$EH_F(\xi, \eta)E^{\top} = \begin{bmatrix} 2 & 0 \\ 0 & 3/2 \end{bmatrix}.$$

As $2, 3/2 > 0$, it follows that $H_F(\xi, \eta)$ is positive semidefinite for all $(\xi, \eta) \in \mathbb{R}^2$. Consequently, F is convex. $\Diamond$

Example 1.57. Let $f : C^1[a, b] \to \mathbb{R}$ be given by

$$f(\mathbf{x}) = \int_a^b \left((\mathbf{x}(t))^2 + \mathbf{x}(t)\mathbf{x}'(t) + (\mathbf{x}'(t))^2 \right) dt, \quad \mathbf{x} \in C^1[a, b].$$

Then f is convex. Let

$$F(\xi, \eta) := \xi^2 + \xi\eta + \eta^2, \quad (\xi, \eta) \in \mathbb{R}^2.$$

Then

$$f(\mathbf{x}) = \int_a^b F(\mathbf{x}(t), \mathbf{x}'(t)) dt.$$

Let $\mathbf{x}_1, \mathbf{x}_2 \in C^1[a, b]$ and $\alpha \in (0, 1)$. Then owing to the convexity of F, we have

$$f((1-\alpha)\mathbf{x}_1 + \alpha\mathbf{x}_2)$$

$$= \int_a^b F\Big((1-\alpha)\mathbf{x}_1(t) + \alpha\mathbf{x}_2(t), (1-\alpha)\mathbf{x}_1'(t) + \alpha\mathbf{x}_2'(t)\Big) dt$$

$$= \int_a^b F\Big((1-\alpha)(\mathbf{x}_1(t), \mathbf{x}_1'(t)) + \alpha(\mathbf{x}_2(t), \mathbf{x}_2'(t))\Big) dt$$

$$\le \int_a^b \Big((1-\alpha)F(\mathbf{x}_1(t), \mathbf{x}_1'(t)) + \alpha F(\mathbf{x}_2(t), \mathbf{x}_2'(t))\Big) dt$$

$$= (1-\alpha)\int_a^b F(\mathbf{x}_1(t), \mathbf{x}_1'(t)) dt + \alpha \int_a^b F(\mathbf{x}_2(t), \mathbf{x}_2'(t)) dt$$

$$= (1-\alpha)f(\mathbf{x}_1) + \alpha f(\mathbf{x}_2).$$

So f is convex.   ◊

More generally, suppose that $F : \mathbb{R}^3 \to \mathbb{R}$ is convex. Let S be a convex subset of $C^1[a, b]$. Then $f : S \to \mathbb{R}$ defined by

$$f(\mathbf{x}) = \int_a^b F(\mathbf{x}(t), \mathbf{x}'(t), t) dt, \quad \mathbf{x} \in S,$$

is convex too. Indeed, we have for all $\mathbf{x}_1, \mathbf{x}_2 \in S$ and $\alpha \in (0, 1)$ that

$$f((1-\alpha)\mathbf{x}_1 + \alpha\mathbf{x}_2)$$

$$= \int_a^b F\Big((1-\alpha)\mathbf{x}_1(t) + \alpha\mathbf{x}_2(t), (1-\alpha)\mathbf{x}_1'(t) + \alpha\mathbf{x}_2'(t), t\Big) dt$$

$$= \int_a^b F\Big((1-\alpha)\mathbf{x}_1(t) + \alpha\mathbf{x}_2(t), (1-\alpha)\mathbf{x}_1'(t) + \alpha\mathbf{x}_2'(t), (1-\alpha)t + \alpha t\Big) dt$$

$$= \int_a^b F\Big((1-\alpha)(\mathbf{x}_1(t), \mathbf{x}_1'(t), t) + \alpha(\mathbf{x}_2(t), \mathbf{x}_2'(t), t)\Big) dt$$

$$\le \int_a^b \Big((1-\alpha)F(\mathbf{x}_1(t), \mathbf{x}_1'(t), t) + \alpha F(\mathbf{x}_2(t), \mathbf{x}_2'(t), t)\Big) dt$$

$$= (1-\alpha)\int_a^b F(\mathbf{x}_1(t), \mathbf{x}_1'(t), t) dt + \alpha \int_a^b F(\mathbf{x}_2(t), \mathbf{x}_2'(t), t) dt$$

$$= (1-\alpha)f(\mathbf{x}_1) + \alpha f(\mathbf{x}_2).$$

It may happen that F is not convex, but f happens to be convex. Here is an example.

Example 1.58. Consider the function used in Example 0.2. Recall that

$$S := \{\mathbf{x} \in C^1[0, T] : \mathbf{x}(0) = 0 \text{ and } \mathbf{x}(T) = Q\},$$

and $f : S \to \mathbb{R}$ was given by

$$f(\mathbf{x}) = \int_0^T (a\mathbf{x}(t) + b\mathbf{x}'(t))\mathbf{x}'(t)dt = \int_0^T F(\mathbf{x}(t), \mathbf{x}'(t))dt, \quad \mathbf{x} \in S,$$

where $a, b, Q > 0$ are constants, and $F : \mathbb{R}^2 \to \mathbb{R}$ is given by $F(\xi, \eta) = (a\xi + b\eta)\eta$, $(\xi, \eta) \in \mathbb{R}^2$. Then F is not convex in $\mathbb{R}^2$: for example,

$$\begin{aligned}
F(2b, -a) &= (a \cdot 2b + b \cdot (-a)) \cdot (-a) = -a^2b, \\
F(-2b, a) &= (a \cdot (-2b) + b \cdot a) \cdot a = -a^2b, \\
F(0, 0) &= 0,
\end{aligned}$$

and

$$F\left(\frac{(2b, -a) + (-2b, a)}{2}\right) = F(0, 0) = 0 \nleq -a^2b = \frac{F(2b, -a) + F(-2b, a)}{2}.$$

Nevertheless, let us check that f is convex. The convexity of the map

$$\mathbf{x} \mapsto b \int_0^T (\mathbf{x}'(t))^2 dt : S \to \mathbb{R}$$

follows from the convexity of $\eta \mapsto \eta^2 : \mathbb{R} \to \mathbb{R}$. The map

$$\mathbf{x} \mapsto a \int_0^T \mathbf{x}(t)\mathbf{x}'(t)dt : S \to \mathbb{R}$$

is constant on S because

$$\int_0^T \mathbf{x}(t)\mathbf{x}'(t)dt = \frac{1}{2} \int_0^T \frac{d}{dt}(\mathbf{x}(t))^2 dt = \frac{(\mathbf{x}(T))^2 - (\mathbf{x}(0))^2}{2} = \frac{Q^2 - 0^2}{2} = \frac{Q^2}{2},$$

and so this map is trivially convex. Hence f, being the sum of two convex functions, is convex too.   ◇

Exercise 1.59. Let $C := \{(\xi, \eta) \in \mathbb{R}^2 : \xi = \eta\}$. Show that C is a convex set. Consider the map $f : C \to \mathbb{R}$ given by $f(\xi, \eta) = \xi \cdot \eta$, $(\xi, \eta) \in \mathbb{R}^2$. Prove that f is convex. Calculate the Hessian $H_f(\xi, \eta)$, and show that $H_f(\xi, \eta)$ is *not* positive semidefinite on C.

We now prove the following result on the sufficiency of the vanishing derivative for a minimizer in the case of convex functions.

Theorem 1.60. *Let X be a normed space and $f : X \to \mathbb{R}$ be convex and differentiable. If $x_* \in X$ is such that $f'(x_*) = \mathbf{0}$, then f has a minimum at x_*.*

Proof. Suppose that $x_0 \in X$ and $f(x_0) < f(x_*)$. Define $\varphi : \mathbb{R} \to \mathbb{R}$ by

$$\varphi(t) = f(tx_0 + (1-t)x_*), \quad t \in \mathbb{R}.$$

The function φ is convex, since if $\alpha \in (0,1)$ and $t_1, t_2 \in \mathbb{R}$, then we have

$$\varphi((1-\alpha)t_1 + \alpha t_2) = f\left(\left((1-\alpha)t_1 + \alpha t_2\right)x_0 + \left(1 - (1-\alpha)t_1 - \alpha t_2\right)x_*\right)$$

$$= f\left((1-\alpha)(t_1 x_0 - t_1 x_*) + \alpha(t_2 x_0 - t_2 x_*) + x_*\right)$$

$$= f\left((1-\alpha)(t_1 x_0 - t_1 x_*) + \alpha(t_2 x_0 - t_2 x_*) + (1-\alpha)x_* + \alpha x_*\right)$$

$$= f\left((1-\alpha)(t_1 x_0 - t_1 x_* + x_*) + \alpha(t_2 x_0 - t_2 x_* + x_*)\right)$$

$$= f\left((1-\alpha)(t_1 x_0 + (1-t_1)x_*) + \alpha(t_2 x_0 + (1-t_2)x_*)\right)$$

$$\leq (1-\alpha)f\left(t_1 x_0 + (1-t_1)x_*\right) + \alpha f\left(t_2 x_0 + (1-t_2)x_*\right)$$

$$= (1-\alpha)\varphi(t_1) + \alpha\varphi(t_2).$$

Also, from Exercise 1.41 on page 32, it follows that φ is differentiable at 0, and

$$\varphi'(0) = f'(x_*)(x_0 - x_*) = \mathbf{0}(x_0 - x_*) = 0.$$

Since $\varphi(1) = f(x_0) < f(x_*) = \varphi(0)$, it follows from the Mean Value Theorem that there exists a $\theta \in (0,1)$ such that

$$\varphi'(\theta) = \frac{\varphi(1) - \varphi(0)}{1 - 0} < 0 = \varphi'(0).$$

But this is a contradiction because φ is convex, and so φ' must be increasing (see the proof of the "only if" part of Theorem 1.50). Thus there cannot exist an $x_0 \in X$ such that $f(x_0) < f(x_*)$. Consequently, f has a minimum at x_*. $\qquad\square$

Exercise 1.61. Consider the function $f : C[0,1] \to \mathbb{R}$ defined by

$$f(\mathbf{x}) = \int_0^1 (\mathbf{x}(t))^2 dt, \quad \mathbf{x} \in C[0,1].$$

Let $\mathbf{x}_0 \in C[0,1]$. From Example 1.36, we know that $f'(\mathbf{x}_0) : C[0,1] \to \mathbb{R}$ is given by

$$(f'(\mathbf{x}_0))(\mathbf{h}) = 2 \int_0^1 \mathbf{x}_0(t)\mathbf{h}(t) dt, \quad \mathbf{h} \in C[0,1].$$

Prove that $f'(\mathbf{x}_0) = \mathbf{0}$ if and only if $\mathbf{x}_0(t) = 0$ for all $t \in [0,1]$. We have also seen that in Example 1.53, that f is a convex function. Find all solutions to the optimization problem

$$\begin{cases} \text{minimize} & f(\mathbf{x}) \\ \text{subject to} & \mathbf{x} \in C[0,1]. \end{cases}$$

1.5. What We Will Do Next

We quickly outline what we will do in the next chapter in order to prepare the reader for what is going to be done there. We will apply the two optimization results we have just learnt to specific scenarios. For example, we will learn the following result.

Theorem 1.62. *Let*

(1) $S = \{\mathbf{x} \in C^1[a, b] : \mathbf{x}(a) = y_a, \ \mathbf{x}(b) = y_b\}$,

(2) $F : \mathbb{R}^3 \to \mathbb{R}$, $(\xi, \eta, \tau) \overset{F}{\mapsto} F(\xi, \eta, \tau)$,

(3) $f : S \to \mathbb{R}$ *be given by* $f(\mathbf{x}) = \int_a^b F(\mathbf{x}(t), \mathbf{x}'(t), t)dt$, $\mathbf{x} \in S$.

Then we have:

(1) *If* $\mathbf{x}_*$ *is a minimizer of* f, *then it satisfies the* Euler-Lagrange equation:

$$\frac{\partial F}{\partial \xi}(\mathbf{x}_*(t), \mathbf{x}'_*(t), t) - \frac{d}{dt}\left(\frac{\partial F}{\partial \eta}(\mathbf{x}_*(t), \mathbf{x}'_*(t), t)\right) = 0 \ \text{for all} \ t \in [a, b].$$

(2) *If* f *is convex, and* $\mathbf{x}_* \in S$ *satisfies the Euler-Lagrange equation, then* $\mathbf{x}_*$ *is a minimizer of* f.

So instead of working in a general normed space X, we consider particular subsets (like S) of a concrete normed space (like $X = C^1[a, b]$), and instead of a general function f, we will work with a function having a specific form (for example, the f specified in statement (3) of above result, where f is given by an integral). For such concrete situations, we will apply Theorems 1.42 and 1.60 to obtain results of the type mentioned above. These special results then cover a wide range of examples that appear in practice. For instance, let us revisit Example 0.2, and solve it by observing that it falls in the class of problems considered in the above result.

Example 1.63. Recall that

$$S := \{\mathbf{x} \in C^1[0, T] : \mathbf{x}(0) = 0 \ \text{and} \ \mathbf{x}(T) = Q\},$$

so that we have $a = 0$, $b = T$, $y_a = 0$ and $y_b = Q$. $f : S \to \mathbb{R}$ was given by

$$f(\mathbf{x}) = \int_0^T (a\mathbf{x}(t) + b\mathbf{x}'(t))\mathbf{x}'(t)dt = \int_0^T F(\mathbf{x}(t), \mathbf{x}'(t), t)dt, \quad \mathbf{x} \in S,$$

where $a, b, Q > 0$ are constants, and $F : \mathbb{R}^3 \to \mathbb{R}$ is given by

$$F(\xi, \eta, \tau) = (a\xi + b\eta)\eta, \quad (\xi, \eta, \tau) \in \mathbb{R}^3.$$

So this problem does fall into the class of problems covered by Theorem 1.62. In order to apply the result to solve this problem, we compute

$$\frac{\partial F}{\partial \xi}(\xi, \eta, \tau) = a\eta,$$

$$\frac{\partial F}{\partial \eta}(\xi, \eta, \tau) = a\xi + 2b\eta.$$

Thus if $x_* \in S$, then

$$\frac{\partial F}{\partial \xi}(\mathbf{x}_*(t), \mathbf{x}'_*(t), t) = a\mathbf{x}'_*(t),$$

$$\frac{\partial F}{\partial \eta}(\mathbf{x}_*(t), \mathbf{x}'_*(t), t) = a\mathbf{x}_*(t) + 2b\mathbf{x}'_*(t).$$

The Euler-Lagrange equation is:

$$\frac{\partial F}{\partial \xi}(\mathbf{x}_*(t), \mathbf{x}'_*(t), t) - \frac{d}{dt}\left(\frac{\partial F}{\partial \eta}(\mathbf{x}_*(t), \mathbf{x}'_*(t), t)\right) = 0$$

that is, $a\mathbf{x}'_*(t) - \frac{d}{dt}(a\mathbf{x}_*(t) + 2b\mathbf{x}'_*(t)) = 0$ for all $t \in [0, T]$. Thus

$$\frac{d}{dt}(2b\mathbf{x}'_*(t)) = 0, \quad t \in [0, T].$$

By the Fundamental Theorem of Calculus, it follows that there is a constant A such that $\mathbf{x}'_*(t) = A$, $t \in [0, T]$, and integrating again, we obtain a constant B such that $\mathbf{x}_*(t) = At + B$, $t \in [0, T]$. But since $\mathbf{x}_* \in S$, we also have that $\mathbf{x}_*(0) = 0$ and $\mathbf{x}_*(T) = Q$, which we can use to find the constants A, B:

$$A \cdot 0 + B = 0,$$
$$A \cdot T + B = Q,$$

so that $B = 0$ and $A = Q/T$. Consequently, by part (1) of the conclusion in Theorem 1.62, we know that *if* $\mathbf{x}_*$ is a minimizer of f, then

$$\mathbf{x}_*(t) = Q\frac{t}{T}, \quad t \in [0, T].$$

On the other hand, we had checked in Example 1.58 that f is convex. And we know that the $\mathbf{x}_*$ given above satisfies the Euler-Lagrange equation. Consequently, by part (2) of the conclusion in Theorem 1.62, we know that this $\mathbf{x}_*$ is a minimizer. So we have shown, using Theorem 1.62, that

$$\boxed{\mathbf{x}_* \text{ is a minimizer of } f} \quad \Leftrightarrow \quad \boxed{\mathbf{x}_*(t) = Q\frac{t}{T}, t \in [0, T].}$$

So we have solved our optimal mining question, and we now know that the optimal mining operation is the humble straight line! See Figure 11. ◊

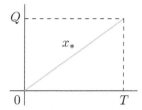

Figure 11. The optimal mining operation.

1.6. A Historical Episode

Loosely speaking, "Calculus of Variations" is the subject where one deals with optimization problems when the domain of the function to be optimized is itself a set of functions. The name comes from the fact that often the procedure involved the calculation of the "variation" in the function when its argument (which was typically a curve) was changed, and then passing limits. The history of calculus of variations is deeply interwoven with the history of mathematics, with problems of this nature dealt with since antiquity. We mention one important milestone, called the Brachistochrone Problem. The word "brachistochrone" means "shortest time" (from the Greek *brachystos*, shortest, and *chronos*, time). The problem was posed by Johann Bernoulli (1667-1748) as a challenge to the mathematical world in June 1696 in *Acta Eruditorum*:

> "**Invitation to all mathematicians to solve a new problem.** If in a vertical plane two points A and B are given, then it is required to specify the path AMB of the movable point M, along which it, starting from A, and under the influence of its own weight, arrives at B in the shortest possible time.
>
> So that those who are keen of such matters will be tempted to solve this problem, is it good to know that it is not, as it may seem, purely speculative and without practical use. Rather it even appears, and this may seem hard to believe, that it is very useful also for other branches of science than mechanics.
>
> In order to avoid a hasty conclusion, it should be remarked that the straight line is certainly the line of shortest distance between A and B, but it is not the one which is travelled in the shortest time.
>
> However, the curve AMB— which I shall divulge if by the end of the year nobody else has found it— is very well known among geometers."

Let us note that the problem is indeed one of the type we are dealing with in this book: corresponding to each shape of the curve (that is, a function), we have an associated number (the time of travel). And the question is: which curve gives the smallest time of travel. The fact that the straight line is not the minimizing curve is also intuitively clear, since if the slope is high at the beginning, the body picks up a high speed and so its plausible that the travel time could be reduced) and it can be verified experimentally by sliding beads down wires in various shapes. Let us make the problem more precise by giving a mathematical formulation. To this end, let us look at Figure 12, which shows the point A, which we choose as the origin, and the point B with coordinates, say (x_0, y_0), in the vertical plane. We will take the y-axis to be oriented positively in the downward direction.

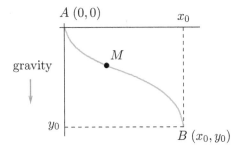

Figure 12. The brachistochrone problem.

As the particle is released from rest at A, conservation of energy gives

$$\frac{1}{2}mv^2 = mgy,$$

where we have taken the zero potential energy level at $y = 0$, and where v denotes the speed of the particle. Thus the speed is given by

$$v = \frac{ds}{dt} = \sqrt{2gy},$$

where s denotes arc length along the curve. From Figure 13, we see using Pythagoras's Theorem that an element of arc length

$$ds = \sqrt{(dx)^2 + (dy)^2} = \sqrt{\left(\frac{dx}{dy}\right)^2 + 1} \cdot dy = \sqrt{1 + (\mathbf{x}'(y))^2}\, dy.$$

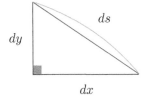

Figure 13. Element of arc length.

Hence the total time of travel from A to B is given by

$$T = \int_{\text{curve}} \frac{ds}{\sqrt{2gy}} = \int_0^{y_0} \sqrt{\frac{1 + (\mathbf{x}'(y))^2}{2gy}} dy.$$

Our problem is to find the path $\mathbf{x}$, $y \mapsto \mathbf{x}(y) : [0, y_0] \to \mathbb{R}$, satisfying $\mathbf{x}(0) = 0$ and $\mathbf{x}(y_0) = x_0$, which minimizes T, that is, to determine the minimizer for the function $f : S \to \mathbb{R}$, where $S = \{\mathbf{x} \in C^1[0, y_0] : \mathbf{x}(0) = 0 \text{ and } \mathbf{x}(y_0) = x_0\}$ and

$$f(\mathbf{x}) = \int_0^{y_0} \sqrt{\frac{1 + (\mathbf{x}'(y))^2}{2gy}} dy, \quad \mathbf{x} \in S.$$

Thus we see that the problem fits in the class of problems considered in Theorem 1.62. Let us see if we can discover what the curve is. With

$$F(\xi, \eta, \tau) = \sqrt{\frac{1 + \eta^2}{2g\tau}},$$

we have that $f(\mathbf{x}) = \int_0^{y_0} F(\mathbf{x}(y), \mathbf{x}'(y), y) dy$. We compute

$$\frac{\partial F}{\partial \xi}(\xi, \eta, \tau) = 0,$$

$$\frac{\partial F}{\partial \eta}(\xi, \eta, \tau) = \frac{1}{\sqrt{2g\tau}} \cdot \frac{\eta}{\sqrt{1 + \eta^2}}.$$

Thus the Euler-Lagrange equation is:

$$0 - \frac{d}{dy}\left(\frac{1}{\sqrt{2gy}} \cdot \frac{\mathbf{x}'(y)}{\sqrt{1 + (\mathbf{x}'(y))^2}}\right) = 0, \quad y \in [0, y_0],$$

which gives, upon integrating,

$$\frac{\mathbf{x}'(y)}{\sqrt{1 + (\mathbf{x}'(y))^2}} \cdot \frac{1}{\sqrt{y}} = C,$$

in $[0, y_0]$, for some constant C. Also, $\mathbf{x}(0) = 0$ and $\mathbf{x}(y_0) = x_0$. It turns out that the solution to this is given (in parametric form) by

$$\mathbf{x}(\mathbf{y}(\Theta)) = r(\Theta - \sin \Theta), \tag{1.18}$$

$$\mathbf{y}(\Theta) = r(1 - \cos \Theta), \tag{1.19}$$

for $\Theta \in [0, \Theta_0]$, where r, Θ_0 are appropriate constants chosen so that

$$x_0 = r(\Theta_0 - \sin \Theta_0),$$

$$y_0 = r(1 - \cos \Theta_0).$$

How to solve the exotic differential equation we got from the Euler-Lagrange equation is something we won't discuss here. But let us at least check that the differential equation is indeed satisfied by the parametric form solutions we have given above. We have

$$\frac{d(\mathbf{x} \circ \mathbf{y})}{d\Theta} = r(1 - \cos\Theta), \quad \text{and} \quad \frac{dy}{d\Theta} = r\sin\Theta,$$

and so with $y = \mathbf{y}(\Theta)$, we obtain

$$\mathbf{x}'(y) = \frac{\dfrac{d(\mathbf{x} \circ \mathbf{y})}{d\Theta}}{\dfrac{d\mathbf{y}}{d\Theta}} = \frac{r(1 - \cos\Theta)}{r\sin\Theta} = \frac{1 - \cos\Theta}{\sin\Theta}.$$

Hence

$$1 + (\mathbf{x}'(y))^2 = 1 + \left(\frac{1 - \cos\Theta}{\sin\Theta}\right)^2 = \frac{(\sin\Theta)^2 + 1 - 2\cos\Theta + (\cos\Theta)^2}{(\sin\Theta)^2} = \frac{2(1 - \cos\Theta)}{(\sin\Theta)^2}.$$

Consequently,

$$\frac{\mathbf{x}'(y)}{\sqrt{1 + (\mathbf{x}'(y))^2}} \cdot \frac{1}{\sqrt{y}} = \frac{\dfrac{1 - \cos\Theta}{\sin\Theta}}{\sqrt{\dfrac{2(1 - \cos\Theta)}{(\sin\Theta)^2}}} \cdot \frac{1}{\sqrt{r(1 - \cos\Theta)}} = \frac{1}{\sqrt{2r}}.$$

So the Euler-Lagrange equation is satisfied by the curve x described in parametric form by (1.18) and (1.19). But what is the geometric description of this curve that Bernoulli alluded to? One year later, in June 1697, Bernoulli announced his solution:

"Thus I conclude that the brachystochrone is the ordinary cycloid."

The *cycloid* is a curve described by a point on a circle that rolls without slipping on a straight line. See Figure 14. If the circle has radius r, and the angle the radius makes with the vertical axis is Θ, then the position of the center of the circle is $(r\Theta, r)$, and so the position (x, y) of the point M on the rim is given by

$$x = r(\Theta - \sin\Theta),$$
$$y = r(1 - \cos\Theta).$$

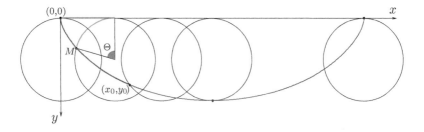

Figure 14. The cycloid through $(0,0)$ and (x_0, y_0).

Chapter 2

Euler-Lagrange Equation

In this chapter, we will give a necessary condition for a minimizer of a function of the type

$$f(\mathbf{x}) = \int_a^b F\left(\mathbf{x}(t), \mathbf{x}'(t), t\right) dt, \quad \mathbf{x} \in S$$

with various types of constraints (that is different classes of S). The necessary condition is in the form of a differential equation that the minimizer should satisfy, and this differential equation is called the *Euler-Lagrange equation*. We begin with the version we had mentioned at the end of Chapter 1.

2.1. Fixed Endpoints

We will prove the following result.

Theorem 2.1. *Suppose that*

(1) $S = \{\mathbf{x} \in C^1[a, b] : \mathbf{x}(a) = y_a,\ \mathbf{x}(b) = y_b\}$,

(2) $F : \mathbb{R}^3 \to \mathbb{R}$, $(\xi, \eta, \tau) \overset{F}{\mapsto} F(\xi, \eta, \tau)$, *has continuous partial derivatives of order 2,*

(3) $f : S \to \mathbb{R}$ *is given by* $f(\mathbf{x}) = \int_a^b F(\mathbf{x}(t), \mathbf{x}'(t), t)dt,\ \mathbf{x} \in S$.

Then we have:

(1) *If* $\mathbf{x}_*$ *is a minimizer of* f, *then it satisfies the* Euler-Lagrange equation:

$$\frac{\partial F}{\partial \xi}(\mathbf{x}_*(t), \mathbf{x}_*'(t), t) - \frac{d}{dt}\left(\frac{\partial F}{\partial \eta}(\mathbf{x}_*(t), \mathbf{x}_*'(t), t)\right) = 0 \text{ for all } t \in [a, b].$$

(2) *If* f *is convex, and* $\mathbf{x}_* \in S$ *satisfies the Euler-Lagrange equation, then* $\mathbf{x}_*$ *is a minimizer of* f.

In other words, the optimization problem consists of finding a minimizer of a function of $f : S \to \mathbb{R}$, where the class of curves in S consists of all smooth curves joining two *fixed* points; see Figure 1.

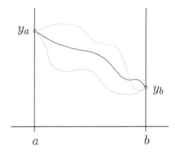

Figure 1. Curves with the fixed endpoints (a, y_a) and (b, y_b).

We will apply the necessary condition for a minimizer (established in Theorem 1.42) to solve the optimization problem described above. But first we will need the following technical result.

Lemma 2.2. *If* $\mathbf{k} \in C[a, b]$ *is such that*

$$\text{for all } \mathbf{h} \in C^1[a, b] \text{ with } \mathbf{h}(a) = \mathbf{h}(b) = 0, \int_a^b \mathbf{k}(t)\mathbf{h}'(t)dt = 0,$$

then there exists a constant c *such that* $\mathbf{k}(t) = c$ *for all* $t \in [a, b]$.

Of course, if $\mathbf{k} \equiv c$, then by the Fundamental Theorem of Calculus,

$$\int_a^b \mathbf{k}(t)\mathbf{h}'(t)dt = \int_a^b c\mathbf{h}'(t)dt = c\int_a^b \mathbf{h}'(t)dt = c(\mathbf{h}(b) - \mathbf{h}(a)) = c(0 - 0) = 0,$$

for all $\mathbf{h} \in C^1[a, b]$ that satisfy $\mathbf{h}(a) = \mathbf{h}(b) = 0$. The remarkable thing is that the converse is true, namely that the special property in the box forces $\mathbf{k}$ to be a constant.

Proof. Set $c := \dfrac{1}{b - a} \displaystyle\int_a^b \mathbf{k}(t)dt.$

(If $\mathbf{k} \equiv c$, then

$$\frac{1}{b - a}\int_a^b \mathbf{k}(t)dt = \frac{1}{b - a} \cdot c \cdot (b - a) = c;$$

so this c is the constant that $\mathbf{k}$ "is supposed to be.")

Define $\mathbf{h}_0 : [a, b] \to \mathbb{R}$ by $\mathbf{h}_0(t) = \displaystyle\int_a^t (\mathbf{k}(\tau) - c)d\tau.$

Then

(1) $\mathbf{h}_0 \in C^1[a, b]$,

(2) $\mathbf{h}_0(a) = \displaystyle\int_a^a (\mathbf{k}(\tau) - c)d\tau = 0$,

(3) $\mathbf{h}_0(b) = \displaystyle\int_a^b (\mathbf{k}(\tau) - c)d\tau = \int_a^b \mathbf{k}(\tau)d\tau - c \cdot (b - a) = 0$ (by definition of c).

Thus $\displaystyle\int_a^b \mathbf{k}(t)\mathbf{h}_0'(t)dt = 0$. Since $\mathbf{h}_0'(t) = \mathbf{k}(t) - c$, $t \in [a, b]$, we obtain

$$
\begin{aligned}
\int_a^b (\mathbf{k}(t) - c)^2 dt &= \int_a^b (\mathbf{k}(t) - c)(\mathbf{k}(t) - c)dt = \int_a^b (\mathbf{k}(t) - c)\mathbf{h}_0'(t)dt \\
&= \int_a^b \mathbf{k}(t)\mathbf{h}_0'(t)dt - \int_a^b c\mathbf{h}_0'(t)dt = 0 - c(\mathbf{h}_0(b) - \mathbf{h}_0(a)) = 0.
\end{aligned}
$$

Thus $\mathbf{k}(t) - c = 0$ for all $t \in [a, b]$, and so $\mathbf{k} \equiv c$. $\qquad\square$

Proof of Theorem 2.1. (1) The proof is long and so we divide it into several steps. We will use the fact that if f is a real-valued function on a normed *space* that has a minimum at $\mathbf{x}_*$, then $f'(\mathbf{x}_*) = \mathbf{0}$. But our S is not a *vector space* (unless $y_a = y_b = 0$; see Exercise 1.3). So we can't apply this result directly.

Step 1. From S to a vector space X. To remedy the problem that S is not a vector space, we introduce a new vector space X, and consider a new function $\tilde{f} : X \to \mathbb{R}$ that is defined in terms of the old function f. Introduce the vector space

$$X = \{\mathbf{h} \in C^1[a, b] : \mathbf{h}(a) = \mathbf{h}(b) = 0\} \ (\subset C^1[a, b]),$$

with the induced norm from $C^1[a, b]$, that is, $\|\mathbf{h}\|_{1,\infty} = \|\mathbf{h}\|_\infty + \|\mathbf{h}'\|_\infty$, $\mathbf{h} \in X$ and where $\|\cdot\|_\infty$ denotes the usual supremum norm. Then X is a normed space, and

$$\mathbf{h} \in X \quad \Leftrightarrow \quad \mathbf{x}_* + \mathbf{h} \in S.$$

(Indeed, if $h \in X$, then

$$
\begin{aligned}
(\mathbf{x}_* + \mathbf{h})(a) &= \mathbf{x}_*(a) + \mathbf{h}(a) = y_a + 0 = y_a, \\
(\mathbf{x}_* + \mathbf{h})(b) &= \mathbf{x}_*(b) + \mathbf{h}(b) = y_b + 0 = y_b,
\end{aligned}
$$

and so $\mathbf{x}_* + \mathbf{h} \in S$. Vice versa, if $\mathbf{x}_* + \mathbf{h} \in S$, then

$$
\begin{aligned}
(\mathbf{x}_* + \mathbf{h})(a) &= y_a \text{ implies } y_a + \mathbf{h}(a) = y_a, \text{ and so } h(a) = 0, \\
(\mathbf{x}_* + \mathbf{h})(b) &= y_b \text{ implies } y_b + \mathbf{h}(b) = y_b \text{ and so } h(b) = 0.
\end{aligned}
$$

Consequently, $\mathbf{h} \in X$.)

Now define $\tilde{f} : X \to \mathbb{R}$ by

$$\tilde{f}(\mathbf{h}) = f(\underbrace{\mathbf{x}_* + \mathbf{h}}_{\in S}), \quad \mathbf{h} \in X.$$

Thus $\widetilde{f}$ is well-defined. We claim that $\widetilde{f}$ has a minimum at $\mathbf{0} \in X$. Indeed, for $\mathbf{h} \in X$, we have

$$\widetilde{f}(\mathbf{h}) = f(\mathbf{x}_* + \mathbf{h}) \geq f(\mathbf{x}_*) = f(\mathbf{x}_* + \mathbf{0}) = \widetilde{f}(\mathbf{0}).$$

So now we are in a position to apply Theorem 1.42! Since $\mathbf{0} \in X$ is a minimizer of $\widetilde{f}$, we must have $\widetilde{f}'(\mathbf{0}) = \mathbf{0}$.

Step 2. What is $\widetilde{f}'(\mathbf{0})$? We now calculate $\widetilde{f}'(\mathbf{0})$. For $\mathbf{h} \in X$, we have

$$
\begin{aligned}
\widetilde{f}(\mathbf{h}) - \widetilde{f}(\mathbf{0}) &= f(\mathbf{x}_* + \mathbf{h}) - f(\mathbf{x}_*) \\
&= \int_a^b \Big(F(\mathbf{x}_*(t) + \mathbf{h}(t), \mathbf{x}_*'(t) + \mathbf{h}'(t), t) - F(\mathbf{x}_*(t), \mathbf{x}_*'(t), t) \Big) dt.
\end{aligned}
$$

By Taylor's Formula for F, we know that

$$
\begin{aligned}
F(\xi_0 + p, \eta_0 + q, \tau_0 + r) &- F(\xi_0, \eta_0, \tau_0) \\
&= p \frac{\partial F}{\partial \xi}(\xi_0, \eta_0, \tau_0) + q \frac{\partial F}{\partial \eta}(\xi_0, \eta_0, \tau_0) + r \frac{\partial F}{\partial \tau}(\xi_0, \eta_0, \tau_0) \\
&\quad + \frac{1}{2!} \begin{bmatrix} p & q & r \end{bmatrix} H_F(\xi_0 + \theta p, \eta_0 + \theta q, \tau_0 + \theta r) \begin{bmatrix} p \\ q \\ r \end{bmatrix}
\end{aligned}
$$

for some θ such that $0 < \theta < 1$. We will apply this for each fixed $t \in [a, b]$, with $\xi_0 := \mathbf{x}_*(t)$, $p := \mathbf{h}(t)$, $\eta_0 := \mathbf{x}_*'(t)$, $q := \mathbf{h}'(t)$, $\tau_0 := t$, $r := 0$, and we will obtain a $\theta \in (0, 1)$ for which the above formula works. If I change the t, then I will get a different $\theta \in (0, 1)$. So we have that the θ depends on $t \in [a, b]$. This gives rise to a function $\Theta : [a, b] \to (0, 1)$ so that

$$
\begin{aligned}
\widetilde{f}(h) &- \widetilde{f}(\mathbf{0}) \\
&= \int_a^b \Big(F(\mathbf{x}_*(t) + \mathbf{h}(t), \mathbf{x}_*'(t) + \mathbf{h}'(t), t) - F(\mathbf{x}_*(t), \mathbf{x}_*'(t), t) \Big) dt \\
&= \int_a^b \Big(\mathbf{h}(t) \frac{\partial F}{\partial \xi}(\mathbf{x}_*(t), \mathbf{x}_*'(t), t) + \mathbf{h}'(t) \frac{\partial F}{\partial \eta}(\mathbf{x}_*(t), \mathbf{x}_*'(t), t) + 0 \cdot \frac{\partial F}{\partial \tau}(\mathbf{x}_*(t), \mathbf{x}_*'(t), t) \\
&\quad + \frac{1}{2!} \begin{bmatrix} \mathbf{h}(t) & \mathbf{h}'(t) & 0 \end{bmatrix} H_F(\mathbf{x}_*(t) + \Theta(t)\mathbf{h}(t), x_*'(t) + \Theta(t)\mathbf{h}'(t), t) \begin{bmatrix} \mathbf{h}(t) \\ \mathbf{h}'(t) \\ 0 \end{bmatrix} \Big) dt \\
&= \int_a^b \Big(\mathbf{A}(t)\mathbf{h}(t) + \mathbf{B}(t)\mathbf{h}'(t) \Big) dt + \int_a^b \frac{1}{2!} \begin{bmatrix} \mathbf{h}(t) & \mathbf{h}'(t) & 0 \end{bmatrix} H_F(\mathbf{P}(t)) \begin{bmatrix} \mathbf{h}(t) \\ \mathbf{h}'(t) \\ 0 \end{bmatrix} \Big) dt,
\end{aligned}
$$

where

$$\mathbf{A}(t) = \frac{\partial F}{\partial \xi}(\mathbf{x}_*(t), \mathbf{x}'_*(t), t),$$

$$\mathbf{B}(t) = \frac{\partial F}{\partial \eta}(\mathbf{x}_*(t), \mathbf{x}'_*(t), t), \text{ and}$$

$$\mathbf{P}(t) = (\mathbf{x}_*(t) + \Theta(t)\mathbf{h}(t), \mathbf{x}'_*(t) + \Theta(t)\mathbf{h}'(t), t)$$

and $H_F(\cdot)$ denotes the Hessian of F. From the above expression, we make a guess for $\widetilde{f}'(0)$. Define $L : X \to \mathbb{R}$ by

$$L(\mathbf{h}) = \int_a^b \Big(\mathbf{A}(t)\mathbf{h}(t) + \mathbf{B}(t)\mathbf{h}'(t) \Big) dt, \quad \mathbf{h} \in X.$$

We have seen that L is a continuous linear transformation in Examples 1.26 and 1.30. For $\mathbf{h} \in X$,

$$|\widetilde{f}(\mathbf{h}) - \widetilde{f}(0) - L(\mathbf{h} - 0)|$$

$$= \left| \frac{1}{2} \int_a^b (\mathbf{h}(t))^2 \frac{\partial^2 F}{\partial \xi^2}(\mathbf{P}(t)) + 2\mathbf{h}(t)\mathbf{h}'(t)\frac{\partial^2 F}{\partial \xi \partial \eta}(\mathbf{P}(t)) + (\mathbf{h}'(t))^2 \frac{\partial^2 F}{\partial \eta^2}(\mathbf{P}(t)) dt \right|$$

$$\leq \frac{1}{2} \int_a^b |\mathbf{h}(t)|^2 \left| \frac{\partial^2 F}{\partial \xi^2}(\mathbf{P}(t)) \right| + 2|\mathbf{h}(t)||\mathbf{h}'(t)| \left| \frac{\partial^2 F}{\partial \xi \partial \eta}(\mathbf{P}(t)) \right| + |\mathbf{h}'(t)|^2 \left| \frac{\partial^2 F}{\partial \eta^2}(\mathbf{P}(t)) \right| dt$$

$$\leq \frac{1}{2} \int_a^b \|\mathbf{h}\|_\infty^2 \left| \frac{\partial^2 F}{\partial \xi^2}(\mathbf{P}(t)) \right| + 2\|\mathbf{h}\|_\infty\|\mathbf{h}'\|_\infty \left| \frac{\partial^2 F}{\partial \xi \partial \eta}(\mathbf{P}(t)) \right| + \|\mathbf{h}'\|_\infty^2 \left| \frac{\partial^2 F}{\partial \eta^2}(\mathbf{P}(t)) \right| dt$$

$$\leq \frac{1}{2} \int_a^b \|\mathbf{h}\|_{1,\infty}^2 \left(\left| \frac{\partial^2 F}{\partial \xi^2}(\mathbf{P}(t)) \right| + 2 \left| \frac{\partial^2 F}{\partial \xi \partial \eta}(\mathbf{P}(t)) \right| + \left| \frac{\partial^2 F}{\partial \eta^2}(\mathbf{P}(t)) \right| \right) dt = M\|\mathbf{h}\|_{1,\infty}^2,$$

where

$$M := \frac{1}{2} \int_a^b \left(\left| \frac{\partial^2 F}{\partial \xi^2}(\mathbf{P}(t)) \right| + 2 \left| \frac{\partial^2 F}{\partial \xi \partial \eta}(\mathbf{P}(t)) \right| + \left| \frac{\partial^2 F}{\partial \eta^2}(\mathbf{P}(t)) \right| \right) dt.$$

We note that for each $t \in [a, b]$, the point

$$\mathbf{P}(t) = (\mathbf{x}_*(t) + \Theta(t)\mathbf{h}(t), \mathbf{x}'_*(t) + \Theta(t)\mathbf{h}'(t), t)$$

in $\mathbb{R}^3$ belongs to a ball with center $(\mathbf{x}_*(t), \mathbf{x}'_*(t), t)$ and radius $\|\mathbf{h}\|_{1,\infty}$. But $\mathbf{x}_*, \mathbf{x}'_*$ are continuous, and so these centers $(\mathbf{x}_*(t), \mathbf{x}'_*(t), t)$, for different values of $t \in [a, b]$, lie inside some big compact[1] set in $\mathbb{R}^3$. And if we look at balls with radius, say 1, around these centers, we get a somewhat bigger compact set, say K, in $\mathbb{R}^3$. Since the partial derivatives

$$\frac{\partial^2 F}{\partial \xi^2}, \quad \frac{\partial^2 F}{\partial \xi \partial \eta}, \quad \frac{\partial^2 F}{\partial \eta^2}$$

are all continuous, it follows that their absolute values are bounded on K, say by m. Hence M is finite.

---

[1] In $\mathbb{R}^n$, a set $K \subset \mathbb{R}^n$ is *compact* if and only if it is closed and bounded.

Let $\epsilon > 0$. Set $\delta := \epsilon/M$. Then if $\mathbf{h} \in X$ satisfies $0 < \|\mathbf{h} - \mathbf{0}\|_{1,\infty} = \|\mathbf{h}\|_{1,\infty} < \delta$, we have

$$\frac{|\widetilde{f}(\mathbf{h}) - \widetilde{f}(\mathbf{0}) - L(\mathbf{h} - \mathbf{0})|}{\|\mathbf{h}\|_{1,\infty}} \leq \frac{M\|\mathbf{h}\|_{1,\infty}^2}{\|\mathbf{h}\|_{1,\infty}} = M\|\mathbf{h}\|_{1,\infty} < M\delta = \epsilon.$$

Consequently, $\widetilde{f}'(\mathbf{0}) = L$.

Step 3. Utilizing $\widetilde{f}'(\mathbf{0}) = \mathbf{0}$. In the previous step, we calculated $\widetilde{f}'(\mathbf{0})$ and found out that it is the continuous linear transformation L. But we know that $\widetilde{f}'(\mathbf{0}) = \mathbf{0}$ (because $\mathbf{0} \in X$ is a minimizer of $\widetilde{f}$). So we must have $L = \mathbf{0}$, that is, for all $\mathbf{h} \in X$, $L\mathbf{h} = 0$, and so

$$\text{for all } \mathbf{h} \in C^1[a,b] \text{ with } \mathbf{h}(a) = \mathbf{h}(b) = 0, \quad \int_a^b \Big(\mathbf{A}(t)\mathbf{h}(t) + \mathbf{B}(t)\mathbf{h}'(t)\Big)dt = 0.$$

We would now like to use the technical result (Lemma 2.2) we had shown. So we rewrite the above integral and convert the term in the integrand which involves $\mathbf{h}$, into a term involving $\mathbf{h}'$, by using integration by parts:

$$
\begin{aligned}
\int_a^b \mathbf{A}(t)\mathbf{h}(t)dt &= \mathbf{h}(t)\int_a^t \mathbf{A}(\tau)d\tau \Big|_a^b - \int_a^b \Big(\mathbf{h}'(t)\int_a^t \mathbf{A}(\tau)d\tau\Big)dt \\
&= 0 - \int_a^b \Big(\mathbf{h}'(t)\int_a^t \mathbf{A}(\tau)d\tau\Big)dt,
\end{aligned}
$$

because $\mathbf{h}(a) = \mathbf{h}(b) = 0$. Hence

$$\text{for all } \mathbf{h} \in C^1[a,b] \text{ with } \mathbf{h}(a) = \mathbf{h}(b) = 0, \quad \int_a^b \Big(-\int_a^t \mathbf{A}(\tau)d\tau + \mathbf{B}(t)\Big)\mathbf{h}'(t)dt = 0.$$

and so by Lemma 2.2, it follows that

$$-\int_a^t \mathbf{A}(\tau)d\tau + \mathbf{B}(t) = c, \quad t \in [a,b],$$

for some constant c. By differentiating with respect to t, we obtain

$$\frac{\partial F}{\partial \xi}(\mathbf{x}_*(t), \mathbf{x}_*'(t), t) - \frac{d}{dt}\Big(\frac{\partial F}{\partial \eta}(\mathbf{x}_*(t), \mathbf{x}_*'(t), t)\Big) = 0 \text{ for all } t \in [a,b].$$

This finishes the proof of (1).

(2) Now suppose that f is convex and that $\mathbf{x}_* \in S$ satisfies the Euler-Lagrange equation. Define X and $\widetilde{f}$ in the same manner as above. By retracing the steps in **Step 3** above, we see that $\widetilde{f}'(\mathbf{0}) = \mathbf{0}$. Also the convexity of f makes $\widetilde{f}$ convex as

well. Indeed, if $\mathbf{h}_1, \mathbf{h}_2 \in X$, and $\alpha \in (0, 1)$, then

$$
\begin{aligned}
\widetilde{f}((1-\alpha)\mathbf{h}_1 + \alpha\mathbf{h}_2) &= f(\mathbf{x}_* + (1-\alpha)\mathbf{h}_1 + \alpha\mathbf{h}_2) \\
&= f((1-\alpha)\mathbf{x}_* + \alpha\mathbf{x}_* + (1-\alpha)\mathbf{h}_1 + \alpha\mathbf{h}_2) \\
&= f((1-\alpha)(\mathbf{x}_* + \mathbf{h}_1) + \alpha(\mathbf{x}_* + \mathbf{h}_2)) \\
&\leq (1-\alpha)f(\mathbf{x}_* + \mathbf{h}_1) + \alpha f(\mathbf{x}_* + \mathbf{h}_2) \\
&= (1-\alpha)\widetilde{f}(\mathbf{h}_1) + \alpha\widetilde{f}(\mathbf{h}_2).
\end{aligned}
$$

Recall that in Theorem 1.60 we showed that for a convex function, the derivative vanishing at a point implies that that point is a minimizer for the function. Since $\widetilde{f}$ is convex, and because $\widetilde{f}'(\mathbf{0}) = \mathbf{0}$, it follows that $\mathbf{0}$ is a minimizer of $\widetilde{f}$. We claim that $\mathbf{x}_*$ is a minimizer of f. Indeed, if $\mathbf{x} \in S$, then

$$
\mathbf{x} = \mathbf{x}_* + \underbrace{(\mathbf{x} - \mathbf{x}_*)}_{=:\mathbf{h}},
$$

and so $\mathbf{h} := \mathbf{x} - \mathbf{x}_* \in X$. Hence

$$
f(\mathbf{x}) = f(\mathbf{x}_* + (\mathbf{x} - \mathbf{x}_*)) = \widetilde{f}(\mathbf{x} - \mathbf{x}_*) \geq \widetilde{f}(\mathbf{0}) = f(\mathbf{x}_* + \mathbf{0}) = f(\mathbf{x}_*).
$$

This completes the proof. $\qquad\square$

Exercise 2.3. Solve the problem

$$
\begin{cases}
\text{minimize} & \displaystyle\int_1^2 t^3 (\mathbf{x}'(t))^2 dt \\[2mm]
\text{subject to} & \mathbf{x} \in C^1[1, 2], \ \mathbf{x}(1) = 5, \mathbf{x}(2) = 2.
\end{cases}
$$

Exercise 2.4. Find the curve that has the shortest length between $(0, 0)$ and $(1, 1)$.

Exercise 2.5. Let $S = \{\mathbf{x} \in C^1[0, 1] : \mathbf{x}(0) = 0 = \mathbf{x}(1)\}$ and $f : S \to \mathbb{R}$ be given by

$$
f(\mathbf{x}) = \int_0^1 (\mathbf{x}(t))^3 dt, \quad \mathbf{x} \in S.
$$

Show that the Euler-Lagrange equation has the unique solution $\mathbf{0} \in C^1[0, 1]$. Is $\mathbf{0}$ a minimizer for f?

Exercise 2.6. Let $S = \{\mathbf{x} \in C^1[0, 1] : \mathbf{x}(0) = 0 \text{ and } \mathbf{x}(1) = 1\}$. Find all solutions in S of the Euler-Lagrange equation associated with the problem

$$
\begin{cases}
\text{minimize} & f(\mathbf{x}) \\
\text{subject to} & \mathbf{x} \in S,
\end{cases}
$$

where the function $f : S \to \mathbb{R}$ is given by

(1) $f(\mathbf{x}) = \displaystyle\int_0^1 \mathbf{x}'(t) dt, \ \mathbf{x} \in S.$

(2) $f(\mathbf{x}) = \displaystyle\int_0^1 \mathbf{x}(t)\mathbf{x}'(t) dt, \ \mathbf{x} \in S.$

(3) $f(\mathbf{x}) = \displaystyle\int_0^1 (\mathbf{x}(t) + t\mathbf{x}'(t)) dt, \ \mathbf{x} \in S.$

Exercise 2.7. Consider the map $(\xi, \eta, \tau) \overset{F}{\mapsto} F(\xi, \eta, \tau)$. Prove that:

(1) If F does not depend on ξ, then the Euler-Lagrange equation becomes

$$\frac{\partial F}{\partial \eta}(\mathbf{x}_*(t), \mathbf{x}'_*(t), t) = C,$$

where C is a constant.

(2) If F does not depend on η, then the Euler-Lagrange equation becomes

$$\frac{\partial F}{\partial \xi}(\mathbf{x}_*(t), \mathbf{x}'_*(t), t) = 0.$$

(3) If F does not depend on τ and if $\mathbf{x}_*$ is twice-differentiable in $[a, b]$, then the Euler-Lagrange equation becomes

$$F(\mathbf{x}_*(t), \mathbf{x}'_*(t), t) - \mathbf{x}'_*(t)\frac{\partial F}{\partial \eta}(\mathbf{x}_*(t), \mathbf{x}'_*(t), t) = C,$$

where C is a constant.

Hint: Find $\dfrac{d}{dt}\left(F(\mathbf{x}_*(t), \mathbf{x}'_*(t), t) - \mathbf{x}'_*(t)\dfrac{\partial F}{\partial \eta}(\mathbf{x}_*(t), \mathbf{x}'_*(t), t)\right).$

Exercise 2.8. Suppose that a is a real number such that $0 \leq a < 1$. Let

$$S := \{\mathbf{x} \in C^1[a, 1] : \mathbf{x}(a) = 0, \ \mathbf{x}(1) = 1\}.$$

Define $f : S \to \mathbb{R}$ by $f(\mathbf{x}) = \displaystyle\int_a^1 t(\mathbf{x}'(t))^2 dt$, $\mathbf{x} \in S$, and consider the optimization problem

$$\begin{cases} \text{minimize} & f(\mathbf{x}) \\ \text{subject to} & \mathbf{x} \in S. \end{cases}$$

(1) Let $a \neq 0$. Show that there is a unique solution in S to the above Euler-Lagrange equation.

(2) Let $a = 0$. Prove that there is no solution in S to the Euler-Lagrange equation. What does this say about the solvability of the optimization problem above?

Exercise 2.9 (Optimization in function spaces versus that in $\mathbb{R}^n$). Let f be given by

$$f(\mathbf{x}) = \int_a^b F(\mathbf{x}(t), \mathbf{x}'(t), t)\, dt \text{ for } \mathbf{x} \in C^1[a, b] \text{ such that } \mathbf{x}(a) = y_a \text{ and } \mathbf{x}(b) = y_b.$$

We obtain a related finite dimensional optimization problem by "discretization" as follows. First we divide the interval $[a, b]$ into n equal parts: $a = t_0, t_1, \cdots, t_{n-1}, t_n = b$. Then we replace the curve $\{\mathbf{x}(t), t \in [a, b]\}$ by the polygonal line joining the points

$$(t_0, y_a), (t_1, \mathbf{x}(t_1)), \cdots, (t_{n-1}, \mathbf{x}(t_{n-1})), (t_n, y_b),$$

and we approximate the function f at $\mathbf{x}$ by the sum

$$f_n(y_1, \cdots, y_{n-1}) = \sum_{k=0}^{n-1} F\left(y_k, \frac{y_{k+1} - y_k}{\frac{b-a}{n}}, t_k\right) \cdot \frac{b-a}{n}, \tag{2.1}$$

where $y_k = \mathbf{x}(t_k)$, $k = 0, \cdots, n$. Each polygonal line is uniquely determined by the ordinates $y_1, \cdots, y_{n-1}$ of its vertices (recall that $y_0 = y_a$ and $y_n = y_b$ are fixed), and the sum (2.1) is therefore a function of the $n-1$ variables $y_1, \cdots, y_{n-1}$. Thus as an approximation, we can replace the given optimization problem by the problem of optimizing the function $f_n(y_1, \cdots, y_{n-1})$. Euler used this "method of finite differences" to solve optimization problems in function spaces. By replacing smooth curves by polygonal lines, he replaced the problem of finding minimizers of a function whose domain is a set of functions to the problem of finding minimizers of a function of n variables, and then he obtained exact solutions by passing to the limit as $n \to \infty$. In this sense, f can be regarded as "functions of infinitely many variables" (that is, the infinitely many free values of $x(t)$ at different points $t \in [a, b]$), and the subject of calculus of variations can be regarded as the corresponding analog of optimization of functions of finitely many real variables. Carry out the above procedure for the problem of finding the curve $\mathbf{x} \in C^1[0,1]$ such that $\mathbf{x}(0) = 0$ and $\mathbf{x}(1) = 1$ that minimizes

$$f(\mathbf{x}) = \int_0^1 \sqrt{1 + (\mathbf{x}'(t))^2} \, dt.$$

2.2. Free Endpoints

Given a real-valued function F on $\mathbb{R}^3$, consider the optimization problem

$$\begin{cases} \text{minimize} \quad f(\mathbf{x}) := \int_a^b F(\mathbf{x}(t), \mathbf{x}'(t), t) \, dt \\ \text{subject to} \quad \mathbf{x} \in C^1[a, b]. \end{cases}$$

Thus as opposed to the problem in the previous section, where $\mathbf{x} \in C^1[a, b]$ had fixed values y_a, y_b at the endpoints a, b, respectively, now the endpoints are "free." See Figure 2.

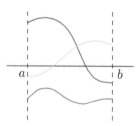

Figure 2. Free endpoints.

How does the result from the previous section change? We will prove the following.

Theorem 2.10. *Suppose that*

(1) *$F : \mathbb{R}^3 \to \mathbb{R}$, $(\xi, \eta, \tau) \overset{F}{\mapsto} F(\xi, \eta, \tau)$, has continuous partial derivatives of order 2,*

(2) *$f : C^1[a, b] \to \mathbb{R}$ is given by $f(\mathbf{x}) = \int_a^b F(\mathbf{x}(t), \mathbf{x}'(t)t) \, dt$, $\mathbf{x} \in C^1[a, b]$.*

Then we have:

(1) *If* $\mathbf{x}_*$ *is a minimizer of* f, *then it satisfies the* Euler-Lagrange equation

$$\frac{\partial F}{\partial \xi}(\mathbf{x}_*(t), \mathbf{x}_*'(t), t) - \frac{d}{dt}\left(\frac{\partial F}{\partial \eta}(\mathbf{x}_*(t), \mathbf{x}_*'(t), t)\right) = 0 \text{ for all } t \in [a, b]$$

and the endpoint conditions

$$\frac{\partial F}{\partial \eta}\left(\mathbf{x}_*(a), \mathbf{x}_*'(a), a\right) = 0 \quad \text{and} \quad \frac{\partial F}{\partial \eta}\left(\mathbf{x}_*(b), \mathbf{x}_*'(b), b\right) = 0. \tag{2.2}$$

(2) *If* f *is convex, and* $\mathbf{x}_* \in C^1[a, b]$ *satisfies the Euler-Lagrange equation and the endpoint conditions, then* $\mathbf{x}_*$ *is a minimizer of* f.

Proof. Step 1.   Calculation of $f'(\mathbf{x}_*)$. We take $X = C^1[a, b]$ and compute $f'(\mathbf{x}_*)$. We have

$$f(\mathbf{x}_* + \mathbf{h}) - f(\mathbf{x}_*)$$

$$= \int_a^b \left(F(\mathbf{x}_*(t) + \mathbf{h}(t), \mathbf{x}_*'(t) + \mathbf{h}'(t), t) - F(\mathbf{x}_*(t), \mathbf{x}_*'(t), t)\right) dt$$

$$= \int_a^b \left(\mathbf{A}(t)\mathbf{h}(t) + \mathbf{B}(t)\mathbf{h}'(t)\right) dt + \frac{1}{2}\int_a^b \begin{bmatrix} \mathbf{h}(t) & \mathbf{h}'(t) & 0 \end{bmatrix} H_F(\mathbf{P}(t)) \begin{bmatrix} \mathbf{h}(t) \\ \mathbf{h}'(t) \\ 0 \end{bmatrix} dt,$$

where

$$\mathbf{A}(t) := \frac{\partial F}{\partial \xi}(\mathbf{x}_*(t), \mathbf{x}_*'(t), t),$$

$$\mathbf{B}(t) := \frac{\partial F}{\partial \eta}(\mathbf{x}_*(t), \mathbf{x}_*'(t), t),$$

$$\mathbf{P}(t) := (\mathbf{x}_*(t) + \mathbf{\Theta}(t)\mathbf{h}(t), \mathbf{x}_*'(t) + \mathbf{\Theta}(t)\mathbf{h}'(t), t).$$

and $H_F(\cdot)$ denotes the Hessian of F. So we guess that $f'(\mathbf{x}_*) = L$, where the map $L : C^1[a, b] \to \mathbb{R}$ is given by

$$L\mathbf{h} = \int_a^b \left(\mathbf{A}(t)\mathbf{h}(t) + \mathbf{B}(t)\mathbf{h}'(t)\right) dt, \quad \mathbf{h} \in C^1[a, b].$$

We have

$$|f(\mathbf{x}_* + \mathbf{h}) - f(\mathbf{x}_*) - L\mathbf{h}|$$

$$= \left|\frac{1}{2}\int_a^b \left((\mathbf{h}(t))^2 \frac{\partial^2 F}{\partial \xi^2}(\mathbf{P}(t)) + 2\mathbf{h}(t)\mathbf{h}'(t)\frac{\partial^2 F}{\partial \xi \partial \eta}(\mathbf{P}(t)) + (\mathbf{h}'(t))^2 \frac{\partial^2 F}{\partial \eta^2}(\mathbf{P}(t))\right) dt\right|$$

$$\leq \frac{1}{2}\int_a^b \|\mathbf{h}\|_{1,\infty}^2 \left(\left|\frac{\partial^2 F}{\partial \xi^2}(\mathbf{P}(t))\right| + 2\left|\frac{\partial^2 F}{\partial \xi \partial \eta}(\mathbf{P}(t))\right| + \left|\frac{\partial^2 F}{\partial \eta^2}(\mathbf{P}(t))\right|\right) dt$$

$$= M\|\mathbf{h}\|_{1,\infty}^2,$$

where

$$M := \frac{1}{2} \int_a^b \left(\left| \frac{\partial^2 F}{\partial \xi^2}(\mathbf{P}(t)) \right| + 2 \left| \frac{\partial^2 F}{\partial \xi \partial \eta}(\mathbf{P}(t)) \right| + \left| \frac{\partial^2 F}{\partial \eta^2}(\mathbf{P}(t)) \right| \right) dt.$$

Let $\epsilon > 0$. Set $\delta := \epsilon/M$. Then if $\mathbf{h} \in C^1[a,b]$ satisfies $0 < \|\mathbf{h}-\mathbf{0}\|_{1,\infty} = \|\mathbf{h}\|_{1,\infty} < \delta$, we have

$$\frac{|f(\mathbf{x}_* + \mathbf{h}) - f(\mathbf{x}_*) - L(\mathbf{h})|}{\|\mathbf{h}\|_{1,\infty}} \leq \frac{M\|\mathbf{h}\|_{1,\infty}^2}{\|\mathbf{h}\|_{1,\infty}} = M\|\mathbf{h}\|_{1,\infty} < M\delta = \epsilon.$$

Consequently, $f'(\mathbf{x}_*) = L$.

Theorem 1.42 implies that $f'(x_*) = \mathbf{0}$, and so for all $\mathbf{h} \in C^1[a,b]$, $(f'(\mathbf{x}_*))\mathbf{h} = 0$, that is,

$$\int_a^b \left(\mathbf{A}(t)\mathbf{h}(t) + \mathbf{B}(t)\mathbf{h}'(t) \right) dt = 0.$$

Step 2. Obtaining the Euler-Lagrange equation. In particular, also for all $\mathbf{h}$ in $C^1[a,b]$ such that $\mathbf{h}(a) = \mathbf{h}(b) = 0$,

$$\int_a^b \left(\mathbf{A}(t)\mathbf{h}(t) + \mathbf{B}(t)\mathbf{h}'(t) \right) dt = 0.$$

We have

$$\int_a^b \mathbf{A}(t)\mathbf{h}(t)dt = \mathbf{h}(t) \int_0^t \mathbf{A}(\tau)d\tau \Big|_a^b - \int_a^b \left(\mathbf{h}'(t) \int_a^t \mathbf{A}(\tau)d\tau \right) dt$$

$$= 0 - \int_a^b \mathbf{h}'(t) \left(\int_a^t \mathbf{A}(\tau)d\tau \right) dt.$$

So for all $\mathbf{h} \in C^1[a,b]$ with $\mathbf{h}(a) = \mathbf{h}(b) = 0$, we have

$$\int_a^b \left(\mathbf{B}(t) - \int_a^t \mathbf{A}(\tau)d\tau \right) \mathbf{h}'(t)dt = 0.$$

By Lemma 2.2, it follows that $\mathbf{B}(t) - \int_a^t \mathbf{A}(\tau)d\tau = C$ for all $t \in [a,b]$, where C is some constant, and so upon differentiating with respect to t, we obtain

$$\mathbf{A}(t) - \frac{d}{dt}\mathbf{B}(t) = \frac{\partial F}{\partial \xi}(\mathbf{x}_*(t), \mathbf{x}_*'(t), t) - \frac{d}{dt}\left(\frac{\partial F}{\partial \eta}(\mathbf{x}_*(t), \mathbf{x}_*'(t), t) \right) = 0 \text{ for all } t \in [a,b].$$

Step 3. Endpoint conditions. We know that $f'(\mathbf{x}_*) = \mathbf{0}$, that is, for all $\mathbf{h} \in C^1[a,b]$, $(f'(\mathbf{x}_*))\mathbf{h} = 0$, that is,

$$\int_a^b \left(\mathbf{A}(t)\mathbf{h}(t) + \mathbf{B}(t)\mathbf{h}'(t) \right) dt = 0.$$

We have $\int_a^b \mathbf{B}(t)\mathbf{h}'(t)dt = \mathbf{B}(t)\mathbf{h}(t)\Big|_a^b - \int_a^b \mathbf{B}'(t)\mathbf{h}(t)dt$. So for all $\mathbf{h} \in C^1[a,b]$,

$$\int_a^b \underbrace{(\mathbf{A}(t) - \mathbf{B}'(t))}_{=0}\mathbf{h}(t)dt + \mathbf{B}(b)\mathbf{h}(b) - \mathbf{B}(a)\mathbf{h}(a) = 0.$$

Since we have already shown that the Euler-Lagrange equation holds, we have $\mathbf{A}(t) - \mathbf{B}'(t) = 0$ for all $t \in [a,b]$. Hence it follows from the above that for all $\mathbf{h} \in C^1[a,b]$, there holds that

$$\mathbf{B}(b)\mathbf{h}(b) - \mathbf{B}(a)\mathbf{h}(a) = 0. \tag{2.3}$$

Take $\mathbf{h} \in C^1[a,b]$ such that $\mathbf{h}(b) = 0$ but $\mathbf{h}(a) \neq 0$. For example, a concrete such as $\mathbf{h}$ is the function $t \mapsto b - t$, $t \in [a,b]$. Then it follows from (2.3) that $\mathbf{B}(a) = 0$, that is,

$$\frac{\partial F}{\partial \eta}\left(\mathbf{x}_*(a), \mathbf{x}_*'(a), a\right) = 0. \tag{2.4}$$

Similarly, taking another $\mathbf{h} \in C^1[a,b]$ satisfying $\mathbf{h}(b) \neq 0$ (for example, $\mathbf{h} = \mathbf{1}$), we obtain (using (2.3) and (2.4)) that also $\mathbf{B}(b) = 0$, that is,

$$\frac{\partial F}{\partial \eta}\left(\mathbf{x}_*(b), \mathbf{x}_*'(b), b\right) = 0.$$

This completes the proof of (1).

(2) We first note that $f'(\mathbf{x}_*) = \mathbf{0}$. Indeed, this is because the Euler-Lagrange equation and the endpoint conditions ensure that for all $\mathbf{h} \in C^1[a,b]$,

$$\int_a^b \underbrace{(\mathbf{A}(t) - \mathbf{B}'(t))}_{=0}\mathbf{h}(t)dt + \mathbf{B}(b)\mathbf{h}(b) - \mathbf{B}(a)\mathbf{h}(a) = 0,$$

that is,

$$(f'(\mathbf{x}_*))\mathbf{h} = L\mathbf{h} = \int_a^b \left(\mathbf{A}(t)\mathbf{h}(t) - \mathbf{B}(t)\mathbf{h}'(t)\right)dt = 0.$$

So $\mathbf{x}_*$ is a minimizer of f by Theorem 1.60. $\qquad\square$

Example 2.11. Let us find out the curves $t \mapsto x(t) : [0,1] \to \mathbb{R}$ of least length whose endpoints lie on the two lines $t = 0$ and $t = 1$. From Figure 3, we see that any curve parallel to the t axis will be a minimizing curve.

So we expect the set of minimizing curves to be precisely all constant curves. Let us check if our result above yields this. We want to

$$\begin{cases} \text{minimize} \quad f(\mathbf{x}) := \int_0^1 \sqrt{1 + (\mathbf{x}'(t))^2}dt \\ \text{subject to} \quad \mathbf{x} \in C^1[a,b]. \end{cases}$$

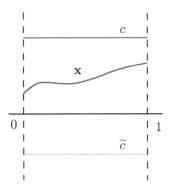

Figure 3. Constant curves are minimizing for the arc length function on the set of all curves between the lines $t = 0$ and $t = 1$.

We have that the integrand is described by F, where $F(\xi, \eta, \tau) = \sqrt{1 + \eta^2}$. Thus

$$\frac{\partial F}{\partial \xi} = 0 \quad \text{and} \quad \frac{\partial F}{\partial \eta} = \frac{\eta}{\sqrt{1 + \eta^2}}.$$

The Euler-Lagrange equation is

$$0 - \frac{d}{dt}\left(\frac{\mathbf{x}'_*(t)}{\sqrt{1 + (\mathbf{x}'_*(t))^2}}\right) = 0, \quad t \in [0, 1],$$

so upon integrating, we obtain

$$\frac{\mathbf{x}'_*(t)}{\sqrt{1 + (\mathbf{x}'_*(t))^2}} = C$$

on $[0, 1]$ for some constant C. By the endpoint condition at $t = 0$, we obtain

$$\frac{\partial F}{\partial \eta}(\mathbf{x}_*(0), \mathbf{x}'_*(0), 0) = 0,$$

that is,

$$\frac{\mathbf{x}'_*(0)}{\sqrt{1 + (\mathbf{x}'_*(0))^2}} = 0.$$

So we conclude that $C = 0$. Hence $\mathbf{x}'_*(t) = 0$ for all $t \in [0, 1]$, and so $\mathbf{x}_*$ is a constant on $[0, 1]$. Also, a constant $\mathbf{x}_*$ also satisfies the endpoint condition at $t = 1$:

$$\frac{\partial F}{\partial \eta}(\mathbf{x}_*(1), \mathbf{x}'_*(1), 1) = \frac{\mathbf{x}'_*(1)}{\sqrt{1 + (\mathbf{x}'_*(1))^2}} = \frac{0}{\sqrt{1 + 0^2}} = 0.$$

In fact, each such constant function $c\mathbf{1}$, where $c \in \mathbb{R}$, is a minimizer, since

$$f(\mathbf{x}) = \int_0^1 \sqrt{1 + (\mathbf{x}'(t))^2} dt \geq \int_0^1 \sqrt{1 + 0^2} dt = 1 = f(c\mathbf{1}),$$

for all $\mathbf{x} \in S$. $\Diamond$

Exercise 2.12. Consider the following problem.

$$\begin{cases} \text{minimize} & \displaystyle\int_0^1 \left(\frac{(\mathbf{x}'(t))^2}{2} + \mathbf{x}(t)\mathbf{x}'(t) + \mathbf{x}'(t) + \mathbf{x}(t)\right) dt \\ \text{subject to} & \mathbf{x} \in C^1[0,1]. \end{cases}$$

Assuming that a minimizer $\mathbf{x}_*$ exists, find it.

What if one endpoint is free and the other is fixed? Let us now consider mixed boundary point conditions, when one endpoint is free and the other is fixed, as illustrated in Figure 4.

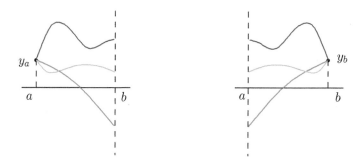

Figure 4. Mixed cases.

In either of these cases, it can again be shown that the Euler-Lagrange equation must be satisfied by a minimizer $\mathbf{x}_*$, and one has the following natural boundary conditions.

In the case when $S = \{\mathbf{x} \in C^1[a,b] : \mathbf{x}(a) = y_a\}$, we have

$$\mathbf{x}_*(a) = y_a \text{ and } \frac{\partial F}{\partial \eta}(\mathbf{x}_*(b), \mathbf{x}'_*(b), b) = 0.$$

On the other hand, if $S = \{\mathbf{x} \in C^1[a,b] : \mathbf{x}(b) = y_b\}$, then

$$\frac{\partial F}{\partial \eta}(\mathbf{x}_*(a), \mathbf{x}'_*(a), a) = 0 \text{ and } \mathbf{x}_*(b) = y_b.$$

In other words, at the free endpoint, we get a free endpoint condition:

$$\frac{\partial F}{\partial \eta}(\mathbf{x}_*(t), \mathbf{x}'_*(t), t)\Big|_{t=\text{free endpoint}} = 0.$$

We won't prove this version because the proof is analogous to the proofs of the previous two theorems.

Summarizing, we have:

S	Endpoint condition for minimizer $\mathbf{x}_*$
$\{\mathbf{x} \in C^1[a,b] : \mathbf{x}(a) = y_a,\ \mathbf{x}(b) = y_b\}$	$\mathbf{x}_*(a) = y_a,$ $\mathbf{x}_*(b) = y_b$
$\{\mathbf{x} \in C^1[a,b] : \mathbf{x}(a) = y_a\}$	$\mathbf{x}_*(a) = y_a,$ $\dfrac{\partial F}{\partial \eta}(\mathbf{x}_*(b), \mathbf{x}_*'(b), b) = 0$
$\{\mathbf{x} \in C^1[a,b] : \mathbf{x}(b) = y_b\}$	$\dfrac{\partial F}{\partial \eta}(\mathbf{x}_*(a), \mathbf{x}_*'(a), a) = 0,$ $\mathbf{x}_*(b) = y_b$
$C^1[a,b]$	$\dfrac{\partial F}{\partial \eta}(\mathbf{x}_*(a), \mathbf{x}_*'(a), a) = 0,$ $\dfrac{\partial F}{\partial \eta}(\mathbf{x}_*(b), \mathbf{x}_*'(b), b) = 0$

Example 2.13. Let us revisit our optimal mining problem from Example 0.2, but now suppose that the company mines in a very big mountain range, and so it isn't the case that there is some amount Q that it can mine in T years, but rather it can mine as much as it wants.

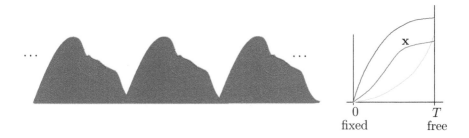

So now the set of admissible mining operations is changed to

$$S = \{\mathbf{x} \in C^1[0,1] : x(0) = 0\},$$

(that is, the endpoint condition $\mathbf{x}(T) = Q$ is dropped). Again the company wants to minimize its cost, but the cost is now modified, because the company sells the final amount of copper at a price P, so that the mining cost associated with $\mathbf{x} \in S$

$$\int_0^T (a\mathbf{x}(t) + b\mathbf{x}'(t))\mathbf{x}'(t)dt \quad (=: f(\mathbf{x}))$$

is reduced by $P \cdot$ (total amount of copper mined) $= P \int_0^T \mathbf{x}'(t)dt.$

(Earlier, the cost was reduced by the fixed amount $P \cdot Q$, which was independent of the mining operation $\mathbf{x}$, and so it did not affect the minimization problem.) So now the problem becomes that of minimizing $\widetilde{f} : S \to \mathbb{R}$, where for $\mathbf{x} \in S$

$$
\begin{aligned}
\widetilde{f}(\mathbf{x}) &= f(\mathbf{x}) - P \int_0^T \mathbf{x}'(t)dt = \int_0^T (a\mathbf{x}(t) + b\mathbf{x}'(t) - P)\mathbf{x}'(t)dt \\
&= \int_0^T F(\mathbf{x}(t), \mathbf{x}'(t), t)dt,
\end{aligned}
$$

and where F is now given by $F(\xi, \eta, \tau) := (a\xi + b\eta - P)\eta$. Thus

$$
\begin{aligned}
\frac{\partial F}{\partial \xi} &= a\eta, \\
\frac{\partial F}{\partial \eta} &= a\xi + 2b\eta - P.
\end{aligned}
$$

The Euler-Lagrange equation is

$$
a\mathbf{x}_*'(t) - \frac{d}{dt}(a\mathbf{x}_*(t) + 2b\mathbf{x}_*'(t) - P) = 0, \quad t \in [0, T].
$$

Hence upon integrating, we obtain $\mathbf{x}_*'(t) = A$, $t \in [0, T]$, and integrating again, we get $\mathbf{x}_*(t) = At + B$, $t \in [0, T]$. Now let us use the endpoint conditions:

(1) $\mathbf{x}_*(0) = 0$ gives $B = 0$, and so $\mathbf{x}_*(t) = At$, $t \in [0, T]$.

(2) The endpoint condition at T is $\dfrac{\partial F}{\partial \eta}(\mathbf{x}_*(T), \mathbf{x}_*'(T), T) = 0$, that is,

$$
aAT + 2bA - P = 0, \text{ and so } A = \frac{P}{aT + 2b}.
$$

Hence

$$
\mathbf{x}_*(t) = \frac{P}{aT + 2b}t, \quad t \in [0, T].
$$

It can be seen again that $\widetilde{f}$ is convex: indeed, from the convex function f we subtract the linear map

$$
\mathbf{x} \mapsto P \int_0^T \mathbf{x}'(t)dt, \quad \mathbf{x} \in S,
$$

giving rise to the convex function $\widetilde{f}$. Consequently, the $\mathbf{x}_*$ given above is the unique minimizer of $\widetilde{f}$. $\Diamond$

Exercise 2.14. Let $S := \{\mathbf{x} \in C^1[0, 1] : \mathbf{x}(0) = 0\}$ and $f : S \to \mathbb{R}$ be defined by

$$
f(\mathbf{x}) = \int_0^1 \cos(\mathbf{x}'(t))dt, \quad \mathbf{x} \in S.
$$

Find all elements in S that minimize f, and also all elements in S that maximize f.

Exercise 2.15. The cost of a manufacturing process in a certain industry is given by

$$f(\mathbf{x}) = \int_0^1 \left(\frac{(\mathbf{x}'(t))^2}{2} + \mathbf{x}(t)\right) dt,$$

for the processes $\mathbf{x} \in C^1[0,1]$.

(1) If $S = \{\mathbf{x} \in C^1[0,1] : \mathbf{x}(0) = 1 \text{ and } \mathbf{x}(1) = 0\}$, then find a minimizer $\mathbf{x}_* \in S$ for $f : S \to \mathbb{R}$. Calculate the minimum cost $f(\mathbf{x}_*)$.

(2) If $\widetilde{S} = \{\mathbf{x} \in C^1[0,1] : \mathbf{x}(0) = 1\}$, then find a minimizer $\widetilde{\mathbf{x}}_*$ for $f : \widetilde{S} \to \mathbb{R}$. Calculate the minimum cost $f(\widetilde{\mathbf{x}}_*)$.

(3) Which of the values $f(\mathbf{x}_*)$ and $f(\widetilde{\mathbf{x}}_*)$ found in parts above is larger? Explain why you would expect this.

Exercise 2.16. Reconsider the problem in Exercise 2.8. Now suppose that $0 < a < 1$, but the endpoint condition $\mathbf{x}(1) = 1$ is dropped. That is, consider the problem

$$\begin{cases} \text{minimize} & \widetilde{f}(\mathbf{x}) \\ \text{subject to} & \mathbf{x} \in \widetilde{S}, \end{cases}$$

where $\widetilde{S} := \{\mathbf{x} \in C^1[a,1] : \mathbf{x}(a) = 0\}$ and $\widetilde{f} : \widetilde{S} \to \mathbb{R}$ is given by

$$\widetilde{f}(\mathbf{x}) = \int_a^1 t(\mathbf{x}'(t))^2 dt, \quad \mathbf{x} \in \widetilde{S}.$$

Find all optimal solutions to this optimization problem.

Remark 2.17. The results in this section can be generalized to the case when the integrand F is a function of more than one independent variable. That is, if we wish to find extremum values of the function

$$f(\mathbf{x}_1, \cdots, \mathbf{x}_n) = \int_a^b F(\mathbf{x}_1(t), \cdots, \mathbf{x}_n(t), \mathbf{x}_1'(t), \cdots, \mathbf{x}_n'(t), t) dt,$$

where $F(\xi_1, \cdots, \xi_n, \eta_1, \cdots, \eta_n, \tau)$ is a function with continuous partial derivatives of order ≤ 2, and $\mathbf{x}_1, \ldots, \mathbf{x}_n$ are continuously differentiable functions of the variable t, then following a similar analysis as before, we obtain n Euler-Lagrange equations to be satisfied by the minimizer:

$$\frac{\partial F}{\partial \xi_k}\left(\mathbf{x}_{1*}(t), \cdots, \mathbf{x}_{n*}(t), \mathbf{x}_{1*}'(t), \cdots, \mathbf{x}_{n*}'(t), t\right)$$

$$-\frac{d}{dt}\left(\frac{\partial F}{\partial \beta_k}\left(\mathbf{x}_{1*}(t), \cdots, \mathbf{x}_{n*}(t), \mathbf{x}_{1*}'(t), \cdots, \mathbf{x}_{n*}'(t), t\right)\right) = 0,$$

for $t \in [a,b]$, $k \in \{1, \ldots, n\}$. Also at any endpoint where $\mathbf{x}_k$ is free,

$$\frac{\partial F}{\partial \eta_k}\left(\mathbf{x}_{1*}(t), \ldots, \mathbf{x}_{n*}(t), \mathbf{x}_{1*}'(t), \ldots, \mathbf{x}_{n*}'(t), t\right) = 0.$$

2.3. Scalar-Valued Integral Constraint

In ordinary calculus, one can solve optimization problems subject to equality constraints using Lagrange multipliers. Consider the optimization problem:

$$(\text{P}) : \begin{cases} \text{minimize} & f(x) \\ \text{subject to} & g(x) = C. \end{cases}$$

Recall that one then has the following result:

Theorem 2.18 (Lagrange Multiplier Theorem). *Let* $C \in \mathbb{R}$,

$$S = \{x \in \mathbb{R}^n : g(x) = C\}$$

and $f : S \to \mathbb{R}$. *If* $x_* \in S$ *is a minimizer of* f, *and* $\nabla g(x_*) \neq \mathbf{0} \in \mathbb{R}^n$, *then there exists a* $\lambda \in \mathbb{R}$ *such that* $\nabla f(x_*) + \lambda \nabla g(x_*) = \mathbf{0}$.

Example 2.19. Suppose we want to find the shape of a triangle with a given fixed perimeter P and given base length $2r$ such that the area A is maximized. See Figure 5.

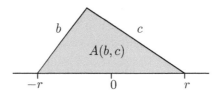

Figure 5. Triangle with given perimeter P and given base length $2r$.

If the other two side lengths are b, c and $P := 2r + b + c$, then the problem is that of maximizing

$$A(b,c) = \sqrt{\frac{P}{2}\left(\frac{P}{2} - 2r\right)\left(\frac{P}{2} - b\right)\left(\frac{P}{2} - c\right)}$$

subject to $b + c = P - 2r$. So we have an optimization problem of the type (P) with $f(b, c) = -A(b, c)$ and $g(b, c) := b + c$. We have

$$\nabla g(b,c) = \begin{bmatrix} 1 & 1 \end{bmatrix} \neq \begin{bmatrix} 0 & 0 \end{bmatrix}.$$

So if (b_*, c_*) is a minimizer for f (that is, a maximizer for A), then there is a $\lambda \in \mathbb{R}$ such that

(1) $\nabla f(b_*, c_*) + \lambda \nabla g(b_*, c_*) = \mathbf{0}$, and

(2) $b_* + c_* = P - 2r$.

(1) gives

$$\sqrt{\frac{P}{2}\left(\frac{P}{2}-2r\right)}\left[\begin{array}{cc}\dfrac{\sqrt{\dfrac{P}{2}}-c_*}{-2\sqrt{\dfrac{P}{2}-b_*}} & \dfrac{\sqrt{\dfrac{P}{2}}-b_*}{-2\sqrt{\dfrac{P}{2}-c_*}}\end{array}\right]+\left[\begin{array}{cc}\lambda & \lambda\end{array}\right]=\left[\begin{array}{cc}0 & 0\end{array}\right],$$

for which we obtain that $b_* = c_*$. The condition in item (2) above then yields that

$$b_* = c_* = \frac{P}{2} - r.$$

Thus the area is maximized when the triangle is isosceles. See Figure 6. $\Diamond$

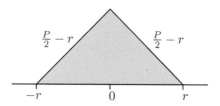

Figure 6. The isosceles triangle maximizes $(b, c) \mapsto A(b, c)$.

Analogous to the problem (P), now let us consider the following.

$$(\text{P}') : \begin{cases} \text{minimize} \quad f(\mathbf{x}) := \displaystyle\int_a^b F(\mathbf{x}(t), \mathbf{x}'(t), t)\,dt \\[2mm] \text{subject to} \quad g(\mathbf{x}) := \displaystyle\int_a^b G(\mathbf{x}(t), \mathbf{x}'(t), t)\,dt = C, \\[2mm] \qquad\qquad \mathbf{x} \in C^1[a, b], \ \mathbf{x}(a) = y_a, \ \mathbf{x}(b) = y_b \end{cases}$$

where $F, G : \mathbb{R}^3 \to \mathbb{R}$, and C is a real constant. Analogous to this method of Lagrange multipliers in finite dimensions, we have the following result.

Theorem 2.20. *If* $\mathbf{x}_*$ *is a solution to* (P'), *and if it is* not *the case that for all* $t \in [a, b]$,

$$\frac{\partial G}{\partial \xi}(\mathbf{x}_*(t), \mathbf{x}'_*(t), t) - \frac{d}{dt}\left(\frac{\partial G}{\partial \eta}(\mathbf{x}_*(t), \mathbf{x}'_*(t), t)\right) = 0,$$

then there exists a $\lambda \in \mathbb{R}$ *such that*
$$\frac{\partial(F + \lambda G)}{\partial \xi}(\mathbf{x}_*(t), \mathbf{x}'_*(t), t) - \frac{d}{dt}\left(\frac{\partial(F + \lambda G)}{\partial \eta}(\mathbf{x}_*(t), \mathbf{x}'_*(t), t)\right) = 0 \text{ for all } t \in [a, b].$$
$$(2.5)$$

One also has similar versions with possibly one/both endpoints being free.

As an illustration of the above theorem, we solve the following isoperimetric problem, this time not restricting ourselves to triangles.

Example 2.21. Look at Figure 7.

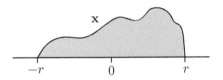

Figure 7. String of length ℓ in the shape of the graph of $\mathbf{x}$.

We are given a string of length $\ell > 2r$ that has a shape described by the function $\mathbf{x} : [-r, r] \to \mathbb{R}$ and the endpoints of the string are fixed so that $\mathbf{x}(-r) = 0 = \mathbf{x}(r)$. The question we ask is this: Which such $\mathbf{x}$ maximizes the area under its graph? So we arrive at the following optimization problem:

$$\begin{cases} \text{maximize} & \displaystyle\int_{-r}^{r} \mathbf{x}(t)dt \\[2mm] \text{subject to} & \displaystyle\int_{-r}^{r} \sqrt{1 + (\mathbf{x}'(t))^2}dt = \ell, \\[2mm] & \mathbf{x} \in C^1[-r, r], \ \mathbf{x}(-r) = 0, \ \mathbf{x}(r) = 0 \end{cases}$$

This is a problem of the type (P'), where

$$F(\xi, \eta, \tau) = -\xi, \quad \text{and} \quad G(\xi, \eta, \tau) = \sqrt{1 + \eta^2}.$$

If for all $t \in [-r, r]$ there holds that

$$\frac{\partial G}{\partial \xi}(\mathbf{x}_*(t), \mathbf{x}_*'(t), t) - \frac{d}{dt}\left(\frac{\partial G}{\partial \eta}(\mathbf{x}_*(t), \mathbf{x}_*'(t), t)\right) = 0 - \frac{d}{dt}\left(\frac{\mathbf{x}_*'(t)}{\sqrt{1 + (\mathbf{x}_*'(t))^2}}\right) = 0,$$

then it follows that

$$\frac{\mathbf{x}_*'(t)}{\sqrt{1 + (\mathbf{x}_*'(t))^2}} = C, \quad t \in [-r, r],$$

for some constant C, and hence $\mathbf{x}_*'(t) = C'$, for some constant C', which in turn implies that $\mathbf{x}_*(t) = C't + D'$, $t \in [-r, r]$ and some constant D'. But then the fact that $\mathbf{x}_*(-r) = 0 = \mathbf{x}_*(r)$ yields $C' = D' = 0$, and so the length of the string is $g(\mathbf{x}_*) = 2r \neq \ell$, a contradiction.

For $\lambda \in \mathbb{R}$, we have $F + \lambda G = -\xi + \lambda\sqrt{1 + \eta^2}$, and so

$$\frac{\partial(F + \lambda G)}{\partial \xi}(\xi, \eta, \tau) = -1, \quad \text{and} \quad \frac{\partial(F + \lambda G)}{\partial \eta}(\xi, \eta, \tau) = \frac{\lambda\eta}{\sqrt{1 + \eta^2}}.$$

So the Euler-Lagrange equation for an $\mathbf{x}_*$ that maximizes the area is given by

$$-1 - \frac{d}{dt}\left(\frac{\lambda\mathbf{x}_*'(t)}{\sqrt{1 + (\mathbf{x}_*'(t))^2}}\right) = 0, \quad t \in [-r, r],$$

and so by integrating, we obtain a constant A such that

$$\frac{\lambda \mathbf{x}_*'(t)}{\sqrt{1 + (\mathbf{x}_*'(t))^2}} = t + A, \quad t \in [-r, r],$$

which yields $(\lambda^2 - (t + A)^2)(\mathbf{x}_*'(t))^2 = (t + A)^2$, and so

$$\mathbf{x}_*'(t) = \pm \frac{t + A}{\sqrt{\lambda^2 - (t + A)^2}}.$$

But $\dfrac{d}{dt} \sqrt{\lambda^2 - (t + A)^2} = -\dfrac{t + A}{\sqrt{\lambda^2 - (t + A)^2}}$, and so

$$\mathbf{x}_*(t) + B = \mp \sqrt{\lambda^2 - (t + A)^2},$$

that is, $(\mathbf{x}_*(t) + B)^2 + (t + A)^2 = \lambda^2$, which describes a circular arc with center at $(-A, -B)$ and radius $|\lambda|$. Also, using $\mathbf{x}_*(-r) = 0 = \mathbf{x}_*(r)$, it follows that $B^2 + (r + A)^2 = \lambda^2 = B^2 + (-r + A)^2$. Thus $(r + A)^2 = (-r + A)^2$. But if $r + A = -r + A$, then $r = 0$, a contradiction. So we must have $r + A = r - A$, and so $A = 0$. Moreover, $B^2 + r^2 = \lambda^2$ gives $\lambda = \pm\sqrt{B^2 + r^2}$, so that $|\lambda| = \sqrt{B^2 + r^2}$. As the arc length of our circular arc is ℓ, we also have

$$\sqrt{B^2 + r^2} \cdot 2 \tan^{-1}\left(\frac{r}{B}\right) = \ell;$$

see Figure 8.

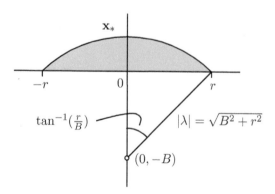

Figure 8. Circular arc maximizing the area.

For example, if $\ell = \pi r$, then we see that $B = 0$ gives a solution:

$$\sqrt{r^2} \cdot 2 \tan^{-1}(+\infty) = r \cdot 2\frac{\pi}{2} = \pi r = \ell,$$

and the shape of $\mathbf{x}$ that maximizes the area is a semicircle. $\diamond$

Exercise 2.22. Solve the optimization problem

$$
\begin{cases}
\text{minimize} & \displaystyle\int_0^1 t\mathbf{x}(t)\,dt \\[2mm]
\text{subject to} & \displaystyle\int_0^1 (\mathbf{x}(t))^2\,dt = \frac{1}{12}, \\[2mm]
& \mathbf{x} \in C^1[0,1].
\end{cases}
$$

Sketch of the proof of Theorem 2.20. Let $\mathbf{h}, \mathbf{k}$ be fixed functions in $C^1[a,b]$ such that

$$
\begin{aligned}
\mathbf{h}(a) &= 0 = \mathbf{h}(b) \text{ and} \\
\mathbf{k}(a) &= 0 = \mathbf{k}(b).
\end{aligned}
$$

Define $\varphi : \mathbb{R}^2 \to \mathbb{R}$ and $\gamma : \mathbb{R}^2 \to \mathbb{R}$ by

$$
\begin{aligned}
\varphi(\epsilon, \delta) &= f(\mathbf{x}_* + \epsilon\mathbf{h} + \delta\mathbf{k}), \\
\gamma(\epsilon, \delta) &= g(\mathbf{x}_* + \epsilon\mathbf{h} + \delta\mathbf{k}).
\end{aligned}
$$

Set $\varphi(0,0) =: m$, the minimum value of f on

$$
S := \{\mathbf{x} \in C^1[a,b] : \mathbf{x}(a) = y_a, \ \mathbf{x}(b) = y_b, \ g(\mathbf{x}) = C\}.
$$

We know that $\gamma(0,0) = C$. Consider the map $\Phi : \mathbb{R}^2 \to \mathbb{R}^2$ that sends $(\epsilon, \delta) \in \mathbb{R}^2$ to $(\varphi(\epsilon, \delta), \gamma(\epsilon, \delta)) \in \mathbb{R}^2$. In particular, the point $(0,0)$ is sent to $(m, C) \in \mathbb{R}^2$. See Figure 9.

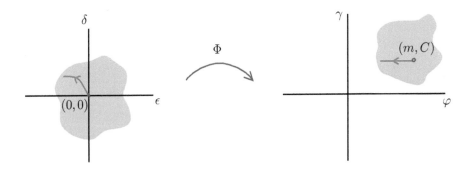

Figure 9. The local behavior of Φ if $\Phi'(0,0)$ is invertible.

The local behavior of Φ around $(0,0)$ is described by its derivative

$$
\Phi'(0,0) = \begin{bmatrix} \dfrac{\partial\varphi}{\partial\epsilon}(0,0) & \dfrac{\partial\varphi}{\partial\delta}(0,0) \\[3mm] \dfrac{\partial\gamma}{\partial\epsilon}(0,0) & \dfrac{\partial\gamma}{\partial\delta}(0,0) \end{bmatrix}.
$$

If $\Phi'(0,0)$ is invertible, then it follows from the Inverse Function Theorem [**R**] that there are open sets U, V in $\mathbb{R}^2$ such that $(0,0) \in U$, $\Phi(0,0) = (m, C) \in V$ and $\Phi(U) = V$. In particular, it follows that we can choose $(\epsilon_0, \delta_0) \in U$ such that $\gamma(\epsilon_0, \delta_0) = C$, and $\varphi(\epsilon_0, \delta_0) < m$. But this contradicts the fact that f has the minimum value m on S. Hence it is not the case that $\Phi'(0,0)$ is invertible. So the rows of $\Phi'(0,0)$ are linearly dependent. Since it is not the case that for all $t \in [a,b]$,

$$\frac{\partial G}{\partial \xi}(\mathbf{x}_*(t), \mathbf{x}'_*(t), t) - \frac{d}{dt}\left(\frac{\partial G}{\partial \eta}(\mathbf{x}_*(t), \mathbf{x}'_*(t), t)\right) = 0,$$

it follows that $\widetilde{g}'(\mathbf{0}) \neq \mathbf{0}$, where $\widetilde{g} : X \to \mathbb{R}$ is given by

$$\widetilde{g}(\mathbf{h}) = g(\mathbf{x}_* + \mathbf{h}), \quad \mathbf{h} \in X := \{\mathbf{h} \in C^1[a,b] : \mathbf{h}(a) = 0 = \mathbf{h}(b)\}.$$

So there exists an $\mathbf{h}_0 \in X$ such that $\widetilde{g}'(\mathbf{0})\mathbf{h}_0 \neq 0$. But

$$\frac{\partial \gamma}{\partial \epsilon}(0,0) = \widetilde{g}'(\mathbf{0})\mathbf{h}_0$$

and so the second row of $\Phi'(0,0)$ is a nonzero vector in $\mathbb{R}^2$. Since the two rows of $\Phi'(0,0)$ are linearly dependent, it follows now that there must exist a $\lambda \in \mathbb{R}$ such that

$$\left[\begin{array}{cc} \dfrac{\partial \varphi}{\partial \epsilon}(0,0) & \dfrac{\partial \varphi}{\partial \delta}(0,0) \end{array}\right] + \lambda \left[\begin{array}{cc} \dfrac{\partial \gamma}{\partial \epsilon}(0,0) & \dfrac{\partial \gamma}{\partial \delta}(0,0) \end{array}\right] = \left[\begin{array}{cc} 0 & 0 \end{array}\right].$$

In particular,

$$\frac{\partial \varphi}{\partial \delta}(0,0) + \lambda \frac{\partial \gamma}{\partial \delta}(0,0) = 0.$$

But

$$\frac{\partial \varphi}{\partial \delta}(0,0) = \widetilde{f}'(\mathbf{0})\mathbf{k} \quad \text{and} \quad \frac{\partial \gamma}{\partial \delta}(0,0) = \widetilde{g}'(\mathbf{0})\mathbf{k},$$

where $\widetilde{f} : X \to \mathbb{R}$ is given by

$$\widetilde{f}(\mathbf{k}) = f(\mathbf{x}_* + \mathbf{k}), \quad \mathbf{k} \in X := \{\mathbf{k} \in C^1[a,b] : \mathbf{k}(a) = 0 = \mathbf{k}(b)\}.$$

Thus the above gives

$$\widetilde{f}'(\mathbf{0})\mathbf{k} + \lambda \widetilde{g}'(\mathbf{0})\mathbf{k} = 0.$$

But the choice of $\mathbf{k} \in X$ was arbitrary, and so the above holds for all $\mathbf{k} \in X$. This yields (2.5). $\qquad \square$

Part 2

Optimal Control

Chapter 3

Differential Equations

In the rest of the book, we will study optimization problems in function spaces where one has a *differential equation constraint*. This is the subject of Optimal Control Theory, and we'll learn to solve the following general sort of a problem:

$$
\begin{cases}
\text{minimize} & \varphi(\mathbf{x}(t_f)) + \displaystyle\int_{t_i}^{t_f} F(t, \mathbf{x}(t), \mathbf{u}(t))dt \\
\text{subject to} & \boxed{\mathbf{x}'(t) = f(t, \mathbf{x}(t), \mathbf{u}(t)), \quad t \in [t_i, t_f]}, \\
& \mathbf{x}(t_i) = x_i \in \mathbb{R}^n, \\
& \mathbf{x}(t_f) \in \mathbb{X}_f \subset \mathbb{R}^n, \\
& \mathbf{u}(t) \in \mathbb{U} \subset \mathbb{R}^m, \quad t \in [t_i, t_f].
\end{cases}
$$

We will specify in detail later what each of the things mean in the above structure of the problem, and also provide motivation for solving these types of problems via examples. And we will learn two methods for solving them, one called the Hamiltonian method, and the other based on Bellman's equation. Right now, we focus on the boxed line in the above problem. This is a differential equation, which appears as a constraint in the above optimization problem. Hence it is clear that we will need some background knowledge of such differential equations, and this is what we will learn now.

3.1. High Order to First Order

Many problems in the applied sciences such as physics, biology, economics, engineering and so on involve rates of change dependent on the interaction of the basic elements; for example:

(1) position or velocities in physics,

(2) population size in biology,

(3) assets, prices, interest rates, etc. in economics.

So relationships between quantities naturally give rise to differential equation models. What is a differential equation? It is an equation involving known and unknown functions of a single variable (t, thought of as time) and the derivatives of (some of) these functions. Such a "definition" doesn't help much, and the best way to become familiar with differential equations is to see examples. For instance,

$$\mathbf{x}'''(t) + (\mathbf{x}(t))^9 = te^{9t}$$

is a differential equation. One can also consider a *system* of differential equations, for example,

$$\begin{cases} \mathbf{x}_1'(t) &= 2\mathbf{x}_1(t) + (\mathbf{x}_2(t))^2, \\ \mathbf{x}_2'(t) &= -3(\sin t)\mathbf{x}_1(t). \end{cases}$$

What we are considering here are Ordinary Differential Equations (ODEs), where we have just one independent variable (namely, time t), and this is in contrast to Partial Differential Equations (PDEs), where the equations involve known and unknown functions of several variables $(t, \xi, \cdots)$ and their partial derivatives; for example,

$$\frac{\partial \mathbf{x}}{\partial t}(\xi, t) = \frac{\partial^2 \mathbf{x}}{\partial \xi^2}(\xi, t) + e^{-\xi - t}.$$

So we will be considering ODEs, and in particular we will only study (systems of) "first order" equations of the following form.

$$\mathbf{x}_1'(t) = f_1(t, \mathbf{x}_1(t), \mathbf{x}_2(t), \ldots, \mathbf{x}_n(t)), \tag{3.1}$$
$$\mathbf{x}_2'(t) = f_2(t, \mathbf{x}_1(t), \mathbf{x}_2(t), \ldots, \mathbf{x}_n(t)), \tag{3.2}$$
$$\vdots$$
$$\mathbf{x}_n'(t) = f_n(t, \mathbf{x}_1(t), \mathbf{x}_2(t), \ldots, \mathbf{x}_n(t)). \tag{3.3}$$

Here the (known) functions $(\tau, \xi_1, \ldots, \xi_n) \mapsto f_i(\tau, \xi_1, \ldots, \xi_n)$ take values in $\mathbb{R}$ (the real numbers) and are defined on a set in $\mathbb{R}^{n+1}$ $(= \mathbb{R} \times \mathbb{R} \times \cdots \times \mathbb{R}, n+1$ times).

Note that the system of equations (3.1)-(3.3) are *first order*, in that the derivatives occurring are of order at most 1. However, in applications, one may end up with a model described by a set of *high order* equations. So why restrict our study only to *first* order systems? Why not also consider *high* order differential equations that involve not just derivatives of the functions as above, but also possibly higher order derivatives (second, third and so on)? The reason is that one can replace high order models by first order models by the following procedure of introducing a "state vector." So throughout the sequel we will consider only a system of first order equations.

Let us consider the system

$$\begin{cases} \mathbf{y}'''(t) + \mathbf{y}''(t) + \sin t + \mathbf{z}(t) = 0, \\ \mathbf{z}''(t) + (\mathbf{z}'(t))^2 + \mathbf{y}'(t)\mathbf{z}(t) = 0. \end{cases}$$

Introduce the new functions $\mathbf{x}_1, \mathbf{x}_2, \cdots$ as follows

$$
\begin{aligned}
\mathbf{x}_1(t) &:= \mathbf{y}(t), \\
\mathbf{x}_2(t) &:= \mathbf{y}'(t), \\
\mathbf{x}_3(t) &:= \mathbf{y}''(t), \\
\mathbf{x}_4(t) &:= \mathbf{z}(t), \\
\mathbf{x}_5(t) &:= \mathbf{z}'(t).
\end{aligned}
$$

So we take $\mathbf{y}$ and its successive derivatives and give them names and stop at one less than the largest order of differentiation of $\mathbf{y}$. Since the order of differentiation of $\mathbf{y}$ is 3, we stop at $\mathbf{y}''$. We do a similar thing with $\mathbf{z}$, and its derivatives. So we now have got all these new functions. What is the big deal? The big deal is that these new functions $\mathbf{x}_1, \mathbf{x}_2, \cdots$ satisfy a first order system!

$$
\begin{aligned}
\mathbf{x}_1'(t) &= \mathbf{y}'(t) = \mathbf{x}_2(t), \\
\mathbf{x}_2'(t) &= \mathbf{y}''(t) = \mathbf{x}_3(t), \\
\mathbf{x}_3'(t) &= \mathbf{y}'''(t) = -\mathbf{y}''(t) - \sin t - \mathbf{z}(t) = -\mathbf{x}_3(t) - \sin t - \mathbf{x}_4(t), \\
\mathbf{x}_4'(t) &= \mathbf{z}'(t) = \mathbf{x}_5(t), \\
\mathbf{x}_5'(t) &= \mathbf{z}''(t) = -(\mathbf{z}'(t))^2 - \mathbf{y}'(t) \cdot \mathbf{z}(t) = -(\mathbf{x}_5(t))^2 - \mathbf{x}_2(t) \cdot \mathbf{x}_4(t).
\end{aligned}
$$

Note that in the third and fifth equality, where we had the largest order of derivatives of $\mathbf{y}$ and $\mathbf{z}$, we used the two differential equations to express these in terms of lower orders of derivatives of $\mathbf{y}$ and $\mathbf{z}$. So we arrive at the following first order system:

$$
\begin{cases}
\mathbf{x}_1'(t) = \mathbf{x}_2(t), \\
\mathbf{x}_2'(t) = \mathbf{x}_3(t), \\
\mathbf{x}_3'(t) = -\mathbf{x}_3(t) - \sin t - \mathbf{x}_4(t), \\
\mathbf{x}_4'(t) = \mathbf{x}_5(t), \\
\mathbf{x}_5'(t) = -(\mathbf{x}_5(t))^2 - \mathbf{x}_2(t) \cdot \mathbf{x}_4(t).
\end{cases}
$$

So if $\mathbf{y}, \mathbf{z}$ satisfy the original pair of differential equations, then the functions $\mathbf{x}_1, \cdots, \mathbf{x}_5$, defined in the above manner (as successive derivatives of $\mathbf{y}$ and $\mathbf{z}$), satisfy the above first order system. Vice versa, if $\mathbf{x}_1, \cdots, \mathbf{x}_5$ satisfy the above first order system, then defining $\mathbf{y} := \mathbf{x}_1$ and $\mathbf{z} := \mathbf{x}_4$, it is easy to check that $\mathbf{y}$ and $\mathbf{z}$ satisfy the high order pair of differential equations considered at the outset.

The auxiliary vector $\mathbf{x}$

$$
\mathbf{x}(t) = \begin{bmatrix} \mathbf{x}_1(t) \\ \vdots \\ \mathbf{x}_5(t) \end{bmatrix}
$$

comprising successive derivatives of the unknown functions in the high order differential equation, is called a *state* and the resulting system of first order differential equations is called a *state equation*.

Exercise 3.1. By introducing appropriate state variables, write a state equation in each case:

> (1) $\mathbf{x}'' + \omega^2 \mathbf{x} = 0$, where ω is a constant.
> (2) $\mathbf{y}'' + \mathbf{y} = 0$, $\mathbf{z}'' + \mathbf{z}' + \mathbf{z} = 0$.
> (3) $\mathbf{x}'' + t \sin \mathbf{x} = 0$.

3.2. Initial Value Problem. Existence and Uniqueness.

Introducing the vector notation

$$
\mathbf{x} := \begin{bmatrix} x_1 \\ \vdots \\ x_n \end{bmatrix}, \quad \mathbf{x}' := \begin{bmatrix} x_1' \\ \vdots \\ x_n' \end{bmatrix}, \quad \text{and} \quad f = \begin{bmatrix} f_1 \\ \vdots \\ f_n \end{bmatrix},
$$

the system of differential equations (3.1)-(3.3) can be abbreviated simply as

$$
\mathbf{x}'(t) = f(t, \mathbf{x}(t)).
$$

Let $t_i \in \mathbb{R}$ (the "initial time") and $x_i \in \mathbb{R}^n$ (the "initial state") be given. We will consider the following "initial value problem":

$$
\begin{cases} \mathbf{x}'(t) = f(t, \mathbf{x}(t)), & t \geq t_i, \\ \mathbf{x}(t_i) = x_i. \end{cases} \tag{3.4}
$$

Here $\mathbf{x}(t), \mathbf{x}'(t) \in \mathbb{R}^n$, for all $t \geq t_i$, and f is an $\mathbb{R}^n$-valued function defined on a subset of $[t_i, \infty) \times \mathbb{R}^n$. As mentioned earlier, the data f, t_i, x_i are given. We seek a function $\mathbf{x} : [t_i, t_f] \to \mathbb{R}^n$, for some $t_f > t_i$, that satisfies (3.4) for each $t \in [t_i, t_f]$.

Example 3.2. Let $\mathbf{x}(t)$ denote the population size of fish in a lake at time t. If the population size is large, then one can ignore the graininess of $\mathbf{x}$, and just imagine it to be continuous-valued. Suppose that the population growth law is given by

$$
\begin{cases} \mathbf{x}'(t) = c\mathbf{x}(t)\left(1 - \dfrac{\mathbf{x}(t)}{M}\right), & t \geq t_i, \\ \mathbf{x}(t_i) = x_i. \end{cases}
$$

where c and M are known constants. Such a law makes sense since the rate $\mathbf{x}'(t)$ of increase of the size of population should be proportional to the current population size (the more fish there are, the more the likelihood of mating), and the term $(1 - \mathbf{x}(t)/M)$ reflects the fact that if the population size is comparable to some maximal population size M, then owing to competition for limited food resources, the population size starts diminishing. x_i is the initial population size at time t_i, which can be measured/estimated. So we see that initial value problems arise naturally in applications.   ◊

What are the questions we are interested in for the initial value problem (3.4)? To begin with, we are interested in existence and uniqueness of solutions:

(1) Does there exist a solution?

(2) Is it unique?

Let us first study some examples which illustrate that these are meaningful and relevant questions.

Example 3.3 (No solution). Suppose that $f : \mathbb{R} \to \mathbb{R}$ is the indicator function of the rational numbers:

$$f(t) = \begin{cases} 1 & \text{if } t \in \mathbb{Q}, \\ 0 & \text{if } t \in \mathbb{R} \setminus \mathbb{Q}. \end{cases}$$

Then for any $x_i, t_i \in \mathbb{R}$, there is no solution to the initial value problem

$$\begin{cases} \mathbf{x}'(t) = f(t), & t \geq t_i, \\ \mathbf{x}(t_i) = x_i. \end{cases}$$

Indeed, if there was a differentiable $\mathbf{x} : [t_i, t_f] \to \mathbb{R}$ for some $t_f > t_i$, then since f is Lebesgue integrable on $[t_i, t_f]$, we have

$$\mathbf{x}(t) - x_i = \int_{t_i}^{t} \mathbf{x}'(\tau) d\tau = \int_{t_i}^{t} f(t) dt = 0, \quad t_i \leq t \leq t_f$$

and so $\mathbf{x} \equiv x_i$. This implies that $\mathbf{x}' = 0 \neq f$, a contradiction.

If a student is not familiar with the Lebesgue integral, then the above can be accepted on faith. Alternatively, here is a different elementary example, relying only on the Riemann integral. Define

$$S = \bigcup_{k=0,1,2,3,\cdots} \left[\frac{1}{2^{2k+1}}, \frac{1}{2^{2k}} \right],$$

and take f to be the indicator function of S. Suppose that the initial value problem with $t_i := 0$ has a differentiable solution $\mathbf{x} : [0, T] \to \mathbb{R}$ for some $T > 0$. Take a large enough k such that $[1/2^{2k+2}, 1/2^{2k}] \subset [0, T]$. Now for all values of t such that $1/2^{2k} > t > 1/2^{2k+1} =: t_0$, we have

$$\mathbf{x}(t) = \mathbf{x}(t_0) + \int_{t_0}^{t} \mathbf{x}'(\tau) d\tau = \mathbf{x}(t_0) + \int_{t_0}^{t} f(\tau) d\tau = \mathbf{x}(t_0) + \int_{t_0}^{t} 1 d\tau = \mathbf{x}(t_0) + t - t_0.$$

Also, for $1/2^{2k+2} < t < t_0 = 1/2^{2k+1}$, we have

$$\mathbf{x}(t_0) - \mathbf{x}(t) = \int_{t}^{t_0} \mathbf{x}'(\tau) d\tau = \int_{t}^{t_0} f(\tau) d\tau = \int_{t}^{t_0} 0 \, d\tau = 0,$$

and so $\mathbf{x}(t) = \mathbf{x}(t_0)$ for these values of t. Thus in the vicinity of t_0, the graph of the function $\mathbf{x}$ has a corner (being flat for $t < t_0$, and linearly increasing with slope 1 for $t > t_0$), and so $\mathbf{x}$ is not differentiable at $t_0 \in [0, T]$, a contradiction! Hence the initial value problem does not have a solution. $\Diamond$

Of course, the above f is rather pathological! In ODE theory, a famous result of Peano guarantees the existence of a solution on some small time interval $[t_i, t_f]$ whenever $(t, x) \mapsto f(t, x)$ is continuous. Then the solution to our initial value problem must necessarily be *continuously* differentiable. However, we will see later on, when we study optimal control, that having f continuous is rather restrictive, and it will prevent us from considering very important practical applications (where one allows "input functions" that may experience step changes in time). We will assume that f is continuous in x and only piecewise continuous in t, in which a solution could only be piecewise continuously differentiable. This will allow us to cover important applications while having a general theory that guarantees the existence of solutions. Let us first explain what we mean by "piecewise continuous."

Definition 3.4. A function $\varphi : [t_i, t_f] \to \mathbb{R}^n$ is called *piecewise continuous* if there exist finitely many points $c_0 := t_i < c_1 < \cdots < c_n < t_f =: c_{n+1}$ such that φ is continuous on each subinterval (c_i, c_{i+1}), and the limits

$$\lim_{t \to t_i+} \varphi(t), \quad \lim_{t \to c_1-} \varphi(t), \lim_{t \to c_1+} \varphi(t), \quad \cdots \quad , \quad \lim_{t \to c_n-} \varphi(t), \lim_{t \to c_n+} \varphi(t), \quad \lim_{t \to t_f-} \varphi(t),$$

all exist. That is, at each point of discontinuity, the left- and right-hand limits exist; see Figure 1.

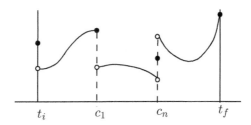

Figure 1. Piecewise continuity.

Similarly, a function $\varphi : [t_i, t_f] \to \mathbb{R}^n$ is called *piecewise continuously differentiable* if it is a continuous function, which is differentiable at all but finitely many points in $[t_i, t_f]$, and whose derivative has left- and right-hand limits at each point where it is not differentiable; see Figure 2.

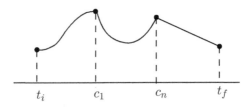

Figure 2. Piecewise differentiability.

Here is an example showing that even if a solution exists, it may not be unique.

Example 3.5 (Nonunique solution). Consider the initial value problem

$$\begin{cases} x'(t) = \sqrt{|x(t)|}, & t \geq 0, \\ x(0) = 0. \end{cases}$$

It is clear that $(t, x) \mapsto \sqrt{|x|} : [0, \infty) \times \mathbb{R} \to \mathbb{R}$ is continuous. Then $x = 0$, is clearly one solution. Yet another solution is given by

$$\tilde{x}(t) = \left(\frac{t}{2}\right)^2, \quad t \geq 0.$$

Indeed, for all $t \geq 0$,

$$\tilde{x}'(t) = 2 \cdot \frac{t}{2} \cdot \frac{1}{2} = \frac{t}{2} = \sqrt{\left|\left(\frac{t}{2}\right)^2\right|} = \sqrt{|\tilde{x}(t)|}$$

and

$$\tilde{x}(0) = \left(\frac{0}{2}\right)^2 = 0.$$

See Figure 3. ◇

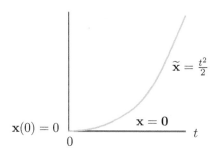

Figure 3. The distinct solutions $x, \tilde{x}$.

Finally, even if there exists a unique solution, it may not exist for *all* times $t \geq t_i$.

Example 3.6 (Finite escape time). Consider the initial value problem

$$\begin{cases} x'(t) = (x(t))^2, & t \geq 0, \\ x(0) = 1. \end{cases}$$

We note that

$$x(t) - x(0) = \int_0^t x'(\tau)d\tau = \int_0^t \underbrace{(x(\tau))^2}_{\geq 0} d\tau,$$

and so $x(t) \geq x(0) = 1$. So $x(t) \neq 0$ for all $t \geq 0$. Now

$$\frac{x'(t)}{(x(t))^2} = \frac{d}{dt}\left(-\frac{1}{x(t)}\right) = 1,$$

and so upon integrating from 0 to t, we obtain $-\dfrac{1}{\mathbf{x}(t)} + \dfrac{1}{\mathbf{x}(0)} = -\dfrac{1}{\mathbf{x}(t)} + 1 = t - 0.$

Hence

$$\mathbf{x}(t) = \frac{1}{1-t}, \quad 0 \le t < 1.$$

So the solution $\mathbf{x}$ "escapes to $+\infty$ in finite time". See Figure 4.   ◇

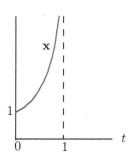

Figure 4. The solution $\mathbf{x}$ escapes to $+\infty$ in finite time.

In order to develop a general theory for optimal control, we would like to have a result which says that

> "If f is such and such, then there exists a unique solution to the initial value problem (3.4)."

Indeed, such a result is available from the theory of ODEs.

Theorem 3.7. *Suppose that $f : [t_i, t_f] \times \mathbb{R}^n \to \mathbb{R}^n$ satisfies the following:*

(1) *For all $x \in \mathbb{R}^n$, $t \mapsto f(t, x) : [t_i, t_f] \to \mathbb{R}^n$ is piecewise continuous.*

(2) *There exists an $L > 0$ such that for all $t \in [t_i, t_f]$ and all $x, y \in \mathbb{R}^n$,*

$$\|f(t, x) - f(t, y)\|_2 \le L\|x - y\|_2.$$

Then for each $x_i \in \mathbb{R}^n$, the initial value problem

$$\begin{cases} \mathbf{x}'(t) = f(t, \mathbf{x}(t)), & t \in [t_i, t_f], \\ \mathbf{x}(t_i) = x_i \end{cases}$$

has a solution and it is unique.

We won't prove this result, and we refer the interested reader to [**K**, Theorem 3.2].

Definition 3.8. A function $\varphi : \mathbb{R}^n \to \mathbb{R}^n$ is said to be *globally Lipschitz* if

$$\text{for all } x, y \in \mathbb{R}^n, \ \|\varphi(x) - \varphi(y)\|_2 \le L\|x - y\|_2.$$

L is then called a *Lipschitz constant*.

"Globally" means the same constant works everywhere in $\mathbb{R}^n$. (There is a local version of this definition, where the above inequality holds in balls of radius r and center 0, with the constant L possibly depending on r.)

The second condition, namely that there exists an $L > 0$ such that for all $t \in [t_i, t_f]$ and all $x, y \in \mathbb{R}^n$,

$$\|f(t, x) - f(t, y)\|_2 \le L\|x - y\|_2,$$

is abbreviated by saying that "f is globally Lipschitz in x uniformly in t."

"Uniformly in t" refers to the fact that the L is independent of which t we choose in $[t_i, t_f]$. Clearly the condition in particular implies that for each fixed $t \in [t_i, t_f]$, the map

$$x \mapsto f(t, x) : \mathbb{R}^n \to \mathbb{R}^n$$

is continuous. Roughly, the condition means the following when $n = 1$. The modulus of the slope of $f(t, \cdot)$ is bounded by L for all t:

$$\left| \frac{f(t, x) - f(t, y)}{x - y} \right| \le L$$

Example 3.9. We consider the three functions $|x|, \sqrt{|x|}, x^2$.

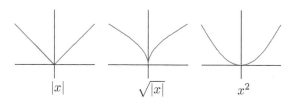

$$|x| \qquad\qquad \sqrt{|x|} \qquad\qquad x^2$$

We expect that $|x|$ is globally Lipschitz in x, since the modulus of the slope is bounded by 1. Indeed it is globally Lipschitz with a Lipschitz constant of 1, and this can be seen using the triangle inequality: for all $x, y \in \mathbb{R}$, $||x| - |y|| \le 1 \cdot |x - y|$. As a consequence, it follows from Theorem 3.7 that the initial value problem

$$\begin{cases} \mathbf{x}'(t) = (\cos t) \cdot |\mathbf{x}(t)|, & t \ge 0, \\ \mathbf{x}(0) = x_0, \end{cases}$$

has a unique solution for all $t \ge 0$.

We have seen that the initial value problem

$$\begin{cases} \mathbf{x}'(t) = \sqrt{|\mathbf{x}(t)|}, & t \geq 0, \\ \mathbf{x}(0) = 0, \end{cases}$$

does not have a unique solution. So in light of Theorem 3.7, we expect f, given by $f(\xi) := \sqrt{|\xi|}$ for $\xi \in \mathbb{R}$, to be not globally Lipschitz, and we prove this below. Suppose on the contrary that f is globally Lipschitz, with a Lipschitz constant L. Then for all $x, y \in \mathbb{R}$, we must have $|\sqrt{|x|} - \sqrt{|y|}| \leq L|x - y|$, and in particular, with $y = 0$, we obtain that $\sqrt{|x|} \leq L|x|$. But if we take $x = 1/n^2$ for $n \in \mathbb{N}$, then this yields that for all $n \in \mathbb{N}$, $n \leq L$, a contradiction. Hence $\sqrt{|x|}$ is not globally Lipschitz.

Finally, x^2 is not globally Lipschitz in x, because if it were, then with $x = n$, $n \in \mathbb{N}$, and $y = 0$, the inequality $|x^2 - y^2| \leq L|x - y|$ yields $n \leq L$ for all $n \in \mathbb{N}$, a contradiction. So it is no surprise that we had discovered that the initial value problem

$$\begin{cases} x'(t) = (x(t))^2, & t \geq 0, \\ x(0) = 1, \end{cases}$$

has a solution which has a finite escape time.   ◊

Exercise 3.10. If the f in the initial value problem (3.4) does not depend on t, that is, it is simply a function defined on some subset of $\mathbb{R}^n$, taking values in $\mathbb{R}^n$, then the differential equation

$$\mathbf{x}'(t) = f(\mathbf{x}(t))$$

is called *autonomous*. Classify the following differential equations as autonomous or as nonautonomous.

(1) $\mathbf{x}'(t) = e^{\mathbf{x}(t)}$.

(2) $\mathbf{x}'(t) = e^t \mathbf{y}(t)$, $\mathbf{y}'(t) = \mathbf{x}(t) + \mathbf{y}(t)$.

(3) $\mathbf{x}'(t) = \mathbf{y}(t)$, $\mathbf{y}'(t) = \mathbf{x}(t)\mathbf{y}(t)$.

Exercise 3.11. Verify that the initial value problem is solved by the given function.

(1) IVP: $\begin{cases} \mathbf{x}'(t) = a\mathbf{x}(t), \ t \geq 0, \\ \mathbf{x}(0) = x_i, \end{cases}$ (a is a constant).
    Solution: $\mathbf{x}(t) = e^{ta}x_i, \ t \geq 0$.

(2) IVP: $\begin{cases} \mathbf{x}_1'(t) = 2\mathbf{x}_2(t), \ t \geq 0, \\ \mathbf{x}_2'(t) = -2\mathbf{x}_1(t), \ t \geq 0, \\ \mathbf{x}_1(0) = 0, \\ \mathbf{x}_2(0) = 1. \end{cases}$
    Solution: $\begin{cases} \mathbf{x}_1(t) = \sin(2t), & t \geq 0 \\ \mathbf{x}_2(t) = \cos(2t), & t \geq 0. \end{cases}$

(3) IVP: $\begin{cases} \mathbf{x}'(t) = 2t(\mathbf{x}(t))^2, \ t \geq 0, \\ \mathbf{x}(0) = 1. \end{cases}$
    Solution: $\mathbf{x}(t) = \dfrac{1}{1 - t^2}$ for $t \in [0, 1)$.

3.3. Linear Systems

Now let us consider a very important special case of ODEs, namely a *linear* first order system:
$$\mathbf{x}'(t) = A(t)\mathbf{x}(t),$$
where for each $t \in \mathbb{R}$, $A(t)$ is a square matrix, with entries denoted by $a_{ij}(t)$, $1 \le i, j \le n$, such that each map $t \mapsto a_{ij}(t) : \mathbb{R} \to \mathbb{R}$ is piecewise continuous.

If we look at the initial value problem
$$(\text{IVP})_{x_i} : \begin{cases} \mathbf{x}'(t) = A(t)\mathbf{x}(t), & t \ge t_i \\ \mathbf{x}(t_i) = x_i, \end{cases}$$

for a linear system, then let us note that this has a unique solution. Indeed, since A is piecewise continuous, the entries of A are bounded on any compact interval $[t_i, t_f]$, and so on this time interval, the solution exists and is unique. But if we take any large time $T_f > t_f$, again the solution exists and is unique. Since the restriction of this new solution to the old time interval $[t_i, t_f]$ also serves as a solution on $[t_i, t_f]$, the new solution is just an extension of the old one to a bigger time interval. As $T_f > t_f$ can be chosen arbitrarily large, it follows that the solution exists for all $t \ge t_i$.

The mapping that sends the initial condition in $\mathbb{R}^n$ to the corresponding unique solution is linear. We explain this below. For the initial condition $v \in \mathbb{R}^n$, let $(\text{IVP})_v$ denote the initial value problem:
$$(\text{IVP})_v : \begin{cases} \mathbf{x}'(t) = A(t)\mathbf{x}(t), & t \ge t_i \\ \mathbf{x}(t_i) = v, \end{cases}$$

and let us denote its unique solution by $\mathbf{x}_v : \mathbb{R} \to \mathbb{R}^n$. Then there holds that
$$\mathbf{x}_{v+w} = \mathbf{x}_v + \mathbf{x}_w$$
$$\mathbf{x}_{\alpha \cdot v} = \alpha \cdot \mathbf{x}_v.$$

How is this observation helpful? We will see that by just finding the solution to n initial conditions, we can find the solution to *every* initial condition of the given linear system. We elaborate on this now. First suppose that

$$e_1 := \begin{bmatrix} 1 \\ 0 \\ \vdots \\ 0 \end{bmatrix}, \quad \cdots \quad, \quad e_n := \begin{bmatrix} 0 \\ \vdots \\ 0 \\ 1 \end{bmatrix},$$

denote the standard basis vectors in $\mathbb{R}^n$, and $t \mapsto \varphi_k(t, t_i)$, $k = 1, \cdots, n$, denote the solutions to
$$(\text{IVP})_{e_k} : \begin{cases} \mathbf{x}'(t) = A(t)\mathbf{x}(t), & t \ge t_i \\ \mathbf{x}(t_i) = e_k. \end{cases}$$
Set
$$\Phi(t, t_i) := \begin{bmatrix} \varphi_1(t, t_i) & \cdots & \varphi_n(t, t_i) \end{bmatrix} \in \mathbb{R}^{n \times n}.$$

Then we have that the initial value problem

$$(\text{IVP})_{x_i} : \begin{cases} \mathbf{x}'(t) = A(t)\mathbf{x}(t), & t \geq t_i \\ \mathbf{x}(t_i) = x_i, \end{cases}$$

has the unique solution $\mathbf{x}(t) = \Phi(t, t_i)x_i$, $t \geq t_i$. Indeed, if $x_i = \alpha_1 \cdot e_1 + \cdots + \alpha_n \cdot e_n$, then it follows from the linearity that

$$\mathbf{x}(t) = \alpha_1 \cdot \varphi_1(t, t_i) + \cdots + \alpha_n \cdot \varphi_n(t, t_i) = \Phi(t, t_1)x_i.$$

If we keep t_i fixed, then clearly the $\Phi(t, t_i)$ is a time varying matrix, but we can also envisage changing the $t_i \in \mathbb{R}$, so that we have a matrix function of two variables:

$$(t, s) \overset{\Phi}{\mapsto} \Phi(t, s), \quad t \geq s.$$

We call Φ the *state transition matrix*. From the result on the existence and uniqueness of solutions to the initial value problem, we now prove the following important properties of Φ, which we will use in the sequel.

Theorem 3.12. *With the notation developed above, there holds that:*

(1) $\Phi(t, t) = I$ *for all $t \in \mathbb{R}$.*

(2) *For all $t_i \in \mathbb{R}$, and all $t_f \geq t_i$, $\Phi(t_f, t_i)$ is invertible.*

(3) $\Phi(T, s)\Phi(s, t) = \Phi(T, t)$ *for all $t \leq s \leq T$.*

Proof.

(1) The solution to the initial value problem

$$(\text{IVP})_{x_i} : \begin{cases} \mathbf{x}'(t) = A(t)\mathbf{x}(t), & t \geq t_i \\ \mathbf{x}(t_i) = x_i, \end{cases}$$

is given by $\mathbf{x}(t) = \Phi(t, t_i)x_i$ for all $t \geq t_i$ and so in particular, with $t = t_i$, we must have $x_i = \Phi(t_i, t_i)x_i$. As the choice of $x_i \in \mathbb{R}^n$ was arbitrary, it follows that $\Phi(t_i, t_i) = I$. But $t_i \in \mathbb{R}$ was arbitrary too, and hence $\Phi(t, t) = I$ for all $t \in \mathbb{R}$.

(2) Since $\Phi(t_f, t_i)$ is square, it is enough to show that it is onto. So let $v \in \mathbb{R}^n$. Consider the initial value problem

$$\begin{cases} \mathbf{y}'(t) = -A(t_f + t_i - t)\mathbf{y}(t), & t \geq t_i \\ \mathbf{y}(t_i) = v, \end{cases}$$

Then $t \mapsto -A(t_f + t_i - t)$ is piecewise continuous, and so this initial value problem has a unique solution. Define $\mathbf{x}(t) = \mathbf{y}(t_f + t_i - t)$, $t \geq t_i$. Then

$$\begin{aligned} \mathbf{x}'(t) &= \mathbf{y}'(t_f + t_i - t) \cdot (-1) \\ &= -A(t_f + t_i - (t_f + t_i - t))\mathbf{y}(t_f + t_i - t) \cdot (-1) \\ &= A(t)\mathbf{x}(t) \end{aligned}$$

for $t \in [t_i, t_f]$, and moreover, $\mathbf{x}(t_i) = \mathbf{y}(t_f)$, and $\mathbf{x}(t_f) = \mathbf{y}(t_i) = v$. But we know that $\mathbf{x}(t_f) = \Phi(t_f, t_i)\mathbf{x}(t_i)$, and so $\Phi(t_f, t_i)\mathbf{y}(t_f) = v$. Thus v belongs to the range of $\Phi(t_f, t_i)$. As the choice of $v \in \mathbb{R}^n$ was arbitrary, it follows that $\Phi(t_f, t_i)$ is onto.

(3) This result is intuitively clear based on Figure 5.

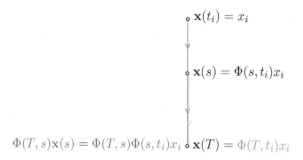

Figure 5. The property $\Phi(T, t_i) = \Phi(T, s)\Phi(s, t_i)$.

Consider the two initial value problems

$$(\text{IVP})_{x_i} : \quad \begin{cases} \mathbf{x}'(t) = A(t)\mathbf{x}(t), & t \geq t_i \\ \mathbf{x}(t_i) = x_i \end{cases} \quad \text{and}$$

$$(\text{IVP})_{\Phi(s, t_i)x_i} : \quad \begin{cases} \mathbf{x}'(t) = A(t)\mathbf{x}(t), & t \geq s \\ \mathbf{x}(s) = \Phi(s, t_i)x_i. \end{cases}$$

It is clear that since at time s, the solution $\mathbf{x}$ to $(\text{IVP})_{x_i}$ is given by

$$\mathbf{x}(s) = \Phi(s, t_i)x_i,$$

which is precisely the initial value at time s of $(\text{IVP})_{\Phi(s, t_i)x_i}$, the restriction of $\mathbf{x}$ for times $t \geq s$ is the solution to $(\text{IVP})_{\Phi(s, t_i)x_i}$. Thus

$$\mathbf{x}(T) = \Phi(T, t_i)x_i = \Phi(T, s)\Phi(s, t_i)x_i \text{ for all } T \geq s.$$

As $x_i \in \mathbb{R}^n$ was arbitrary, it follows from here that $\Phi(T, t_i) = \Phi(T, s)\Phi(s, t_i)$. $\quad\square$

Exercise 3.13. We have seen above that the solution to the initial value problem

$$(\text{IVP})_{x_i} : \begin{cases} \mathbf{x}'(t) = A(t)\mathbf{x}(t), & t \geq t_i \\ \mathbf{x}(t_i) = x_i \end{cases}$$

is given by $\mathbf{x}(t) = \Phi(t, t_i)x_i$, $t \geq t_i$. Now consider the related "final" value problem

$$(\text{FVP})_{\lambda_f} : \begin{cases} \boldsymbol{\lambda}'(t) = -(A(t))^\top \boldsymbol{\lambda}(t), & t \in [t_i, t_f] \\ \boldsymbol{\lambda}(t_f) = \lambda_f. \end{cases}$$

Prove that the unique solution to $(\text{IVP})_{\lambda_f}$ is given by $\boldsymbol{\lambda}(t) = \Phi(t_f, t)\lambda_f$, $t \in [t_i, t_f]$.
Hint: Differentiate $(\boldsymbol{\lambda}(t))^\top \mathbf{x}(t)$.

3.3.1. Scalar case. A special case of the linear system is when $n = 1$, so that we have a scalar system

$$(\text{IVP})_{x_i} : \begin{cases} \mathbf{x}'(t) = a(t)\mathbf{x}(t), & t \geq t_i \\ \mathbf{x}(t_i) = x_i \in \mathbb{R}. \end{cases}$$

In this case, it is easy to see directly that the solution is given by

$$\mathbf{x}(t) = \exp\left(\int_{t_i}^{t} a(\tau)d\tau\right)x_i, \quad t \geq t_i.$$

(And so, $\Phi(t, t_i) = \exp\left(\int_{t_i}^{t} a(\tau)d\tau\right)$.) Indeed, we have for $t \geq t_i$,

$$
\begin{aligned}
\mathbf{x}'(t) &= \exp\left(\int_{t_i}^{t} a(\tau)d\tau\right) \cdot \frac{d}{dt}\left(\int_{t_i}^{t} a(\tau)d\tau\right) \cdot x_i \\
&= \exp\left(\int_{t_i}^{t} a(\tau)d\tau\right) \cdot a(t) \cdot x_i = a(t) \cdot \exp\left(\int_{t_i}^{t} a(\tau)d\tau\right)x_i \\
&= a(t)\mathbf{x}(t)
\end{aligned}
$$

and moreover $\mathbf{x}(t_i) = \exp\left(\int_{t_i}^{t_i} a(\tau)d\tau\right)x_i = (\exp 0) \cdot x_i = x_i$.

Exercise 3.14. Show that in the special case when $a(t) \equiv a$ (autonomous/time-invariant) case, the solution to

$$(\text{IVP})_{x_i} : \begin{cases} \mathbf{x}'(t) = a(t)\mathbf{x}(t), & t \geq t_i \\ \mathbf{x}(t_i) = x_i \in \mathbb{R} \end{cases}$$

is given by $\mathbf{x}(t) = e^{(t-t_i)a}x_i, \ t \geq t_i$.

3.3.2. Autonomous/time-invariant case. Another special case of linear systems is the autonomous case, that is, when $A(t) \equiv A$ for all $t \in \mathbb{R}$.

It turns out that for the initial value problem

$$(\text{IVP})_{x_i} : \begin{cases} \mathbf{x}'(t) = A\mathbf{x}(t), & t \geq t_i \\ \mathbf{x}(t_i) = x_i, \end{cases}$$

just like the scalar case when $A = a \in \mathbb{R}$ discussed in Exercise 3.14 above, the solution in the general case is notationally the same:

$$\mathbf{x}(t) = e^{(t-t_i)A}x_i, \quad t \geq t_i,$$

with the little 'a' now replaced by the matrix 'A'! But what do we mean by the exponential of a matrix, e^{tA}? We first introduce this concept, and then show how one can compute it.

Just like for $a \in \mathbb{R}$, one defines e^a or $\exp a$ as the sum of the series

$$e^a = 1 + \frac{1}{1!}a + \frac{1}{2!}a^2 + \frac{1}{3!}a^3 + \cdots,$$

when $A \in \mathbb{R}^{n \times n}$, we define e^A or $\exp A$ by

$$e^A = I + \frac{1}{1!}A + \frac{1}{2!}A^2 + \frac{1}{3!}A^3 + \cdots .$$

(Here I denotes the identity matrix in $\mathbb{R}^{n \times n}$.) It can be shown that this series converges for all $A^{n \times n}$. What does this mean? We can form the partial sum s_k for $k \in \mathbb{N}$:

$$s_k := I + \frac{1}{1!}A + \frac{1}{2!}A^2 + \frac{1}{3!}A^3 + \cdots + \frac{1}{k!}A^k.$$

If for $1 \leq i, j \leq n$, we denote the entry in the ith row and jth column of the square matrix s_k by $s_k(i, j)$, then by the convergence of e^A, we mean that for each i, j, the sequence $(s_k(i, j))_{k \in \mathbb{N}}$ converges, and the limit is taken as the entry in the ith row and jth column of e^A.

For example, if $\mathbf{0} \in \mathbb{R}^{n \times n}$ denotes the zero matrix with all entries equal to 0, then

$$e^{\mathbf{0}} = I + \mathbf{0} + \mathbf{0} + \cdots = I.$$

This already tells us that computing e^A is not[1] the same as simply taking exponents of the entries of A. (After all, if that was the case, then $e^{\mathbf{0}}$ would have been the matrix with all entries equal to 1, but the off-diagonal entries of I are zeros!) We will learn a method to calculate the exponential of a matrix at the end of this section. But first, let us note that if A is a fixed matrix, then for $t \in \mathbb{R}$, we have

$$e^{tA} = I + \frac{t}{1!}A + \frac{t^2}{2!}A^2 + \frac{t^3}{3!}A^3 + \cdots .$$

We note that we can think of $t \mapsto e^{tA}$ as an $n \times n$ matrix, each entry of which is a function $t \mapsto e^{tA}(i, j)$ of time. So we can imagine differentiating each entry of e^{tA} with respect to time, and can form the matrix of derivatives

$$\frac{d}{dt}e^{tA}.$$

Just like we know in the scalar case that $\frac{d}{dt}e^{ta} = ae^{ta}$, one can show that

$$\frac{d}{dt}e^{tA} = Ae^{tA} = e^{tA}A.$$

(Since we are dealing with matrices, the last equality above is not superfluous.) One can see the plausibility of the above if one differentiates the series for e^{tA}

---

[1]For $n > 1$.

termwise with respect to t:

$$\frac{d}{dt}e^{tA} = \frac{d}{dt}\left(I + \frac{t}{1!}A + \frac{t^2}{2!}A^2 + \frac{t^3}{3!}A^3 + \cdots\right)$$

$$= 0 + \frac{1}{1!}A + \frac{2t}{2!}A^2 + \frac{3t^2}{3!}A^3 + \cdots$$

$$= A\left(I + \frac{t}{1!}A + \frac{t^2}{2!}A^2 + \frac{t^3}{3!}A^3 + \cdots\right) = \left(I + \frac{t}{1!}A + \frac{t^2}{2!}A^2 + \frac{t^3}{3!}A^3 + \cdots\right)A$$

$$= Ae^{tA} = e^{tA}A.$$

Also, although $e^{A+B} \neq e^A e^B$ in general for matrices $A, B \in \mathbb{R}^{n \times n}$ (see Exercise 3.23), it turns out that whenever $AB = BA$, this does hold. Note that it follows from here that e^{tA} is always invertible and in fact its inverse is e^{-tA}: Indeed $I = e^0 = e^{tA - tA} = e^{tA}e^{-tA}$. Using these properties of the matrix exponential, let us now show the following result relating the matrix exponential to differential equations.

Theorem 3.15. *The initial value problem*

$$(\text{IVP})_{x_i} : \begin{cases} \mathbf{x}'(t) = A\mathbf{x}(t), & t \geq t_i \\ \mathbf{x}(t_i) = x_i, \end{cases} \tag{3.5}$$

has the unique solution $\mathbf{x}(t) = e^{(t-t_i)A}x_i,\ t \geq t_i.$

Proof. We have

$$\frac{d}{dt}(e^{(t-t_i)A}x_i) = Ae^{(t-t_i)A}x_i,$$

and so $t \mapsto e^{(t-t_i)A}x_i$ solves $\mathbf{x}'(t) = A\mathbf{x}(t)$ for $t \geq t_i$. Furthermore,

$$\mathbf{x}(t_i) = e^{0A}x_i = e^0 x_i = Ix_i = x_i.$$

Finally we show that the solution is unique. Let $\mathbf{x}$ be a solution to (3.5). Differentiation of the matrix product $e^{-tA}\mathbf{x}(t)$ yields:

$$\frac{d}{dt}(e^{-tA}\mathbf{x}(t)) = -Ae^{-tA}\mathbf{x}(t) + e^{-tA}A\mathbf{x}(t) = 0.$$

Therefore, $e^{-tA}\mathbf{x}(t)$ is a constant column vector, say C, and so $\mathbf{x}(t) = e^{tA}C$. As $\mathbf{x}(t_i) = x_i$, we obtain that $x_i = e^{t_iA}C$, that is, $C = e^{-t_iA}x_i$. Consequently, $\mathbf{x}(t) = e^{tA}e^{-t_iA}x_i = e^{(t-t_i)A}x_i,\ t \geq t_i.$ $\qquad\square$

Thus the matrix exponential enables us to solve the differential equation (3.5). Below, we will learn a method to compute e^{tA} by using Laplace transforms.

Exercise 3.16. Show that for a diagonal matrix D with diagonal entries $\lambda_1, \cdots, \lambda_n$, e^D is a diagonal matrix with entries $e^{\lambda_1}, \cdots, e^{\lambda_n}$. Moreover, show that if A is *diagonalizable*, that is, if there exists a diagonal matrix D and an invertible matrix P such that $A = PDP^{-1}$, then $e^A = Pe^DP^{-1}$.

3.3.3. How to compute e^{tA}. In Exercise 3.16, we saw that the computation of e^{tA} is easy if the matrix A is diagonalizable. However, not all matrices are diagonalizable. For example, consider the matrix

$$A = \begin{bmatrix} 0 & 1 \\ 0 & 0 \end{bmatrix}.$$

The eigenvalues of this matrix are both 0, and so if it were diagonalizable, then the diagonal form will be the zero matrix, but then if there did exist an invertible P such that $P^{-1}AP$ is this zero matrix, then clearly A should be zero, which it is not!

In general, however, every matrix has what is called a *Jordan canonical form*, that is, there exists an invertible P such that $P^{-1}AP = D+N$, where D is diagonal, N is *nilpotent* (that is, there exists an integer $n > 0$ such that $N^n = 0$), and D and N commute. Then one can compute the exponential of A:

$$e^A = Pe^D \left(I + N + \frac{1}{2!}N^2 + \cdots + \frac{1}{n!}N^n \right) P^{-1}.$$

However, the algorithm for computing the P taking A to the Jordan form requires some sophisticated linear algebra. So we give a different procedure for calculating e^{tA} below, using Laplace transforms. This method is based on the following result on what happens when we take the entrywise Laplace transform of the elements of e^{tA}.

Theorem 3.17. *For large enough s,* $\int_0^\infty e^{-st}e^{tA}dt = (sI - A)^{-1}$.

In the above, the integral of a matrix whose elements are functions of t is defined entrywise. From the above result, we see that by taking the inverse Laplace transform of the entries of $(sI - A)^{-1}$, we would get the entries of e^{tA}. But first let us outline why this result holds, skipping all technical details.

Proof. (Sketch.) For all large enough real s, we have

$$\int_0^\infty e^{-ts}e^{tA}dt = \int_0^\infty e^{-t(sI-A)}dt = \int_0^\infty (sI-A)^{-1}(sI-A)e^{-t(sI-A)}dt$$

$$= (sI-A)^{-1}\int_0^\infty (sI-A)e^{-t(sI-A)}dt$$

$$= (sI-A)^{-1}\int_0^\infty -\frac{d}{dt}e^{-t(sI-A)}dt$$

$$= (sI-A)^{-1}\left(-e^{-ts}e^{tA}\Big|_{t=0}^{t=\infty} \right) = (sI-A)^{-1}(0+I) = (sI-A)^{-1}.$$

$\square$

If s is not an eigenvalue of A, then $sI - A$ is invertible, and by *Cramer's Rule*,[2]

$$(sI - A)^{-1} = \frac{1}{\det(sI - A)} \mathrm{adj}(sI - A).$$

Here $\mathrm{adj}(sI - A)$ denotes the *classical adjoint* or *adjugate* of the matrix $sI - A$, which is defined as follows: its (i,j)th entry is obtained by multiplying $(-1)^{i+j}$ and the determinant of the matrix obtained by deleting the jth row and ith column of $sI - A$. Thus we see that each entry of $\mathrm{adj}(sI - A)$ is a polynomial in s whose degree is at most $n-1$. (Here n denotes the size of A, that is, A is an $n \times n$ matrix.)

Consequently, each entry m_{ij} of $(sI - A)^{-1}$ is a rational function: in other words, it is a ratio of two polynomials (in s) p_{ij} and $q := \det(sI - A)$:

$$m_{ij} = \frac{p_{ij}(s)}{q(s)}.$$

Also from the above, we see that $\deg(p_{ij}) \leq \deg(q) - 1$. From the Fundamental Theorem of Algebra, we know that the monic polynomial q can be factored as

$$q(s) = (s - \lambda_1)^{m_1} \ldots (s - \lambda_k)^{m_k},$$

where $\lambda_1, \ldots, \lambda_k$ are the distinct eigenvalues of $q(s) = \det(sI - A)$, with the algebraic multiplicities $m_1, \ldots, m_k$.

By the "partial fraction expansion"[3] one learns in calculus, it follows that one can find suitable coefficients for a decomposition of each rational entry of $(sI - A)^{-1}$ as follows:

$$m_{ij} = \sum_{\ell=1}^{k} \sum_{r=1}^{m_\ell} \frac{C_{\ell,r}}{(s - \lambda_\ell)^r}.$$

Thus if $f_{ij}(t)$ denotes the (i,j)th entry of e^{tA}, then its Laplace transform will be an expression of the type m_{ij} given above. Now it turns out that this determines the f_{ij}, and this is the content of the following result. (For a proof, see for example [**A**, p.348].)

Theorem 3.18. *Let $a \in \mathbb{C}$ and $n \in \mathbb{N}$. If f is a continuous function defined on $[0, \infty)$, and if there exists an s_0 such that for all $s > s_0$,*

$$F(s) := \int_0^\infty e^{-st} f(t) dt = \frac{1}{(s - a)^n},$$

then

$$f(t) = \frac{1}{(n-1)!} t^{n-1} e^{ta} \quad \text{for all } t \geq 0.$$

---

[2]For a proof, see for instance Artin [**Ar**].
[3]See for example Sasane [**S**, Exercise 3.51, p.123-124].

So we have a procedure for computing e^{tA}: form the matrix $sI - A$, compute its inverse (as a rational matrix), perform a partial fraction expansion of each of its entries, and take the inverse Laplace transform of each elementary fraction. Sometimes, the partial fraction expansion may be avoided, by making use of the following corollary (which can be obtained from Theorem 3.18, by a partial fraction expansion!).

Corollary 3.19. *Let f be a continuous function defined on $[0, \infty)$, and let there exist an $s_0 \in [0, \infty)$ such that for all $s > s_0$, F defined by*

$$F(s) := \int_0^\infty e^{-st} f(t)dt,$$

is one of the functions given in the first column below. Then f is given by the corresponding entry in the second column.

F	f
$\dfrac{b}{(s-a)^2 + b^2}$	$e^{ta} \sin(bt)$
$\dfrac{s-a}{(s-a)^2 + b^2}$	$e^{ta} \cos(bt)$
$\dfrac{b}{(s-a)^2 - b^2}$	$e^{ta} \sinh(bt)$
$\dfrac{s-a}{(s-a)^2 - b^2}$	$e^{ta} \cosh(bt)$

Example 3.20. If $A = \begin{bmatrix} 0 & 1 \\ 0 & 0 \end{bmatrix}$, then $sI - A = \begin{bmatrix} s & -1 \\ 0 & s \end{bmatrix}$, and so

$$(sI - A)^{-1} = \frac{1}{s^2} \begin{bmatrix} s & 1 \\ 0 & s \end{bmatrix} = \begin{bmatrix} 1/s & 1/s^2 \\ 0 & 1/s \end{bmatrix}.$$

By using Theorem 3.18 ("taking the inverse Laplace transform"), we obtain

$$e^{tA} = \begin{bmatrix} 1 & t \\ 0 & 1 \end{bmatrix}. \qquad \diamond$$

Exercise 3.21. Let A be a 2×2 matrix so that $e^{tA} = \begin{bmatrix} \cosh t & \sinh t \\ \sinh t & \cosh t \end{bmatrix}$, $t \in \mathbb{R}$. Find A.

Exercise 3.22. Compute e^{tA}, for the "Jordan block" $A = \begin{bmatrix} \lambda & 1 & 0 \\ 0 & \lambda & 1 \\ 0 & 0 & \lambda \end{bmatrix}$.

In general, it can be shown that for $A = \begin{bmatrix} \lambda & 1 & & \\ & \ddots & \ddots & \\ & & & 1 \\ & & & \lambda \end{bmatrix}$, one has

$$e^{tA} = e^{\lambda t} \begin{bmatrix} 1 & t & \dfrac{t^2}{2!} & \cdots & \dfrac{t^{n-1}}{(n-1)!} \\ & \ddots & \ddots & & \vdots \\ & & & & \vdots \\ & & & \ddots & \dfrac{t^2}{2!} \\ & & & \ddots & t \\ & & & & 1 \end{bmatrix}.$$

Exercise 3.23. Give an example of 2×2 matrices A and B such that $e^{A+B} \neq e^A e^B$.

Exercise 3.24. If $A(\cdot)$ is an $\mathbb{R}^{n \times n}$-valued function on $\mathbb{R}$, then formally we have

$$\frac{d}{dt} \exp\left(\int_{t_i}^t A(\tau)d\tau \right) = \frac{d}{dt}\left(I + \frac{1}{1!}\int_{t_i}^t A(\tau)d\tau + \frac{1}{2!}\int_{t_i}^t A(\tau)d\tau \int_{t_i}^t A(\tau)d\tau + \cdots \right)$$

$$= A(t) + \frac{1}{2}\left(A(t)\int_{t_i}^t A(\tau)d\tau + \int_{t_i}^t A(\tau)d\tau\, A(t) \right) + \cdots,$$

and since it may not necessarily hold that

$$A(t)\left(\int_{t_i}^t A(\tau)d\tau \right) = \left(\int_{t_i}^t A(\tau)d\tau \right) A(t),$$

one may not have the pleasant situation where

$$\frac{d}{dt} \exp\left(\int_{t_i}^t A(\tau)d\tau \right) = A(t) \exp\left(\int_{t_i}^t A(\tau)d\tau \right).$$

This is what goes wrong when one tries to prove that in analogy with the $n = 1$ case, the solution to the initial value problem

$$(\text{IVP})_{x_i} : \begin{cases} \mathbf{x}'(t) = A(t)\mathbf{x}(t), & t \geq t_i \\ \mathbf{x}(t_i) = x_i \end{cases}$$

is given by the formula

$$\mathbf{x}(t) = \exp\left(\int_{t_i}^t A(\tau)d\tau \right) x_i, \quad t \geq t_i.$$

Show that the above formula does not yield a solution to the initial value problem when $t_i = 0$, $x_i = e_2$ (the standard basis vector $(0,1) \in \mathbb{R}^2$) and

$$A(t) = \begin{bmatrix} 1 & 2t \\ 0 & 0 \end{bmatrix}, \quad t \in \mathbb{R}.$$

3.4. Underdetermined ODEs and Control Theory

Before looking at underdetermined *differential* equations, we illustrate the main ideas of the setup by looking at an analogous easy *algebraic* equation.

Example 3.25. Consider the following *underdetermined* algebraic equation

$$x = 10 - u,$$

where $x, u \in \mathbb{N}$. We call this an underdetermined equation, because the solution pair (x, u) is *not uniquely* determined by the equation, and instead, we have many different possibilities for solutions:

$$
\begin{aligned}
(x, u) &= (9, 1), \\
(x, u) &= (5, 5), \\
(x, u) &= (2, 8), \text{ etc.}
\end{aligned}
$$

We can think of the u as being a "free" variable, that is something we can "input," and once we make a choice of a particular u, a unique x is determined; see Figure 6.

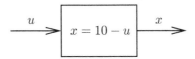

Figure 6. An underdetermined algebraic equation.

Since the value of x is influenced by the choice of our input u, one may ask the question of whether we can *control* the value of x in some desirable way by some suitable choice of u. For example, we could ask: Can one make $x \geq 5$? And the answer is: "Yes, by choosing the input $u \leq 5$!" Note that one has several possible choices of u that do this job of controlling the x in the manner specified: indeed, we can take $u = 1, 2, 3, 4, 5$. These are all inputs that ensure that the corresponding $x \geq 5$. One can then also ask further if there is a "best" possible choice of u among these possibilities based on some criterion. For example, suppose we want u to be the least possible number. So we are now looking at the optimization problem

$$
\begin{cases}
\text{minimize} & u \\
\text{subject to} & x = 10 - u, \\
& x \geq 5, \\
& x, u \in \mathbb{N}.
\end{cases}
$$

It is clear that there is a unique solution to this "optimal control" problem, which is $u_* = 1$. ◊

We are going to ask and solve similar questions, but not for such elementary *algebraic* equations, but rather *differential* equations that are underdetermined.

The basic objects of study in control theory are *underdetermined* differential equations. This means that there is some *freeness* in the functions appearing in the differential equation.

Definition 3.26. A *control system* is a differential equation with an initial condition:

$$\begin{cases} \mathbf{x}'(t) = f(t, \mathbf{x}(t), u(t)), \quad t \in [t_i, t_f], \\ \mathbf{x}(t_i) = x_i. \end{cases} \tag{3.6}$$

Here $f : [t_i, t_f] \times \mathbb{R}^n \times \mathbb{R}^m \to \mathbb{R}^n$.

Thus $\mathbf{u}$, the free variable, has m components, and is called the *input* function. Once a choice has been made for the $\mathbf{u}$ (which will be assumed to be piecewise continuous on $[t_i, t_f]$), then the right-hand side becomes a function of t and the state $\mathbf{x}(t)$, and given the initial condition x_i, a unique $\mathbf{x}$ is then determined by this $\mathbf{u}$. If we change the $\mathbf{u}$, a different corresponding $\mathbf{x}$ is produced. So we may think of a control system as a box, in which we can input $\mathbf{u}$, and the box then produces a corresponding $\mathbf{x}$. See Figure 7.

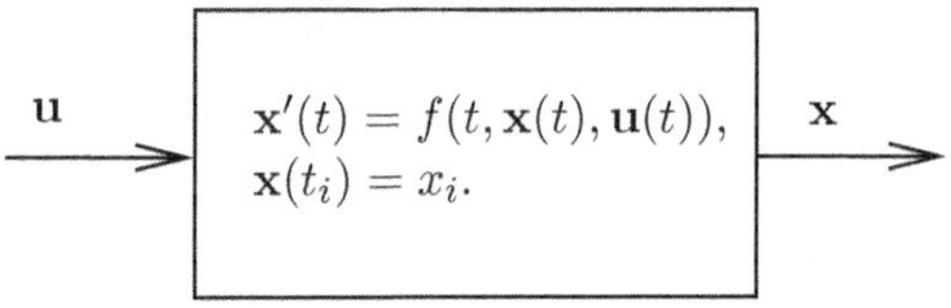

$$\mathbf{u} \longrightarrow \boxed{\begin{array}{l} \mathbf{x}'(t) = f(t, \mathbf{x}(t), \mathbf{u}(t)), \\ \mathbf{x}(t_i) = x_i. \end{array}} \longrightarrow \mathbf{x}$$

Figure 7. A control system.

Example 3.27. Let us revisit Example 3.2. Now suppose that we harvest the fish at a harvesting rate $\mathbf{h}$. Then the population evolution is described by

$$\mathbf{x}'(t) = c\mathbf{x}(t)\left(1 - \frac{\mathbf{x}(t)}{M}\right) - \mathbf{h}(t).$$

But the harvesting rate depends on the harvesting effort $\mathbf{u}$ (which is free, and something we can decide, that is, $\mathbf{u}$ is an input):

$$\mathbf{h}(t) = \mathbf{x}(t)\mathbf{u}(t).$$

(The harvesting effort can be thought in terms of the amount of time used for fishing, or the number of fishing nets used, and so on. Then the above equation makes sense, as the harvesting rate is clearly proportional to the number of fish:

the more fish in the lake, the better the catch.) Hence we arrive at the new model, which is an underdetermined differential equation

$$\mathbf{x}'(t) = c\mathbf{x}(t)\left(1 - \frac{\mathbf{x}(t)}{M}\right) - \mathbf{x}(t)\mathbf{u}(t).$$

This equation is underdetermined, since the $\mathbf{u}$ can be decided by the fisherman. This is the input, and once this has been chosen, then the population evolution is determined by the above equation, given some initial population level x_i of the fish at time t_i, for times $t \geq t_i$. So we see that control systems arise naturally in applications. Depending on how one harvests fish, that is, depending on what $\mathbf{u}$ we choose, the population $\mathbf{x}$ of the fish evolves in different possible ways. A natural question is then if we can influence or *control* $\mathbf{x}$, or say the endpoint value $\mathbf{x}(t_f)$ of $\mathbf{x}$ at the end of some fishing regime in the time interval $[t_i, t_f]$ in some desirable way. For example, an ecologist might recommend that the fishing should be done in such a manner that the chosen $\mathbf{u}$ guarantees that the final population level $\mathbf{x}(t_f)$ is maybe above some critical value, or belongs to a set $\mathbb{X}_f$. There may be many different input functions $\mathbf{u}$ that guarantee this. Then a related question which the fishing company can ask itself is: among these input functions $\mathbf{u}$, which one(s) maximize its profit? For example, suppose that the profit associated with the fishing harvest over a time interval $[t_i, t_f]$ is given by

$$f(\mathbf{u}) = \int_{t_i}^{t_f} e^{-rt}\left(p\mathbf{x}(t)\mathbf{u}(t) - c\mathbf{u}(t)\right) dt.$$

(Here p is the profit per unit harvest, so that p multiplied by the actual harvest $\mathbf{x}(t)\mathbf{u}(t)dt$ gives the profit at time t, and c is the cost per unit effort, so that c times the harvesting effort $\mathbf{u}(t)dt$ gives the cost incurred at time t. The factor e^{-rt} is a discounting factor.) The problem of deciding how to harvest so that the above profit is maximized now arises: Which $\mathbf{u}$ maximizes f? This is an optimal control problem:

$$\begin{cases} \text{minimize} & \int_{t_i}^{t_f} e^{-rt}\left(p\mathbf{x}(t)\mathbf{u}(t) - c\mathbf{u}(t)\right) dt \\ \\ \text{subject to} & \mathbf{x}'(t) = c\mathbf{x}(t)\left(1 - \frac{\mathbf{x}(t)}{M}\right) - \mathbf{x}(t)\mathbf{u}(t), \quad t \in [t_i, t_f] \\ \\ & \mathbf{x}(t_i) = x_i, \\ \\ & \mathbf{x}(t_f) \in \mathbb{X}_f. \end{cases}$$

So we see that optimal control problems arise naturally in applications. $\Diamond$

3.4.1. Linear control systems. Let $A \in \mathbb{R}^{n \times n}$, $B \in \mathbb{R}^{n \times m}$. A *linear control system*[4] is an equation of the form

$$\begin{cases} x'(t) = Ax(t) + B\mathbf{u}(t), \quad t \in [t_i, t_f], \\ \mathbf{x}(t_i) = x_i. \end{cases}$$

It turns out that in the case of linear control systems, we can give an explicit formula for the solution $\mathbf{x}$ in terms of the input chosen, and this is the content of the following result.

Theorem 3.28. *If* $\mathbf{u}$ *is a piecewise continuous* $\mathbb{R}^m$*-valued function on* $[t_i, t_f]$*, then the differential equation*

$$\begin{cases} \mathbf{x}'(t) = A\mathbf{x}(t) + B\mathbf{u}(t), \quad t \in [t_i, t_f], \\ \mathbf{x}(t_i) = x_i \end{cases} \tag{3.7}$$

has the unique solution $\mathbf{x}$ *given by*

$$\mathbf{x}(t) = e^{(t-t_i)A} x_i + \int_{t_i}^t e^{(t-\tau)A} B\mathbf{u}(\tau) d\tau. \tag{3.8}$$

Proof. We have

$$\frac{d}{dt}\left(e^{(t-t_i)A} x_i + \int_{t_i}^t e^{(t-\tau)A} B\mathbf{u}(\tau) d\tau \right)$$

$$= \frac{d}{dt}\left(e^{(t-t_i)A} x_i + e^{tA} \int_{t_i}^t e^{-\tau A} B\mathbf{u}(\tau) d\tau \right)$$

$$= A e^{(t-t_i)A} x_i + A e^{tA} \int_{t_i}^t e^{-\tau A} B\mathbf{u}(\tau) d\tau + e^{tA} e^{-tA} B\mathbf{u}(t)$$

$$= A\left(e^{(t-t_i)A} x_i + e^{tA} \int_{t_i}^t e^{-\tau A} B\mathbf{u}(\tau) d\tau \right) + e^{tA - tA} B\mathbf{u}(t)$$

$$= A\left(e^{(t-t_i)A} x_i + e^{tA} \int_{t_i}^t e^{-\tau A} B\mathbf{u}(\tau) d\tau \right) + B\mathbf{u}(t),$$

and so it follows that $\mathbf{x}(\cdot)$ given by (3.8) satisfies $\mathbf{x}'(t) = A\mathbf{x}(t) + B\mathbf{u}(t)$ on $[t_i, t_f]$. Furthermore,

$$e^{(t_i - t_i)A} x_i + \int_{t_i}^{t_i} e^{(t_i - \tau)A} B\mathbf{u}(\tau) d\tau = e^0 x_i + 0 = I x_i = x_i.$$

---

[4]We only consider the *time-invariant* case, that is, we assume A, B are fixed matrices, which do not vary with time.

Finally we show uniqueness. If $\mathbf{x}_1, \mathbf{x}_2$ are both solutions to (3.7), then $\mathbf{x} := \mathbf{x}_1 - \mathbf{x}_2$ satisfies

$$\begin{cases} \mathbf{x}'(t) = A\mathbf{x}(t), & t \in [t_i, t_f], \\ \mathbf{x}(t_i) = 0 \end{cases}$$

and so from Theorem 3.15 it follows that $\mathbf{x}(t) = 0$ for all $t \in [t_i, t_f]$, that is $\mathbf{x}_1 = \mathbf{x}_2$. $\qquad\square$

Exercise 3.29 (Riccati equation).

(1) Let $\alpha, \beta, \gamma \in \mathbb{R}$. Suppose that $\mathbf{p} \in C^1[0, T]$ is such that for all $t \in [0, T]$, $\mathbf{p}(t) + \alpha \neq 0$, and it satisfies the *Riccati Equation*

$$\mathbf{p}'(t) = \gamma(\mathbf{p}(t) + \alpha)(\mathbf{p}(t) + \beta).$$

Prove that $\mathbf{q}$ given by $\mathbf{q}(t) := \dfrac{1}{\mathbf{p}(t) + \alpha}$, $t \in [0, T]$, satisfies

$$\mathbf{q}'(t) = \gamma(\alpha - \beta)\mathbf{q}(t) - \gamma, \quad t \in [0, T].$$

(2) Find $\mathbf{p} \in C^1[0, 1]$ such that $\mathbf{p}'(t) = (\mathbf{p}(t))^2 - 1$, $t \in [0, 1]$, $\mathbf{p}(1) = 0$.

We end this section by considering a simple control problem for a linear control system. We will see in this example that there are many possible inputs that accomplish the control task. We will then envisage various possible scenarios in which we choose the "best" possible input by considering many different objective/cost functions that arise naturally in this application. In this manner we will arrive at the most general version of the optimal control problem that we gave at the beginning of this chapter (and in the next two chapters, we will learn ways to solve this optimization problem).

Example 3.30 (Rocket car). Consider a "rocket car," with mass m, on which there are engines/thrusters that can exert a force $\mathbf{F}$ in either direction. See Figure 8.

Figure 8. A rocket car moving in a straight line.

We consider the following problem: Drive the rocket car (by a suitable choice of the force $\mathbf{F}$) from its initial position z_i, where it is initially at rest at time $t = 0$, to the final desired position 0, where it should come to rest, say at time T. Let us first write the equations of motion and see that in fact this is a control problem for a linear system.

By Newton's law of motion, the force is equal to the mass of the body times its acceleration, and so if $\mathbf{z}(t)$ denotes the position at time t, we have

$$m\mathbf{z}''(t) = \mathbf{F}(t).$$

We can rewrite this as a first order system. To this end, we introduce the state variables

$$\mathbf{x}_1 := \mathbf{z},$$
$$\mathbf{x}_2 := \mathbf{z}',$$

and we set $\mathbf{u} := \mathbf{F}/m$, which is the input. Hence we obtain

$$\begin{bmatrix} \mathbf{x}_1' \\ \mathbf{x}_2' \end{bmatrix} = \begin{bmatrix} \mathbf{z}' \\ \mathbf{z}'' \end{bmatrix} = \begin{bmatrix} \mathbf{x}_2 \\ \mathbf{F}/m \end{bmatrix} = \begin{bmatrix} \mathbf{x}_2 \\ \mathbf{u} \end{bmatrix} = \underbrace{\begin{bmatrix} 0 & 1 \\ 0 & 0 \end{bmatrix}}_{A} \underbrace{\begin{bmatrix} \mathbf{x}_1 \\ \mathbf{x}_2 \end{bmatrix}}_{\mathbf{x}} + \underbrace{\begin{bmatrix} 0 \\ 1 \end{bmatrix}}_{B} \mathbf{u},$$

and

$$\mathbf{x}(0) = \begin{bmatrix} \mathbf{z}(0) \\ \mathbf{z}'(0) \end{bmatrix} = \underbrace{\begin{bmatrix} z_i \\ 0 \end{bmatrix}}_{=:x_i}.$$

The final desired state is

$$\mathbf{x}(T) = \begin{bmatrix} \mathbf{z}(T) \\ \mathbf{z}'(T) \end{bmatrix} = \underbrace{\begin{bmatrix} 0 \\ 0 \end{bmatrix}}_{=:x_f}.$$

The control problem is that of finding a (piecewise continuous) $\mathbf{u} : [0, T] \to \mathbb{R}$ such that the solution $\mathbf{x}$ to

$$\begin{cases} \mathbf{x}'(t) = A\mathbf{x}(t) + B\mathbf{u}(t), & t \in [0, T], \\ \mathbf{x}(0) = x_i \end{cases}$$

satisfies $\mathbf{x}(T) = x_f$.

Our expression (3.8) for the solution established earlier tells us that the $\mathbf{u}$ must satisfy

$$0 = \mathbf{x}(T) = e^{TA}x_i + \int_0^T e^{(T-\tau)A}B\mathbf{u}(\tau)d\tau.$$

How do we find such input functions $\mathbf{u}$? Here is a way. Let us suppose that

$$W := \int_0^T e^{(T-\tau)A}BB^\top e^{(T-\tau)A^\top}d\tau$$

is invertible. In our case the particular A, B guarantee this, since a calculation reveals that

$$W = \begin{bmatrix} T^3/3 & T^2/2 \\ T^2/2 & T \end{bmatrix},$$

which is indeed invertible. Then we claim that $\mathbf{u}$ defined by

$$\mathbf{u}(t) := -B^\top e^{(T-t)A^\top} W^{-1} e^{TA} x_i + \mathbf{u}_0(t), \quad t \in [0, T]$$

does the job, where $\mathbf{u}_0$ satisfies

$$\int_0^T e^{-A\tau} B\mathbf{u}_0(\tau)d\tau = 0. \tag{3.9}$$

To check this, we simply substitute this $\mathbf{u}$ in (3.8)! We have

$$e^{TA}x_i + \int_0^T e^{(T-\tau)A} B\mathbf{u}(\tau)d\tau$$

$$= e^{TA}x_i + \int_0^T e^{(T-\tau)A} B\left(-B^\top e^{(T-\tau)A^\top} W^{-1}e^{TA}x_i + \mathbf{u}_0(\tau)\right)d\tau$$

$$= e^{TA}x_i - WW^{-1}e^{TA}x_i + e^{TA}\int_0^T e^{-A\tau}B\mathbf{u}_0(\tau)d\tau$$

$$= e^{TA}x_i - e^{TA}x_i + 0 = 0 = \mathbf{x}(T).$$

But there are infinitely many $\mathbf{u}_0$'s which satisfy (3.9). Indeed, in our case (3.9) is equivalent to

$$\int_0^T \tau\mathbf{u}_0(\tau)d\tau = 0 \text{ and } \int_0^T \mathbf{u}_0(\tau)d\tau = 0.$$

For example, this holds for $\mathbf{u}_0$ defined by

$$\mathbf{u}_0(t) = \frac{2(n-1)T^n}{(n+1)(n+2)} - \frac{6nT^{n-1}}{(n+1)(n+2)}t + t^n, \quad t \in [0,T], \quad n \in \mathbb{N}.$$

(What follows now is a natural motivation for considering *optimal* control problems.) Given the choice in picking a $\mathbf{u}$ that accomplishes the control task, it makes sense in practice to choose a $\mathbf{u}$ that is in some sense the "best." So it is natural to associate a cost with each admissible $\mathbf{u}$, which allows us to compare the "bestness" of each $\mathbf{u}$, and then we will take the $\mathbf{u}_*$ that minimizes this cost. We choose which cost to use, depending on the situation at hand. Here are a few examples.

(1) (Linear Quadratic=LQ) Suppose that $T > 0$ is fixed. In order to see the minimum "energy" control input, one considers the following optimization problem:

$$\begin{cases} \text{minimize} & \int_0^T (\mathbf{u}(t))^2 dt \\ \text{subject to} & \mathbf{x}'(t) = A\mathbf{x}(t) + B\mathbf{u}(t), \quad t \in [0,T] \\ & \mathbf{x}(0) = x_i, \\ & \mathbf{x}(T) = 0. \end{cases}$$

(2) (Time optimal control) Suppose we don't care about the energy expended, but only want to accomplish the control task in the least amount of time ("Time is money!"), then we are led to the following optimization problem:

$$
\begin{cases}
\text{minimize} & T \\
\text{subject to} & \mathbf{x}'(t) = A\mathbf{x}(t) + B\mathbf{u}(t), \quad t \in [0, T] \\
& \mathbf{x}(0) = x_i, \\
& \mathbf{x}(T) = 0, \\
& \boxed{T \text{ is free}}.
\end{cases}
$$

Thus in this time optimal problem, the variables are T (≥ 0) *and* a function $\mathbf{u}$ defined on $[0, T]$, and we are seeking an optimal pair $(T_*, \mathbf{u}_*)$.

(3) (Constrained input, time optimal control) Now suppose that we want to accomplish the control task in the least amount of time, but there is also a size constraint on the input at any instant of time, say $|\mathbf{u}(t)| \leq 1$ for all t. (Imagine that there is some maximum acceleration that the rocket car can be subjected to in either direction.) Then we arrive at the following problem:

$$
\begin{cases}
\text{minimize} & T \\
\text{subject to} & \mathbf{x}'(t) = A\mathbf{x}(t) + B\mathbf{u}(t), \quad t \in [0, T] \\
& \mathbf{x}(0) = x_i, \\
& \mathbf{x}(T) = 0, \\
& \boxed{|\mathbf{u}(t)| \leq 1 \text{ for all } t \in [0, T],} \\
& T \text{ is free.}
\end{cases}
$$

(4) (Approximate solution) Suppose that $T > 0$ is fixed. If in addition to expending the smallest energy in our control effort (LQ optimal control), we don't care much about being exactly at the origin and about being perfectly at rest at the final time, then it is natural to consider the optimization problem

$$
\begin{cases}
\text{minimize} & \boxed{\|\mathbf{x}(T)\|_2^2} + \int_0^T (\mathbf{u}(t))^2 dt \\
\text{subject to} & \mathbf{x}'(t) = A\mathbf{x}(t) + B\mathbf{u}(t), \quad t \in [0, T] \\
& \mathbf{x}(0) = x_i.
\end{cases}
$$

Thus we have removed the condition "$\mathbf{x}(T) = 0$" from our list of constraints, and instead added the extra term "$\|\mathbf{x}(T)\|_2^2$" to the cost, which penalizes the deviation of $\mathbf{x}(T)$ from 0. $\diamondsuit$

Based on the above example, it is now not so strange that we consider the type of problem we mentioned at the beginning of this chapter. Let us now recall this general version of the optimal control problem we will be studying in the sequel, and pay close attention to the various constituents in the problem:

$$
\left\{
\begin{aligned}
&\text{minimize} && \varphi(\mathbf{x}(t_f)) + \int_{t_i}^{t_f} F(t, \mathbf{x}(t), \mathbf{u}(t)) dt \\
&\text{subject to} && \mathbf{x}'(t) = f(t, \mathbf{x}(t), \mathbf{u}(t)), \quad t \in [t_i, t_f], \\
& && \mathbf{x}(t_i) = x_i \in \mathbb{R}^n, \\
& && \mathbf{x}(t_f) \in \mathbb{X}_f \subset \mathbb{R}^n, \\
& && \mathbf{u}(t) \in \mathbb{U} \subset \mathbb{R}^m, \quad t \in [t_i, t_f].
\end{aligned}
\right.
$$

The given data are

$$
\begin{aligned}
&\varphi : \mathbb{R}^n \to \mathbb{R}, \\
&F : [t_i, t_f] \times \mathbb{R}^n \times \mathbb{R}^m \to \mathbb{R}, \\
&f : [t_i, t_f] \times \mathbb{R}^n \times \mathbb{R}^m \to \mathbb{R}^n, \\
&x_i \in \mathbb{R}^n, \\
&\mathbb{X}_f \subset \mathbb{R}^n, \\
&\mathbb{U} \subset \mathbb{R}^m, \\
&\boxed{t_f \geq t_i}.
\end{aligned}
$$

In time optimal control problems, the final time t_f is not specified, and is not fixed, but rather it is a variable that takes values in the interval $[t_i, +\infty)$. In other words, it is free. Let us elaborate on the given data.

The term

$$
\Phi(\mathbf{x}(t_f)) + \int_{t_i}^{t_f} F(t, \mathbf{x}(t), \mathbf{u}(t)) dt
$$

denotes the cost to be minimized. Here the integral term is to be thought of as the "running cost," while the term "$\varphi(\mathbf{x}(t_f))$" is a cost on the final state, which may be a penalty on the deviation of the final state from some desired state, for example $\varphi(x) = \|x\|_2^2$, which we had considered above in the approximate solution to the rocket car optimal control problem in Example 3.30 under item (4).

The equation $\mathbf{x}'(t) = f(t, \mathbf{x}(t), \mathbf{u}(t))$, $t \in [t_i, t_f]$, is the *state equation* (our given control system), and $\mathbf{x}(t_i) = x_i$ is the *initial condition*. Thus given an input $\mathbf{u}$, the initial value problem

$$
\begin{cases}
\mathbf{x}'(t) = f(t, \mathbf{x}(t), \mathbf{u}(t)), & t \in [t_i, t_f], \\
\mathbf{x}(t_i) = x_i
\end{cases}
$$

determines a corresponding unique $\mathbf{x}$. The constraint $\mathbf{x}(t_f) \in \mathbb{X}_f$ is called the *final state constraint*, and it says that we are only allowed to use those inputs that achieve this control task that the final state is in $\mathbb{X}_f$. In principle, the set $\mathbb{X}_f$ could conceivably be any subset of $\mathbb{R}^n$, but in the sequel, we will assume that $\mathbb{X}_f$ is in fact a "manifold" that is the common zero set of a bunch of continuously differentiable functions $g_k : \mathbb{R}^n \to \mathbb{R}$, $k = 1, \cdots, p$.

$$\mathbb{X}_f = \bigcap_{k=1}^{p} \{x \in \mathbb{R}^n : g_k(x) = 0\}.$$

A special example of this is when $\mathbb{X}_f$ is a singleton set: $\mathbf{x}(t_f) = x_f$. (For instance, in the rocket car problem discussed in Example 3.30, x_f was 0.) We will elaborate on this later, but it is useful to keep the picture shown in Figure 9 in mind as an example.

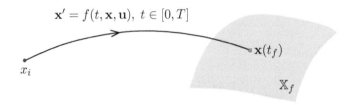

Figure 9. The final state constraint $\mathbf{x}(t_f) \in \mathbb{X}_f$.

3.5. A Useful Perturbation Result

We end our discussion on differential equations by establishing a useful perturbation result, which we will use when we justify something known as the "Pontryagin Minimum Principle" for solving optimal control problems. What do we mean by a "perturbation result"? Consider the initial value problem

$$\begin{cases} \mathbf{x}'(t) = f(t, \mathbf{x}(t), \mathbf{u}(t)), & t \in [t_i, t_f], \\ \mathbf{x}(t_i) = x_i \end{cases}$$

We can think of this as a box, which given an initial condition x_i and an input $\mathbf{u}$, produces a corresponding $\mathbf{x}$; see Figure 10.

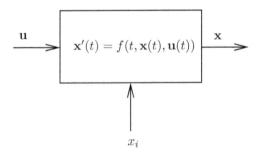

Figure 10. Control system.

So given a pair $(\mathbf{u}, x_i)$ of a control input function $\mathbf{u}$ and an initial condition x_i, we have a corresponding unique state function $\mathbf{x}$. Suppose that the state corresponding to $(\mathbf{u}_*, x_i)$ is denoted by $\mathbf{x}_*$. We ask the following question: if we perturb the initial condition x_i to $x_i + \delta x_i$, and perturb the control input $\mathbf{u}_*$ to $\mathbf{u}_* + \delta \mathbf{u}$, then how does the new state $\mathbf{x}$ compare with $\mathbf{x}_*$? In other words, instead of using $(\mathbf{u}_*, x_i)$ and getting $\mathbf{x}_*$, if we instead use $(\mathbf{u}_* + \delta \mathbf{u}, x_i + \delta x_i)$ and hence obtain a new $\mathbf{x}$, then how big is $\mathbf{x} - \mathbf{x}_*$? We will show the following result, which gives an estimate on $\mathbf{x} - \mathbf{x}_*$, and this will be used later when we prove results on solving optimal control problems.

Theorem 3.31 (Perturbation Result). *Let $f : [t_i, t_f] \times \mathbb{R}^n \times \mathbb{R}^m \to \mathbb{R}^n$ satisfy the following:*

(1) *for all $x \in \mathbb{R}^n$, $u \in \mathbb{R}^m$, $t \mapsto f(t, x, u) : [t_i, t_f] \to \mathbb{R}^n$ is piecewise continuous,*

(2) *there exists an $L > 0$ such that for all $t \in [t_i, t_f]$, $x, y \in \mathbb{R}^n$, $u, v \in \mathbb{R}^m$,*

$$\|f(t, x, u) - f(t, y, v)\|_2 \leq L(\|x - y\|_2 + \|u - v\|_2).$$

Let $\mathbf{u}_* : [t_i, t_f] \to \mathbb{R}^m$ be piecewise continuous, $x_i \in \mathbb{R}^n$, and let $\mathbf{x}_*$ be such that

$$\begin{cases} \mathbf{x}'_*(t) = f(t, \mathbf{x}_*(t), \mathbf{u}_*(t)), & t \in [t_i, t_f], \\ \mathbf{x}_*(t_i) = x_i. \end{cases}$$

Let $\delta x_i \in \mathbb{R}^n$, and $\delta \mathbf{u} : [t_i, t_f] \to \mathbb{R}^m$ be piecewise continuous such that

$$\text{for all } t \in [t_i, t_f], \quad \|\delta \mathbf{u}(t)\|_2 \leq M.$$

If $\mathbf{x}$ satisfies

$$\begin{cases} \mathbf{x}'(t) = f(t, \mathbf{x}(t), \mathbf{u}_*(t) + \delta \mathbf{u}(t)), & t \in [t_i, t_f], \\ \mathbf{x}(t_i) = x_i + \delta x_i, \end{cases}$$

then

$$\text{for all } t \in [t_i, t_f], \quad \|\mathbf{x}(t) - \mathbf{x}_*(t)\|_2 \leq e^{L(t-t_i)}\|\delta x_i\|_2 + M(e^{L(t-t_i)} - 1).$$

We note that if $\delta \mathbf{u}$ is small, then M can be small as well, and if δx_i is small too, then the right-hand side estimate is small, so that pointwise $\mathbf{x} \approx \mathbf{x}_*$ in the time interval $[t_i, t_f]$. More importantly, the above estimate says that the change $\mathbf{x} - \mathbf{x}_*$ produced by perturbing the input and/or the initial condition is of the same order as the perturbations in the input/initial condition. See Figure 11.

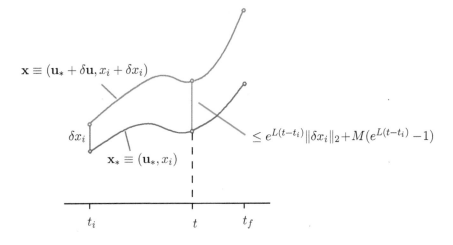

Figure 11. The Perturbation Result.

To prove this result, we will first show the following lemma, which is interesting in its own right.

Lemma 3.32 (Gronwall). *If $\alpha, \varphi : [t_i, t_f] \to \mathbb{R}$ are such that there is a $\beta > 0$ so that*

$$\text{for all } t \in [t_i, t_f], \quad \varphi(t) \leq \alpha(t) + \int_{t_i}^{t} \beta\varphi(\tau)d\tau,$$

then

$$\text{for all } t \in [t_i, t_f], \quad \varphi(t) \leq \alpha(t) + \int_{t_i}^{t} \beta\alpha(\tau)e^{\beta(t-\tau)}d\tau.$$

We note that in the given first inequality, we don't know anything very explicit about the φ, as it appears on both sides of the inequality, while in the latter inequality, we have an estimate on the size of φ purely in terms of α. So the result is giving something meaningful, and let us now see the proof.

Proof of Gronwall's Lemma. Set $\mathbf{F}(t) = \int_{t_i}^{t} \beta\varphi(\tau)d\tau$. Then

$$\mathbf{F}'(t) = \beta\varphi(t) \leq \beta\alpha(t) + \beta \int_{t_i}^{t} \beta\varphi(\tau)d\tau = \beta\alpha(t) + \beta\mathbf{F}(t).$$

Thus $\mathbf{F}' - \beta\mathbf{F} \leq \beta\alpha$, and so $\dfrac{d}{dt}(e^{-\beta t}\mathbf{F}(t)) = e^{-\beta t}(\mathbf{F}'(t) - \beta\mathbf{F}(t)) \leq e^{-\beta t}\beta\alpha(t)$. Hence

$$e^{-\beta t}\mathbf{F}(t) - e^{-\beta t_i}\underbrace{\mathbf{F}(t_i)}_{=0} \leq \int_{t_i}^{t} \beta e^{-\beta\tau}\alpha(\tau)d\tau,$$

and so $\mathbf{F}(t) \le \int_{t_i}^{t} \beta e^{-\beta(t-\tau)} \boldsymbol{\alpha}(\tau) d\tau$. Consequently,

$$\boldsymbol{\varphi}(t) \le \boldsymbol{\alpha}(t) + \underbrace{\int_{t_i}^{t} \beta \boldsymbol{\varphi}(\tau) d\tau}_{\mathbf{F}(t)} \le \boldsymbol{\alpha}(t) + \int_{t_i}^{t} \beta e^{-\beta(t-\tau)} \boldsymbol{\alpha}(\tau) d\tau.$$

This completes the proof. $\square$

We are now ready to prove the Perturbation Result.

Proof of Theorem 3.31. We have

$$\mathbf{x}_*(t) = x_i + \int_{t_i}^{t} f(\tau, \mathbf{x}_*(\tau), \mathbf{u}_*(\tau)) d\tau,$$

$$\mathbf{x}(t) = x_i + \delta x_i + \int_{t_i}^{t} f(\tau, \mathbf{x}(\tau), \mathbf{u}_*(\tau) + \delta\mathbf{u}(\tau)) d\tau.$$

Thus $\mathbf{x}(t) - \mathbf{x}_*(t) = \delta x_i + \int_{t_i}^{t} \Big(f(\tau, \mathbf{x}(\tau), \mathbf{u}_*(\tau) + \delta\mathbf{u}(\tau)) - f(\tau, \mathbf{x}_*(\tau), \mathbf{u}_*(\tau)) \Big) d\tau.$ So

$$\|\mathbf{x}(t) - \mathbf{x}_*(t)\|_2 \le \|\delta x_i\|_2 + \int_{t_i}^{t} \|f(\tau, \mathbf{x}(\tau), \mathbf{u}_*(\tau) + \delta\mathbf{u}(\tau)) - f(\tau, \mathbf{x}_*(\tau), \mathbf{u}_*(\tau))\|_2 d\tau$$

$$\le \|\delta x_i\|_2 + \int_{t_i}^{t} L\Big(\|\mathbf{x}(\tau) - \mathbf{x}_*(\tau)\|_2 + \|\delta\mathbf{u}(\tau)\|_2 \Big) d\tau$$

$$\le \|\delta x_i\|_2 + \int_{t_i}^{t} L\Big(\|\mathbf{x}(\tau) - \mathbf{x}_*(\tau)\|_2 + M \Big) d\tau$$

$$= \underbrace{\|\delta x_i\|_2 + LM(t - t_i)}_{=:\boldsymbol{\alpha}(t)} + \int_{t_i}^{t} \underbrace{L}_{\beta} \underbrace{\|\mathbf{x}(\tau) - \mathbf{x}_*(\tau)\|_2}_{=:\boldsymbol{\varphi}(\tau)} d\tau.$$

Applying Gronwall's Lemma, we obtain

$$\|\mathbf{x}(t) - \mathbf{x}_*(t)\|_2 \le \|\delta x_i\|_2 + LM(t - t_i) + \int_{t_i}^{t} Le^{L(t-\tau)}(\|\delta x_i\|_2 + LM(\tau - t_i)) d\tau.$$

But a calculation reveals that

$$\int_{t_i}^{t} Le^{L(t-\tau)} d\tau = -e^{L(t-\tau)}\Big|_{t_i}^{t} = e^{L(t-t_i)} - 1,$$

while

$$\int_{t_i}^{t} Le^{L(t-\tau)}(\tau - t_i) d\tau = -(\tau - t_i)e^{L(t-\tau)}\Big|_{t_i}^{t} + \int_{t_i}^{t} 1 \cdot e^{L(t-\tau)} d\tau$$

$$= -(t - t_i) + 0 + \frac{e^{L(t-\tau)}}{-L}\Big|_{t_i}^{t} = -(t - t_i) + \frac{e^{L(t-t_i)} - 1}{L}.$$

Thus

$$
\begin{aligned}
\|\mathbf{x}(t) - \mathbf{x}_*(t)\|_2 \ &\leq \ \|\delta x_i\|_2 + LM(t - t_i) + (e^{L(t-t_i)} - 1)\|\delta x_i\|_2 \\
&\quad -LM(t - t_i) + M(e^{L(t-t_i)} - 1) \\
&= \ e^{L(t-t_i)}\|\delta x_i\|_2 + M(e^{L(t-t_i)} - 1).
\end{aligned}
$$

This completes the proof. $\square$

Chapter 4

Pontryagin's Minimum Principle

In this chapter we will learn a result, known as the Pontryagin Minimum Principle, which gives a *necessary* condition that the solution of an optimal control problem must satisfy.

Let us first recall the general version of the optimal control problem that we discussed at the end of the previous chapter:

$$
\begin{cases}
\text{minimize} & \varphi(\mathbf{x}(t_f)) + \displaystyle\int_{t_i}^{t_f} F(t, \mathbf{x}(t), \mathbf{u}(t)) dt \\[2mm]
\text{subject to} & \mathbf{x}'(t) = f(t, \mathbf{x}(t), \mathbf{u}(t)), \quad t \in [t_i, t_f], \\[1mm]
& \mathbf{x}(t_i) = x_i \in \mathbb{R}^n, \\[1mm]
& \mathbf{x}(t_f) \in \mathbb{X}_f \subset \mathbb{R}^n, \\[1mm]
& \mathbf{u}(t) \in \mathbb{U} \subset \mathbb{R}^m, \quad t \in [t_i, t_f].
\end{cases}
$$

We will begin the chapter by looking at the *special case* when

(1) $\mathbb{U} = \mathbb{R}^m$ (unconstrained control),

(2) $\mathbb{X}_f = \mathbb{R}^n$ (no endpoint condition on the state).

4.1. Unconstrained Control with no Endpoint Condition on the State

So the optimization problem is given by:

$$
\begin{cases}
\text{minimize} & \varphi(\mathbf{x}(t_f)) + \displaystyle\int_{t_i}^{t_f} F(t, \mathbf{x}(t), \mathbf{u}(t)) dt \\[2mm]
\text{subject to} & \mathbf{x}'(t) = f(t, \mathbf{x}(t), \mathbf{u}(t)), \quad t \in [t_i, t_f], \\[1mm]
& \mathbf{x}(t_i) = x_i \in \mathbb{R}^n.
\end{cases}
$$

Our standing assumption is that

$$\varphi, f, F \in C^1,$$

that is, these functions are continuously differentiable. Here is the statement of the result we will prove in this section.

Theorem 4.1 (Baby Pontryagin Minimum Principle).

Let $\mathbf{u}_* : [t_i, t_f] \to \mathbb{R}^m$ *be a piecewise smooth optimal solution to*

$$\begin{cases} \text{minimize} & \varphi(\mathbf{x}(t_f)) + \displaystyle\int_{t_i}^{t_f} F(t, \mathbf{x}(t), \mathbf{u}(t))dt \\ \text{subject to} & \mathbf{x}'(t) = f(t, \mathbf{x}(t), \mathbf{u}(t)), \quad t \in [t_i, t_f], \\ & \mathbf{x}(t_i) = x_i \in \mathbb{R}^n, \end{cases}$$

and let $\mathbf{x}_*$ *be the corresponding state.*
Define the Hamiltonian $H : [t_i, t_f] \times \mathbb{R}^n \times \mathbb{R}^m \times \mathbb{R}^n \to \mathbb{R}$ *by*

$$H(t, x, u, \lambda) := F(t, x, u) + \lambda^\top f(t, x, u).$$

Then there exists a continuous and piecewise continuously differentiable function $\boldsymbol{\lambda} : [t_i, t_f] \to \mathbb{R}^n$ *such that*

(1) $$\begin{cases} \boldsymbol{\lambda}'(t) = -\left(\dfrac{\partial H}{\partial x}(t, \mathbf{x}_*(t), \mathbf{u}_*(t), \boldsymbol{\lambda}(t)) \right)^\top, \quad t \in [t_i, t_f], \\ \boldsymbol{\lambda}(t_f) = (\nabla\varphi(\mathbf{x}_*(t_f)))^\top. \end{cases}$$

(2) $\dfrac{\partial H}{\partial u}(t, \mathbf{x}_*(t), \mathbf{u}_*(t), \boldsymbol{\lambda}(t)) = 0, \; t \in [t_i, t_f].$

(3) $H_*(t) := H(t, \mathbf{x}_*(t), \mathbf{u}_*(t), \boldsymbol{\lambda}(t)), \; t \in [t_i, t_f]$ *satisfies*

$$H_*(t_f) - H_*(t) = \int_{t_i}^{t_f} \frac{\partial H}{\partial t}(\tau, \mathbf{x}_*(\tau), \mathbf{u}_*(\tau), \boldsymbol{\lambda}(\tau))d\tau, \quad t \in [t_i, t_f].$$

In particular, if F, f do not depend on t (the autonomous case), then item (3) above becomes

$$\frac{\partial H}{\partial t} = \frac{\partial}{\partial t}(F(x, u) + \lambda^\top f(x, u)) = 0,$$

and so $H_*(t_f) = H_*(t)$ for all $t \in [t_i, t_f]$, that is, $H(t, \mathbf{x}_*(t), \mathbf{u}_*(t), \boldsymbol{\lambda}(t)) = \text{constant}$ on $[t_i, t_f]$.

Proof. Step 1. If $\boldsymbol{\lambda} : [t_i, t_f] \to \mathbb{R}^n$ is an arbitrary (!) function, then by multiplying the constraint

$$f(t, \mathbf{x}(t), \mathbf{u}(t)) - \mathbf{x}'(t) = 0, \quad t \in [t_i, t_f]$$

pointwise by this $\boldsymbol{\lambda}$, we obtain

$$\boldsymbol{\lambda}(t)^\top \left(f(t, \mathbf{x}(t), \mathbf{u}(t)) - \mathbf{x}'(t) \right) = 0, \quad t \in [t_i, t_f],$$

and so by integrating from t_i to t_f,

$$\int_{t_i}^{t_f} \boldsymbol{\lambda}(t)^\top \Big(f(t, \mathbf{x}(t), \mathbf{u}(t)) - \mathbf{x}'(t) \Big) dt = 0.$$

So the cost of u is

$$
\begin{aligned}
I(u) &:= \varphi(\mathbf{x}(t_f)) + \int_{t_i}^{t_f} F(t, \mathbf{x}(t), \mathbf{u}(t)) dt \\
&= \varphi(\mathbf{x}(t_f)) + \int_{t_i}^{t_f} F(t, \mathbf{x}(t), \mathbf{u}(t)) dt + 0 \\
&= \varphi(\mathbf{x}(t_f)) + \int_{t_i}^{t_f} \Big(F(t, \mathbf{x}(t), \mathbf{u}(t)) + \boldsymbol{\lambda}(t)^\top f(t, \mathbf{x}(t), \mathbf{u}(t)) \Big) dt - \int_{t_i}^{t_f} \boldsymbol{\lambda}(t)^\top \mathbf{x}'(t) dt \\
&= \varphi(\mathbf{x}(t_f)) + \int_{t_i}^{t_f} H(t, \mathbf{x}(t), \mathbf{u}(t), \boldsymbol{\lambda}(t)) dt - \int_{t_i}^{t_f} \boldsymbol{\lambda}(t)^\top \mathbf{x}'(t) dt.
\end{aligned}
$$

(This finishes Step 1, in which we expressed the cost of any input u in terms of the Hamiltonian and an arbitrary $\boldsymbol{\lambda}$.)

Step 2. Suppose $\mathbf{u}_*$ is optimal. Let us perturb $\mathbf{u}_*$ to $\mathbf{u} := \mathbf{u}_* + \delta\mathbf{u}$, where $\delta\mathbf{u} : [t_i, t_f] \to \mathbb{R}^m$ is a piecewise continuous function that satisfies the pointwise estimate $\|\delta\mathbf{u}(t)\|_2 < \epsilon$, $t \in [t_i, t_f]$. Let $\mathbf{x}$ denote the state corresponding to $(\mathbf{u}, x_i)$. If we define $\delta\mathbf{x} := \mathbf{x} - \mathbf{x}_*$, then we know from Theorem 3.31 that this $\delta\mathbf{x}$ satisfies $\|\delta\mathbf{x}(t)\|_2 \leq \epsilon(e^{L(t-t_i)} - 1)$, $t \in [t_i, t_f]$, (where L is the Lipschitz constant associated with f). So if we choose ϵ to be small enough at the outset, then we can make $\delta\mathbf{x}$ as small as we want pointwise in the interval $[t_i, t_f]$.

(This finishes Step 2, where we noted that by choosing the perturbation $\delta\mathbf{u}$ of the optimal solution $\mathbf{u}_*$ small enough, we can make the resulting perturbation $\delta\mathbf{x}$ in the optimal state $\mathbf{x}_*$ as small as we please.)

Step 3. Using the result in Step 1, we have

$$
\begin{aligned}
&I(\mathbf{u}) - I(\mathbf{u}_*) \\
&= I(\mathbf{u}_* + \delta\mathbf{u}) - I(\mathbf{u}_*) \\
&= \varphi(\mathbf{x}_*(t_f) + \delta\mathbf{x}(t_f)) - \varphi(\mathbf{x}(t_f)) \\
&\quad + \int_{t_i}^{t_f} \Big(H(t, \mathbf{x}_*(t) + \delta\mathbf{x}(t), \mathbf{u}_*(t) + \delta\mathbf{u}(t), \boldsymbol{\lambda}(t)) - H(t, \mathbf{x}_*(t), \mathbf{u}_*(t), \boldsymbol{\lambda}(t)) \Big) dt \\
&\quad - \int_{t_i}^{t_f} \boldsymbol{\lambda}(t)^\top (\cancel{\mathbf{x}'_*(t)} + \delta\mathbf{x}'(t) - \cancel{\mathbf{x}'_*(t)}) dt.
\end{aligned}
$$

Using Taylor's Formula, we have

$$\varphi(\mathbf{x}_*(t_f) + \delta\mathbf{x}(t_f)) - \varphi(\mathbf{x}(t_f)) = \nabla\varphi(\mathbf{x}_*(t_f)) \cdot (\delta\mathbf{x})(t_f) + o(\epsilon),$$

and (occasionally suppressing the time argument)

$$H(t, \mathbf{x}_* + \delta\mathbf{x}, \mathbf{u}_* + \delta\mathbf{u}, \boldsymbol{\lambda}) - H(t, \mathbf{x}_*, \mathbf{u}_*, \boldsymbol{\lambda})$$
$$= \frac{\partial H}{\partial x}(t, \mathbf{x}_*, \mathbf{u}_*, \boldsymbol{\lambda}) \cdot (\delta\mathbf{x}) + \frac{\partial H}{\partial u}(t, \mathbf{x}_*, \mathbf{u}_*, \boldsymbol{\lambda}) \cdot (\delta\mathbf{u}) + o(\epsilon).$$

By integrating by parts, we have using $(\delta\mathbf{x})(t_i) = x_i - x_i = 0$ that

$$-\int_{t_i}^{t_f} \boldsymbol{\lambda}(t)^\top (\delta\mathbf{x})'(t)dt = -\boldsymbol{\lambda}(t)^\top (\delta\mathbf{x})(t)\Big|_{t_i}^{t_f} + \int_{t_i}^{t_f} \boldsymbol{\lambda}'(t)^\top (\delta\mathbf{x})(t)dt$$

$$= -\boldsymbol{\lambda}(t_f)^\top (\delta\mathbf{x})(t_f) + \int_{t_i}^{t_f} \boldsymbol{\lambda}'(t)^\top (\delta\mathbf{x})(t)dt.$$

Thus

$$I(\mathbf{u}) - I(\mathbf{u}_*) = \left(\nabla\varphi(\mathbf{x}_*(t_f)) - \boldsymbol{\lambda}(t_f)^\top\right)(\delta\mathbf{x})(t_f)$$
$$+ \int_{t_i}^{t_f} \left(\frac{\partial H}{\partial x}(t, \mathbf{x}_*(t), \mathbf{u}_*(t), \boldsymbol{\lambda}(t)) + \boldsymbol{\lambda}'(t)^\top\right)(\delta\mathbf{x})(t)dt$$
$$+ \int_{t_i}^{t_f} \frac{\partial H}{\partial u}(t, \mathbf{x}_*(t), \mathbf{u}_*(t), \boldsymbol{\lambda}(t)) \cdot (\delta\mathbf{u})(t)dt + o(\epsilon).$$

Step 4. So far the $\boldsymbol{\lambda}$ was arbitrary, but now we will make a special choice of $\boldsymbol{\lambda}$! Let $\boldsymbol{\lambda}$ be the solution to:

$$\begin{cases} \boldsymbol{\lambda}'(t) = -\left(\frac{\partial H}{\partial x}(t, \mathbf{x}_*(t), \mathbf{u}_*(t), \boldsymbol{\lambda}(t))\right)^\top, & t \in [t_i, t_f], \\ \boldsymbol{\lambda}(t_f) = (\nabla\varphi(\mathbf{x}_*(t_f)))^\top. \end{cases}$$

Then with this special $\boldsymbol{\lambda}$, we obtain from the previous step that:

$$I(\mathbf{u}) - I(\mathbf{u}_*)$$
$$= (0) \cdot (\delta\mathbf{x})(t_f) + \int_{t_i}^{t_f} 0 \cdot (\delta\mathbf{x})(t)dt + \int_{t_i}^{t_f} \frac{\partial H}{\partial u}(t, \mathbf{x}_*(t), \mathbf{u}_*(t), \boldsymbol{\lambda}(t)) \cdot (\delta\mathbf{u})(t)dt$$
$$+ o(\epsilon).$$

Let $\mathbf{u} := \mathbf{u}_* + \epsilon\mathbf{v}$, where $\mathbf{v}$ is a fixed function and ϵ is small enough. (So $\delta\mathbf{u} = \epsilon\mathbf{v}$.) Thanks to the optimality of $\mathbf{u}_*$, we have $I(\mathbf{u}) \geq I(\mathbf{u}_*)$, and so

$$0 \leq I(\mathbf{u}) - I(\mathbf{u}_*) = \int_{t_i}^{t_f} \frac{\partial H}{\partial u}(t, \mathbf{x}_*(t), \mathbf{u}_*(t), \boldsymbol{\lambda}(t)) \cdot \epsilon\mathbf{v}(t)dt + o(\epsilon).$$

Divide by ϵ:

$$0 \leq \int_{t_i}^{t_f} \frac{\partial H}{\partial u}(t, \mathbf{x}_*(t), \mathbf{u}_*(t), \boldsymbol{\lambda}(t)) \cdot \mathbf{v}(t)dt + \frac{o(\epsilon)}{\epsilon}.$$

Hence by passing the limit as $\epsilon \to 0$, since $\dfrac{o(\epsilon)}{\epsilon} \to 0$, we obtain

$$0 \le \int_{t_i}^{t_f} \frac{\partial H}{\partial u}(t, \mathbf{x}_*(t), \mathbf{u}_*(t), \boldsymbol{\lambda}(t)) \cdot \mathbf{v}(t) dt.$$

The choice of $\mathbf{v}$ was arbitrary, and so the same conclusion holds with $-\mathbf{v}$ instead of $\mathbf{v}$, and so

$$\int_{t_i}^{t_f} \frac{\partial H}{\partial u}(t, \mathbf{x}_*(t), \mathbf{u}_*(t), \boldsymbol{\lambda}(t)) \cdot \mathbf{v}(t) dt = 0.$$

As $\mathbf{v}$ was free, it follows (by choosing $\mathbf{v}$ as

$$\mathbf{v}(t)^\top := \frac{\partial H}{\partial u}(t, \mathbf{x}_*(t), \mathbf{u}_*(t), \boldsymbol{\lambda}(t)),$$

$t \in [t_i, t_f]$) that

$$\frac{\partial H}{\partial u}(t, \mathbf{x}_*(t), \mathbf{u}_*(t), \boldsymbol{\lambda}(t)) = 0, \quad t \in [t_i, t_f].$$

Consequently, we have so far established the conclusion (1) (which was for free by our choice of $\boldsymbol{\lambda}$), and the conclusion (2).

Step 5. It remains to show conclusion (3). Instead of proving this in full generality, we will just prove this under the simplifying assumption that $\mathbf{u}_* \in C^1$. We have

$$H_*(t_f) - H_*(t) = \int_t^{t_f} \frac{dH_*}{dt}(\tau) d\tau,$$

and

$$\frac{dH_*}{dt}(\tau) = \frac{\partial H}{\partial t}(\tau, \mathbf{x}_*(\tau), \mathbf{u}_*(\tau), \boldsymbol{\lambda}(\tau)) \cdot 1 + \underbrace{\frac{\partial H}{\partial u}(\tau, \mathbf{x}_*(\tau), \mathbf{u}_*(\tau), \boldsymbol{\lambda}(\tau)) \cdot \mathbf{u}'_*(\tau)}_{=0}$$

$$+ \underbrace{\frac{\partial H}{\partial x}(\tau, \mathbf{x}_*(\tau), \mathbf{u}_*(\tau), \boldsymbol{\lambda}(\tau)) \cdot \mathbf{x}'_*(\tau)}_{=-\boldsymbol{\lambda}'(\tau)^\top} + \frac{\partial H}{\partial \lambda}(\tau, \mathbf{x}_*(\tau), \mathbf{u}_*(\tau), \boldsymbol{\lambda}(\tau)) \cdot \boldsymbol{\lambda}'(\tau).$$

But we have

$$\frac{\partial H}{\partial \lambda}(\tau, \mathbf{x}_*(\tau), \mathbf{u}_*(\tau), \boldsymbol{\lambda}(\tau)) = (f(\tau, \mathbf{x}_*(\tau), \mathbf{u}_*(\tau)))^\top = (\mathbf{x}'_*(\tau))^\top,$$

and so from the above we obtain

$$\frac{dH_*}{dt}(\tau) = \frac{\partial H}{\partial t}(\tau, \mathbf{x}_*(\tau), \mathbf{u}_*(\tau), \boldsymbol{\lambda}(\tau)) - \cancel{\boldsymbol{\lambda}'(\tau)^\top \mathbf{x}'_*(\tau)} + \cancel{\mathbf{x}'_*(\tau)^\top \boldsymbol{\lambda}'(\tau)}.$$

Consequently,

$$H_*(t_f) - H_*(t) = \int_t^{t_f} \frac{\partial H}{\partial t}(\tau, \mathbf{x}_*(\tau), \mathbf{u}_*(\tau), \boldsymbol{\lambda}(\tau)) d\tau, \quad t \in [t_i, t_f].$$

This completes the proof. $\square$

Remark 4.2.

(1) **Why is this result called the "Minimum" Principle?** Actually the condition

$$\frac{\partial H}{\partial u}(t, \mathbf{x}_*(t), \mathbf{u}_*(t), \boldsymbol{\lambda}(t)) = 0, \quad t \in [t_i, t_f]$$

can be replaced with a stronger condition. Fix $t \in [t_i, t_f]$. Look at the map

$$u \mapsto H(t, \mathbf{x}_*(t), u, \boldsymbol{\lambda}(t)) : \mathbb{R}^m \to \mathbb{R}.$$

Then we will prove in the next section that in fact $\mathbf{u}_*(t)$ is the *minimizer* of this function, that is, for all $t \in [t_i, t_f]$, and for all $u \in \mathbb{R}^m$,

$$H(t, \mathbf{x}_*(t), u, \boldsymbol{\lambda}(t)) \geq H(t, \mathbf{x}_*(t), \mathbf{u}_*(t), \boldsymbol{\lambda}(t)).$$

In other words,

$$H(t, \mathbf{x}_*(t), \mathbf{u}_*(t), \boldsymbol{\lambda}(t)) = \min_{u \in \mathbb{R}^m} H(t, \mathbf{x}_*(t), u, \boldsymbol{\lambda}(t)), \quad t \in [t_i, t_f],$$

that is,

$$\mathbf{u}_*(t) = \arg \min_{u \in \mathbb{R}^m} H(t, \mathbf{x}_*(t), u, \boldsymbol{\lambda}(t)), \quad t \in [t_i, t_f].$$

This is the reason behind the terminology Pontryagin *Minimum* Principle.

(2) The function $\boldsymbol{\lambda}$ is called the *co-state*, since it satisfies the following equation, called the *co-state equation*, namely

$$\boldsymbol{\lambda}'(t) = -\left(\frac{\partial H}{\partial x}(t, \mathbf{x}_*(t), \mathbf{u}_*(t), \boldsymbol{\lambda}(t))\right)^{\top}, \quad t \in [t_i, t_f],$$

which looks similar to the equation satisfied by the state:

$$\mathbf{x}_*'(t) = \frac{\partial H}{\partial \lambda}(t, \mathbf{x}_*(t), \mathbf{u}_*(t), \boldsymbol{\lambda}(t)), \quad t \in [t_i, t_f].$$

(Indeed, $\mathbf{x}_*'(t) = f(t, \mathbf{x}_*(t), \mathbf{u}_*(t)) = \frac{\partial H}{\partial \lambda}(t, \mathbf{x}_*(t), \mathbf{u}_*(t), \boldsymbol{\lambda}(t)), \quad t \in [t_i, t_f].$)

Example 4.3 (LQ problem). Let q_0, q, a, x_i be fixed real numbers such that $q_0 \geq 0$ and $q > 0$. Consider the optimization problem given by

$$\begin{cases} \text{minimize} & \frac{1}{2}q_0(\mathbf{x}(1))^2 + \frac{1}{2}\int_0^1 \left(q(\mathbf{x}(t))^2 + (\mathbf{u}(t))^2\right) dt \\ \text{subject to} & \mathbf{x}'(t) = a\mathbf{x}(t) + \mathbf{u}(t), \quad t \in [0, 1], \\ & \mathbf{x}(0) = x_i \in \mathbb{R}^n. \end{cases}$$

So we have

$$\varphi(x) = \frac{1}{2}q_0 x^2,$$

$$F(t, x, u) = \frac{1}{2}(qx^2 + u^2),$$

$$f(t, x, u) = ax + u.$$

Since F, f don't depend on t, we are in the autonomous case. Let us see what the Pontryagin Minimum Principle says about the optimal control u_*, assuming that it exists. We have

$$H(x, u, \lambda) = \frac{1}{2}(qx^2 + u^2) + \lambda(ax + u),$$

and so

$$\frac{\partial H}{\partial u} = u + \lambda.$$

Thus we obtain

$$\mathbf{u}_*(t) + \boldsymbol{\lambda}(t) = 0, \quad t \in [0, 1],$$

that is, $\mathbf{u}_* = -\boldsymbol{\lambda}$. Using $\dfrac{\partial H}{\partial x} = qx + a\lambda$, we obtain

$$\mathbf{x}' = a\mathbf{x} - \boldsymbol{\lambda},$$
$$\boldsymbol{\lambda}' = -q\mathbf{x} - a\boldsymbol{\lambda},$$

on $[0, 1]$, with the boundary conditions

$$\mathbf{x}(0) = x_i,$$

$$\boldsymbol{\lambda}(1) = \varphi'(\mathbf{x}(1)) = q_0\mathbf{x}(1).$$

Hence we have the first order system

$$\begin{bmatrix} \mathbf{x}' \\ \boldsymbol{\lambda}' \end{bmatrix} = \begin{bmatrix} a & -1 \\ -q & -a \end{bmatrix} \begin{bmatrix} \mathbf{x} \\ \boldsymbol{\lambda} \end{bmatrix}, \quad t \in [0, 1], \begin{cases} \mathbf{x}(0) = x_i \\ \boldsymbol{\lambda}(1) = q_0\mathbf{x}(1). \end{cases}$$

We note that the endpoint conditions are at *different* times.

Let us suppose now that $a := 0$, $q := 1$, $q_0 := 0$. Then one can compute that

$$\exp\begin{bmatrix} a & -1 \\ -q & -a \end{bmatrix} = \exp\begin{bmatrix} 0 & -1 \\ -1 & 0 \end{bmatrix} = \begin{bmatrix} \cosh t & -\sinh t \\ -\sinh t & \cosh t \end{bmatrix},$$

and so

$$\begin{bmatrix} \mathbf{x} \\ \boldsymbol{\lambda} \end{bmatrix} = \begin{bmatrix} \cosh t & -\sinh t \\ -\sinh t & \cosh t \end{bmatrix} \begin{bmatrix} x_i \\ \boldsymbol{\lambda}(0) \end{bmatrix}.$$

Hence $\boldsymbol{\lambda}(t) = -(\sinh t)x_i + (\cosh t)\boldsymbol{\lambda}(0)$. As $\boldsymbol{\lambda}(1) = q_0\mathbf{x}(1) = 0 \cdot \mathbf{x}(1) = 0$, we obtain

$$0 = -(\sinh 1)x_i + (\cosh 1)\boldsymbol{\lambda}(0),$$

and so $\boldsymbol{\lambda}(0) = \dfrac{\sinh 1}{\cosh 1} x_i$. Consequently, $\boldsymbol{\lambda}(t) = \left(-\sinh t + (\cosh t)\dfrac{\sinh 1}{\cosh 1}\right)x_i$, and

$$\mathbf{u}_*(t) = \left(\sinh t - (\cosh t)\frac{\sinh 1}{\cosh 1}\right)x_i = \frac{\sinh(t-1)}{\cosh 1}x_i, \quad t \in [0,1].$$

So if $\mathbf{u}_*$ is an optimal solution, then it has to be equal to $\dfrac{\sinh(\cdot - 1)}{\cosh 1} x_i$ for $t \in [0,1]$.
$\diamond$

Henceforth we will use the "minimum principle" formulation (which we will prove later on) to solve problems.

Example 4.4 (Co-state = "shadow price"). Consider the optimization problem

$$\begin{cases} \text{minimize} & \displaystyle\int_0^T \Big((\mathbf{u}(t))^2 + t\mathbf{x}(t) - 1\Big)dt \\ \text{subject to} & \mathbf{x}'(t) = \mathbf{u}(t), \ t \in [0,T], \\ & \mathbf{x}(0) = x_i \in \mathbb{R}^n. \end{cases}$$

The Hamiltonian is given by $H(t,x,u,\lambda) = u^2 + tx - 1 + \lambda u$. By the Pontryagin Minimum Principle, $\mathbf{u}_*$ minimizes $u \mapsto H(t,\mathbf{x}_*(t),u,\boldsymbol{\lambda}(t)) : \mathbb{R} \to \mathbb{R}$, and so

$$\mathbf{u}_*(t) = \arg\min_{u\in\mathbb{R}} \left(u^2 + t\mathbf{x}_*(t) - 1 + \boldsymbol{\lambda}(t)u\right) = -\frac{1}{2}\boldsymbol{\lambda}(t),$$

since the quadratic function $u \mapsto u^2 + t\mathbf{x}_*(t) - 1 + \boldsymbol{\lambda}(t)u$ on the right-hand side is minimized when $2u + \boldsymbol{\lambda}(t) = 0$. We have

$$\frac{\partial H}{\partial x} = t,$$

and so $\boldsymbol{\lambda}'(t) = -t$. Also, since $\varphi \equiv 0$, $\boldsymbol{\lambda}(T) = \varphi'(\mathbf{x}_*(T)) = 0$, so that

$$\boldsymbol{\lambda}(t) = \boldsymbol{\lambda}(t) - 0 = \boldsymbol{\lambda}(t) - \boldsymbol{\lambda}(T) = -\int_t^T -\tau\, d\tau = \frac{T^2 - t^2}{2}.$$

So if $\mathbf{u}_*$ is an optimal control, then

$$\mathbf{u}_*(t) = \frac{t^2 - T^2}{4}, \quad t \in [0,T].$$

Let us now find the optimal cost. First we find the state $\mathbf{x}_*$ using the state equation and the initial condition

$$\mathbf{x}_*(t) - x_i = \int_0^t \mathbf{x}_*'(\tau)d\tau = \int_0^t \mathbf{u}_*(\tau)d\tau = \int_0^t \frac{\tau^2 - T^2}{4}d\tau = \frac{t^3 - 3T^2 t}{12}, \quad t \in [0,T].$$

Thus the optimal cost is

$$\begin{aligned} V(x_i,T) &:= \int_0^T \Big((\mathbf{u}_*(t))^2 + t\mathbf{x}_*(t) - 1\Big)dt \\ &= \int_0^T \left(\frac{(t^2 - T^2)^2}{16} + t\Big(x_i + \frac{t^3 - 3T^2 t}{12}\Big) - 1\right)dt, \end{aligned}$$

and so

$$\frac{\partial V}{\partial x_i}(x_i, T) = \int_0^T t\, dt = \frac{T^2}{2} = \lambda(0).$$

This is no coincidence! The number $\lambda(0)$ is the marginal change in the optimal cost per unit change in x_i. If we think of the cost in economic terms, and x_i as the capital stock, then $\lambda(0)$ is a "shadow price" of the capital stock, since it measures the marginal change in cost per unit change in the capital stock. So one can think of the co-state as the shadow price. ◊

Example 4.5. Consider the optimization problem

$$\begin{cases} \text{maximize} & \displaystyle\int_0^T \Big(qf(\mathbf{x}(t)) - c(\mathbf{u}(t))\Big)\, dt \\[2mm] \text{subject to} & \mathbf{x}'(t) = \mathbf{u}(t) - \delta \cdot \mathbf{x}(t), \quad t \in [0, T], \\[2mm] & \mathbf{x}(0) = x_i \in \mathbb{R}^n. \end{cases}$$

(Here is an economic interpretation:

$\mathbf{x}(t)$	is the capital stock of a firm,
$f(\cdot)$	is the production function,
q	is the price per unit of output,
$c(\cdot)$	is the cost function of investment,
δ	is the rate of depreciation of capital,
x_i	is the initial capital stock, and
T	is the fixed planning horizon.

We note that it is a *maximization* problem, and so the Hamiltonian is given by

$$H(t, x, u, \lambda) := -qf(x) + c(u) + \lambda(u - \delta \cdot x).$$

Let $\mathbf{u}_*$ be optimal. As $\dfrac{\partial H}{\partial x} = -qf'(x) - \delta \cdot \lambda$, the co-state λ must satisfy

$$\begin{aligned} \lambda'(t) &= qf'(\mathbf{x}_*(t)) + \delta \cdot \lambda(t), \quad t \in [0, T], \\ \lambda(T) &= \varphi'(\mathbf{x}(T)) = 0. \end{aligned}$$

Also, $\mathbf{u}_*(t) = \arg\min_{u \in \mathbb{R}} \Big(-qf(\mathbf{x}_*(t)) + c(u) + \lambda(t)(-\delta \cdot \mathbf{x}_*(t) + u)\Big)$, $t \in [0, T]$.

Suppose that we are given

$$\begin{array}{ll} f(x) = x - 0.03x^2 & q = 1 \\ c(u) = u^2 & \delta = 0.1 \\ x_i = 10 & T = 10. \end{array}$$

Then $f'(x) = 1 - 0.06x$, and so the co-state equation becomes

$$\lambda'(t) = 1 - 0.06\mathbf{x}_*(t) + 0.1\lambda(t), \ t \in [0, 10],$$
$$\lambda(T) = \lambda(10) = 0.$$

Also

$$\mathbf{u}_*(t) = \arg\min_{u \in \mathbb{R}} \left(-\mathbf{x}_*(t) + 0.03(\mathbf{x}_*(t))^2 + u^2 + \lambda(t)(-0.1\mathbf{x}_*(t) + u) \right)$$
$$= -\frac{\lambda(t)}{2}, \ t \in [0, 10].$$

Hence we obtain $\mathbf{x}'_*(t) = \mathbf{u}_*(t) - 0.1\mathbf{x}_*(t) = -\dfrac{\lambda(t)}{2} - 0.1\mathbf{x}_*(t)$, and so

$$\mathbf{x}''_*(t) = -\frac{\lambda'(t)}{2} - 0.1\mathbf{x}'_*(t)$$
$$= -\frac{1}{2}(1 - 0.06\mathbf{x}_*(t) + 0.1\lambda(t)) - 0.1\mathbf{x}'_*(t) = -0.5 + 0.04\mathbf{x}_*(t), \ t \in [0, 10].$$

Consequently, $\mathbf{x}_*(t) = Ae^{0.2t} + Be^{-0.2t} + 12.5$ for some constants A, B, which can be determined using $\mathbf{x}_*(0) = 10$ and $\lambda(10) = 0$. $\qquad\Diamond$

Exercise 4.6. Suppose that u_T is an optimal solution to

$$\begin{cases} \text{minimize} & \dfrac{1}{2}\displaystyle\int_0^T \left(3(\mathbf{x}(t))^2 + (\mathbf{u}(t))^2\right)dt \\ \text{subject to} & \mathbf{x}'(t) = \mathbf{x}(t) + \mathbf{u}(t), \ t \in [0, T], \\ & \mathbf{x}(0) = x_i. \end{cases}$$

and let x_T be the corresponding state. Find u_T and x_T, and show that there is a constant k such that for each fixed $t \in \mathbb{R}$,

$$\lim_{T \to \infty} u_T(t) = k \cdot \lim_{T \to \infty} x_T(t).$$

Exercise 4.7. Suppose that $\mathbf{u}_*$ is an optimal solution to

$$\begin{cases} \text{maximize} & 2\mathbf{x}(1) + 3 + \displaystyle\int_0^1 \left(1 - t\mathbf{u}(t) - (\mathbf{u}(t))^2\right)dt \\ \text{subject to} & \mathbf{x}'(t) = \mathbf{u}(t), \ t \in [0, 1], \\ & \mathbf{x}(0) = 1. \end{cases}$$

Show that $\mathbf{u}_*(t) = 1 - \dfrac{t}{2}, \ t \in [0, 1]$.

Exercise 4.8. Suppose that $\mathbf{u}_*$ is an optimal solution to

$$\begin{cases} \text{maximize} & \sqrt{\mathbf{x}(1)} - \dfrac{1}{2}\displaystyle\int_0^1 (\mathbf{u}(t))^2 dt \\ \text{subject to} & \mathbf{x}'(t) = \mathbf{x}(t) + \mathbf{u}(t), \ t \in [0, 1], \\ & \mathbf{x}(0) = 0. \end{cases}$$

Show that $\mathbf{u}_*(t) = \dfrac{e^{1-t}}{\sqrt[3]{2(e^2 - 1)}}, \ t \in [0, 1]$.

Exercise 4.9. Consider the optimal control problem given by

$$
\begin{cases}
\text{maximize} & \displaystyle\int_0^1 \left(1 - (\mathbf{u}(t))^2\right) \cdot (\mathbf{x}(t))^2 dt \\
\text{subject to} & \mathbf{x}'(t) = \mathbf{u}(t) \cdot \mathbf{x}(t), \quad t \in [0,1], \\
& \mathbf{x}(0) = 1.
\end{cases}
$$

Suppose that $\mathbf{u}_*$ is an optimal solution. Show that, in particular,

$$
0 = -2\mathbf{u}_*(t)(\mathbf{x}_*(t))^2 + \boldsymbol{\lambda}(t)\mathbf{x}_*(t), \quad t \in [0,1], \tag{4.1}
$$

where $\boldsymbol{\lambda}$ denotes the co-state and $\mathbf{x}_*$ the state corresponding to $\mathbf{u}_*$. Assuming that the state $\mathbf{x}_*$ is such that $\mathbf{x}_*(t) \neq 0$ for all $t \in [0,1]$, show that $\mathbf{u}_*$ satisfies

$$
\mathbf{u}_*'(t) = -1 - (\mathbf{u}_*(t))^2, \quad t \in [0,1], \quad \mathbf{u}_*(1) = 0.
$$

Hint: Differentiate both sides of (4.1) with respect to t.

Find $\mathbf{u}_*$ and $\mathbf{x}_*$.

4.2. Constrained Input, Fixed Final State, Autonomous

In this section, we will look at a more general problem than that considered in the previous section. In particular, we will add the following two complications:

(1) $\mathbf{x}(t_f) = x_f$,

(2) $\mathbf{u}(t) \in \mathbb{U} \subset \mathbb{R}^m$ for all $t \in [t_i, t_f]$.

We may assume that the cost is just given by an integral, rather than having added an extra term $\varphi(\mathbf{x}(t_f))$ as the terminal state cost. This is because we are assuming that the control problem demands that our admissible inputs guarantee that the final state $\mathbf{x}(t_f)$ is necessarily x_f, so that $\varphi(\mathbf{x}(t_f))$ is just the constant $\varphi(x_f)$, and this constant additive term in the cost, which is superfluous for the optimization problem, may be ignored.

Also, we will consider the autonomous case first: that is, F, f do not depend on t. We will now show the following result.

Theorem 4.10. *Suppose that $\mathbf{u}_*$ is an optimal solution to*

$$
\begin{cases}
\text{minimize} & \displaystyle\int_{t_i}^{t_f} F(\mathbf{x}(t), \mathbf{u}(t))dt \\
\text{subject to} & \mathbf{x}'(t) = f(\mathbf{x}(t), \mathbf{u}(t)), \quad t \in [t_i, t_f], \\
& \mathbf{x}(t_i) = x_i \in \mathbb{R}^n, \\
& \mathbf{x}(t_f) = x_f \subset \mathbb{R}^n, \\
& \mathbf{u}(t) \in \mathbb{U} \subset \mathbb{R}^m, \quad t \in [t_i, t_f].
\end{cases}
$$

Define the Hamiltonian $H : \mathbb{R}^n \times \mathbb{U} \times \mathbb{R} \times \mathbb{R}^n \to \mathbb{R}$ by

$$
H(x, u, \lambda_0, \lambda) := \lambda_0 F(x, u) + \lambda^\top f(x, u).
$$

Then there exists a nonnegative number λ_0, and a continuous and piecewise continuously differentiable function $\boldsymbol{\lambda} : [t_i, t_f] \to \mathbb{R}^n$ such that:

(P1) $\boldsymbol{\lambda}'(t) = -\left(\dfrac{\partial H}{\partial x}(\mathbf{x}_*(t), \mathbf{u}_*(t), \lambda_0, \boldsymbol{\lambda}(t))\right)^{\top}, \ t \in [t_i, t_f].$

(P2) *There exists a constant C such that*

$$H(\mathbf{x}_*(t), \mathbf{u}_*(t), \lambda_0, \boldsymbol{\lambda}(t)) = \min_{u \in \mathbb{U}} H(\mathbf{x}_*(t), u, \lambda_0, \boldsymbol{\lambda}(t)) = C, \ \ t \in [t_i, t_f].$$

(P3) $(\lambda_0, \boldsymbol{\lambda}(t)^{\top}) \neq 0, \ t \in [t_i, t_f].$

Remark 4.11. Only in some "bizarre" problems will we need $\lambda_0 = 0$. We will see an example later on, but one may think of this as being a pathological case since the optimality criteria does not depend on the cost being optimized.

Also, in the case when λ_0 is positive, there is no loss of generality in assuming that $\lambda_0 = 1$. This will be apparent from the sketch of the proof below. So when solving exercises, we may always assume that $\lambda_0 = 1$.

We will outline the proof of this result, but we will not justify all the technical details. Some of these will be replaced by a convincing plausibility argument. We give the proof as there are many nice geometric ideas involved in the proof.

We begin by recasting the problem in a new form by introducing an extra state variable. Then the optimization problem has a simple geometric interpretation, which can be exploited to prove the Minimum Principle.

Besides the given state equation

$$\mathbf{x}'(t) = f(\mathbf{x}(t), \mathbf{u}(t)), \quad t \in [t_i, t_f],$$

with the initial condition $\mathbf{x}(t_i) = x_i$, suppose we introduce the new state variable $\mathbf{x}_0$ defined by

$$\mathbf{x}_0(t) = \int_{t_i}^{t} F(\mathbf{x}(\tau), \mathbf{u}(\tau))d\tau, \quad t \in [t_i, t_f].$$

Note that this can be interpreted at the "intermediate cost at time t." Also, by the Fundamental Theorem of Calculus,

$$\mathbf{x}_0'(t) = F(\mathbf{x}(t), \mathbf{u}(t)), \quad t \in [t_i, t_f],$$

and $\mathbf{x}_0(t_i) = 0$. If we define the extended state $\widetilde{\mathbf{x}}$ by

$$\widetilde{\mathbf{x}} := \begin{bmatrix} \mathbf{x}_0 \\ \mathbf{x} \end{bmatrix}, \ \text{and set } \widetilde{f}(\mathbf{x}, \mathbf{u}) := \begin{bmatrix} F(\mathbf{x}, \mathbf{u}) \\ f(\mathbf{x}, \mathbf{u}) \end{bmatrix},$$

then we obtain

$$\begin{aligned} \widetilde{\mathbf{x}}' &= \widetilde{f}(\mathbf{x}, \mathbf{u}), \ t \in [t_i, t_f], \\ \widetilde{\mathbf{x}}(t_i) &= \begin{bmatrix} 0 \\ x_i \end{bmatrix}. \end{aligned}$$

If u is "admissible,"[1] then

$$\widetilde{\mathbf{x}}(t_f) = \begin{bmatrix} \mathbf{x}_0(t_f) \\ x_f \end{bmatrix}.$$

We know that the cost of an admissible u is $\mathbf{x}_0(t_f)$, and the optimization task is to minimize this. So if we view the extended state $\widetilde{\mathbf{x}}$ in $\mathbb{R}^{n+1}$ dimensions, thinking of $\mathbb{R}^n$ as the plane, with the x_0 axis emanating perpendicular to it, then the optimization task is to choose the input $\mathbf{u}$ among admissible inputs in such a manner that the 0th-component of the final extended state $\widetilde{\mathbf{x}}$ is as small as possible. See Figure 1.

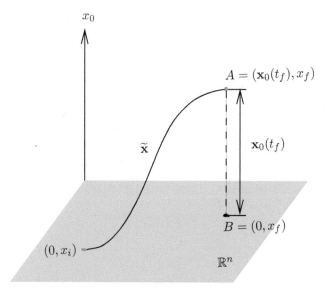

Figure 1. Geometric interpretation of the optimization problem: the closer the point A is to the point B, the smaller the cost.

Let $\mathbf{u}_*$ be optimal and let $\widetilde{\mathbf{x}}_*$ be the corresponding state in the extended state space ending at A_*. Then if we take any other $\mathbf{u}$ and look at the corresponding extended state $\widetilde{\mathbf{x}}$, then this must terminate at a point A higher up than A_*. See Figure 2.

Thus no extended state corresponding to an admissible input can end up on the segment A_*B. This will be the key fact that delivers the Pontryagin Minimum Principle.

---

[1]Rather than defining this painfully, we just understand this to mean that all the constraints are satisfied: so in the case we are considering, this means that $\mathbf{x}(t_f) = x_f$ and $\mathbf{u}(t) \in \mathbb{U}$ for all $t \in [t_i, t_f]$.

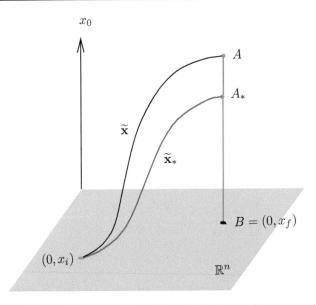

Figure 2. Geometric interpretation of the optimization problem: any admissible input gives an extended state that cannot end on the segment A_*B.

Sketch of the proof of Theorem 4.10. Among the admissible controls, suppose that $\mathbf{u}_*$ is optimal. Thus if $\mathbf{x}_*$ is the extended state, then $\mathbf{u}_*$ takes

$$\widetilde{\mathbf{x}}_*(t_i) = \left[\begin{array}{c} 0 \\ x_i \end{array} \right]$$

to

$$\widetilde{\mathbf{x}}_*(t_f) = \left[\begin{array}{c} \mathbf{x}_{0,*}(t_f) \\ x_f \end{array} \right]$$

and the cost $\mathbf{x}_{0,*}(t_f)$ is the least possible one. Then we need to show that (P1)-(P3) hold.

Step 1. Effect of a simple perturbation of the optimal control input.

We will make a small change in $\mathbf{u}_*$, and find out the effect on the extended state at time t_f. Let $v \in \mathbb{U}$, $\Delta > 0$, $\tau \in (t_i, t_f]$ such that $\tau - \Delta \geq t_i$. Define the perturbed input $\mathbf{u}$ by

$$\mathbf{u}(t) = \left\{ \begin{array}{ll} \mathbf{u}_*(t) & \text{for } t_i < t < \tau - \Delta, \\ v & \text{for } \tau - \Delta \leq t \leq \tau, \\ \mathbf{u}_*(t) & \text{for } \tau < t < t_f. \end{array} \right.$$

See Figure 3.

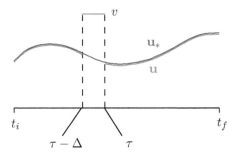

Figure 3. Perturbation $\mathbf{u}$ of the optimal control input $\mathbf{u}_*$: a needle variation.

We remark that Δ is to be imagined to be "small," but v can be very much different from $\mathbf{u}_*(\tau)$. We now ask: How much change is produced by this perturbation in the final cost $x_0(t_f)$? Up to time $\tau - \Delta$, $\mathbf{u}$ coincides with $\mathbf{u}_*$, and by the uniqueness of solutions, also $\widetilde{\mathbf{x}}$ coincides with $\widetilde{\mathbf{x}}_*$ until that time. However, between the time $\tau - \Delta$ to τ, the inputs differ. Let us first see what happens in the little interval $[\tau - \Delta, \tau]$. We have

$$
\begin{aligned}
\widetilde{\mathbf{x}}(\tau) - \widetilde{\mathbf{x}}_*(\tau) &= \int_{t_i}^{\tau} \widetilde{f}(\mathbf{x}(t), \mathbf{u}(t)) dt - \int_{t_i}^{\tau} \widetilde{f}(\mathbf{x}_*(t), \mathbf{u}_*(t)) dt \\
&= \int_{t_i}^{\tau} \Big(\widetilde{f}(\mathbf{x}(t), \mathbf{u}(t)) - \widetilde{f}(\mathbf{x}_*(t), \mathbf{u}_*(t)) \Big) dt \\
&= \Big(\widetilde{f}(\mathbf{x}_*(\tau), v) - \widetilde{f}(\mathbf{x}_*(\tau), \mathbf{u}_*(\tau)) \Big) \cdot \Delta + o(\Delta). \quad (4.2)
\end{aligned}
$$

In the above, we have used

$$
\widetilde{f}(\mathbf{x}(t), v) = \widetilde{f}(\mathbf{x}_*(t), v) + O(\Delta), \quad t \in [\tau - \Delta, \tau]. \quad (4.3)
$$

This follows because to a first order approximation $\mathbf{x}(t)$ can be replaced by $\mathbf{x}_*(t)$ in this time interval by the Perturbation Result in Theorem 3.31. Indeed, the initial condition is the same: $\mathbf{x}(\tau - \Delta) = \mathbf{x}_*(\tau - \Delta)$, and the final time τ is close to the initial time $\tau - \Delta$ (that is, Δ is small) and so $M(e^{L\Delta} - 1)$ is of order $O(\Delta)$.

But now for times $t \in [\tau, t_f]$, the two inputs coincide again. But at the initial time τ, there is a perturbation in the initial condition: instead of the initial condition being $\widetilde{\mathbf{x}}_*(\tau)$, it is now $\widetilde{\mathbf{x}}(\tau)$. Now (4.2) in the calculation done above shows that this perturbation $\widetilde{\mathbf{x}}(\tau) - \widetilde{\mathbf{x}}_*(\tau)$ is of order $O(\Delta)$:

$$
\widetilde{\mathbf{x}}(\tau) - \widetilde{\mathbf{x}}_*(\tau) = \Big(\widetilde{f}(\mathbf{x}_*(\tau), v) - \widetilde{f}(\mathbf{x}_*(\tau), \mathbf{u}_*(\tau)) \Big) \cdot \Delta + o(\Delta).
$$

By the Perturbation Result in Theorem 3.31, it follows that

$$
\begin{aligned}
\|\widetilde{\mathbf{x}}(t) - \widetilde{\mathbf{x}}_*(t)\| &\leq \|\widetilde{\mathbf{x}}(\tau) - \widetilde{\mathbf{x}}_*(\tau)\| = O(\Delta), \text{ for } t \in [\tau, t_f], \text{ and in particular,} \\
\|\widetilde{\mathbf{x}}(t_f) - \widetilde{\mathbf{x}}_*(t_f)\| &\leq \|\widetilde{\mathbf{x}}(\tau) - \widetilde{\mathbf{x}}_*(\tau)\| = O(\Delta).
\end{aligned}
$$

We have for $t \in [\tau, t_f]$, using Taylor's Formula, that

$$
\begin{aligned}
\widetilde{\mathbf{x}}'(t) - \widetilde{\mathbf{x}}'_*(t) &= \widetilde{f}(\mathbf{x}(t), \mathbf{u}_*(t)) - \widetilde{f}(\mathbf{x}_*(t), \mathbf{u}_*(t)) \\
&= \frac{\partial \widetilde{f}}{\partial x}(\mathbf{x}_*(t), \mathbf{u}_*(t)) \cdot (\mathbf{x}(t) - \mathbf{x}_*(t)) + o(\mathbf{x}(t) - \mathbf{x}_*(t)) \\
&= \frac{\partial \widetilde{f}}{\partial x}(\mathbf{x}_*(t), \mathbf{u}_*(t)) \cdot (\mathbf{x}(t) - \mathbf{x}_*(t)) + o(\mathbf{x}(\tau) - \mathbf{x}_*(\tau)). \quad (4.4)
\end{aligned}
$$

Let $\delta\widetilde{\mathbf{x}}$ be the solution to the initial value problem

$$
\delta\widetilde{\mathbf{x}}'(t) = \underbrace{\frac{\partial \widetilde{f}}{\partial x}(\mathbf{x}_*(t), \mathbf{u}_*(t))}_{=:A(t)} \cdot \delta\widetilde{\mathbf{x}}(t), \ t \in [\tau, t_f],
$$

$$
\delta\widetilde{\mathbf{x}}(\tau) = \left(\widetilde{f}(\mathbf{x}_*(\tau), v) - \widetilde{f}(\mathbf{x}_*(\tau), \mathbf{u}_*(\tau)) \right) \cdot \Delta.
$$

The solution to the above is given by

$$
\delta\widetilde{\mathbf{x}}(t) = \Phi(t_f, \tau) \cdot \delta\widetilde{\mathbf{x}}(\tau) = \Phi(t_f, \tau) \left(\widetilde{f}(\mathbf{x}_*(\tau), v) - \widetilde{f}(\mathbf{x}_*(\tau), \mathbf{u}_*(\tau)) \right) \cdot \Delta,
$$

where Φ is the state transition matrix for linear system $\delta\widetilde{\mathbf{x}}'(t) = A(t)\delta\widetilde{\mathbf{x}}(t)$. Then (4.4) gives us that

$$
\begin{aligned}
\widetilde{\mathbf{x}}(t) - \widetilde{\mathbf{x}}_*(t) &= \delta\widetilde{\mathbf{x}}(t) + o(\Delta) \\
&= \Phi(t_f, \tau) \left(\widetilde{f}(\mathbf{x}_*(\tau), v) - \widetilde{f}(\mathbf{x}_*(\tau), \mathbf{u}_*(\tau)) \right) \cdot \Delta + o(\Delta), \ t \in [\tau, t_f].
\end{aligned}
$$

Step 2. Effect of multiple needle perturbations of the optimal control input.

Suppose we make several perturbations of the type described in the previous step, as depicted in Figure 4.

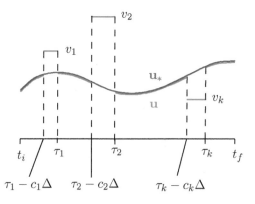

Figure 4. Multiple needle variations of the optimal control input $\mathbf{u}_*$.

Let $I_\ell := [\tau_\ell - c_\ell \Delta, \tau_\ell]$, $\ell = 1, \cdots, k$, and $\mathbf{u}$ be given by

$$\mathbf{u}(t) = \begin{cases} \mathbf{u}_*(t) & \text{if } t \notin I_1 \cup \cdots \cup I_k \\ v_\ell & \text{if } t \in I_\ell, \ \ell = 1, \cdots, k. \end{cases}$$

Suppose for the moment that $k = 2$. Let us first look at the interval $[t_i, \tau_2 - c_2\Delta]$, thinking of $\tau_2 - c_2\Delta$ as the final time. Then from Step 1, we know that

$$\widetilde{\mathbf{x}}(\tau_2 - c_2\Delta) - \widetilde{\mathbf{x}}_*(\tau_2 - c_2\Delta)$$
$$= \Phi(\tau_2 - c_2\Delta, \tau_1)\Big(\widetilde{f}(\mathbf{x}_*(\tau_1), v_1) - \widetilde{f}(\mathbf{x}_*(\tau_1), \mathbf{u}_*(\tau_1))\Big) \cdot c_1\Delta + o(\Delta).$$

On the other hand, we have (by using (4.3))

$$\widetilde{\mathbf{x}}(\tau_2) - \widetilde{\mathbf{x}}(\tau_2 - c_2\Delta) - \Big(\widetilde{\mathbf{x}}_*(\tau_2) - \widetilde{\mathbf{x}}_*(\tau_2 - c_2\Delta)\Big)$$
$$= \Big(\widetilde{f}(\mathbf{x}_*(\tau_2), v_2) - \widetilde{f}(\mathbf{x}_*(\tau_2), \mathbf{u}_*(\tau_2))\Big) \cdot c_2\Delta + o(\Delta),$$

and so

$$\widetilde{\mathbf{x}}(\tau_2) - \widetilde{\mathbf{x}}_*(\tau_2)$$
$$= \Big(\widetilde{f}(\mathbf{x}_*(\tau_2), v_2) - \widetilde{f}(\mathbf{x}_*(\tau_2), \mathbf{u}_*(\tau_2))\Big) \cdot c_2\Delta$$
$$\quad + \Big(\widetilde{\mathbf{x}}(\tau_2 - c_2\Delta) - \widetilde{\mathbf{x}}_*(\tau_2 - c_2\Delta)\Big) + o(\Delta)$$
$$= \Big(\widetilde{f}(\mathbf{x}_*(\tau_2), v_2) - \widetilde{f}(\mathbf{x}_*(\tau_2), \mathbf{u}_*(\tau_2))\Big) \cdot c_2\Delta$$
$$\quad + \Phi(\tau_2 - c_2\Delta, \tau_1)\Big(\widetilde{f}(\mathbf{x}_*(\tau_1), v_1) - \widetilde{f}(\mathbf{x}_*(\tau_1), \mathbf{u}_*(\tau_1))\Big) \cdot c_1\Delta + o(\Delta)$$
$$= \Big(\widetilde{f}(\mathbf{x}_*(\tau_2), v_2) - \widetilde{f}(\mathbf{x}_*(\tau_2), \mathbf{u}_*(\tau_2))\Big) \cdot c_2\Delta$$
$$\quad + \Phi(\tau_2, \tau_1)\Big(\widetilde{f}(\mathbf{x}_*(\tau_1), v_1) - \widetilde{f}(\mathbf{x}_*(\tau_1), \mathbf{u}_*(\tau_1))\Big) \cdot c_1\Delta + o(\Delta).$$

Consequently,

$$\widetilde{\mathbf{x}}(\tau_f) - \widetilde{\mathbf{x}}_*(\tau_f)$$
$$= \Phi(t_f, \tau_2)(\widetilde{\mathbf{x}}(\tau_2) - \widetilde{\mathbf{x}}_*(\tau_2)) + o(\delta)$$
$$= \Phi(t_f, \tau_2)\Big(\widetilde{f}(\mathbf{x}_*(\tau_2), v_2) - \widetilde{f}(\mathbf{x}_*(\tau_2), \mathbf{u}_*(\tau_2))\Big) \cdot c_2\Delta$$
$$\quad + \Phi(t_f, \tau_2)\Phi(\tau_2, \tau_1)\Big(\widetilde{f}(\mathbf{x}_*(\tau_1), v_1) - \widetilde{f}(\mathbf{x}_*(\tau_1), \mathbf{u}_*(\tau_1))\Big) \cdot c_1\Delta + o(\Delta)$$
$$= \Phi(t_f, \tau_2)\Big(\widetilde{f}(\mathbf{x}_*(\tau_2), v_2) - \widetilde{f}(\mathbf{x}_*(\tau_2), \mathbf{u}_*(\tau_2))\Big) \cdot c_2\Delta$$
$$\quad + \Phi(t_f, \tau_1)\Big(\widetilde{f}(\mathbf{x}_*(\tau_1), v_1) - \widetilde{f}(\mathbf{x}_*(\tau_1), \mathbf{u}_*(\tau_1))\Big) \cdot c_1\Delta + o(\Delta)$$

where in order to get the last equality we used $\Phi(t_f, \tau_2)\Phi(\tau_2, \tau_1) = \Phi(t_f, \tau_1)$; see Theorem 3.12. So we see that the perturbation in the final state is the "superposition" of the effects of the individual needle perturbations. If $k > 2$, then the analysis is similar, and we have in the general case that

$$\widetilde{\mathbf{x}}(\tau_f) - \widetilde{\mathbf{x}}_*(\tau_f) = \Delta \cdot \sum_{\ell=1}^{k} c_\ell \cdot \underbrace{\Phi(t_f, \tau_\ell)\Big(\widetilde{f}(\mathbf{x}_*(\tau_\ell), v_\ell) - \widetilde{f}(\mathbf{x}_*(\tau_\ell), \mathbf{u}_*(\tau_\ell))\Big)}_{=:w_\ell}.$$

Note that the numbers c_ℓ are nonnegative, and so we see that

$$\widetilde{\mathbf{x}}(\tau_f) - \widetilde{\mathbf{x}}_*(\tau_f) = \Delta \cdot \sum_{\ell=1}^{k} c_\ell \cdot w_\ell \in C,$$

where C denotes the positive cone given by:

$$C = \left\{ \sum_{\ell=1}^{k} c_\ell \cdot w_\ell : \begin{array}{l} k \geq 0, \ \tau_1, \cdots, \tau_k \in (t_i, t_f], \ c_1, \cdots, c_\ell \geq 0, \\ w_\ell := \Phi(t_f, \tau_\ell)\Big(\widetilde{f}(\mathbf{x}_*(\tau_\ell), v_\ell) - \widetilde{f}(\mathbf{x}_*(\tau_\ell), \mathbf{u}_*(\tau_\ell))\Big), \ \ell = 1, \cdots, k \end{array} \right\}.$$

See Figure 5.

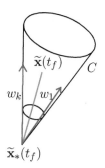

Figure 5. $\widetilde{\mathbf{x}}(\tau_f) - \widetilde{\mathbf{x}}_*(\tau_f)$ belongs to the positive cone C.

Step 3. Separating hyperplane.

Let us look at Figure 2 again. It is intuitively clear that C cannot intersect AB, because otherwise a perturbation of $\mathbf{u}_*$ reduces the cost, contradicting the optimality of $\mathbf{u}_*$. So we have a convex cone C in $\mathbb{R}^{n+1}$ with apex at $\widetilde{\mathbf{x}}_*(t_f)$, which does not intersect the ray $\overrightarrow{A_*B}$ starting from A_* toward B, consisting of the set of all points below A_*:

$$\overrightarrow{A_*B} = \{(r, x_f) : r \leq \mathbf{x}_{0,*}(t_f)\}.$$

This means that there must exist a separating hyperplane $\mathbb{H}$ that separates $\overrightarrow{A_*B}$ and C. In other words, there exists a $d \in \mathbb{R}^{n+1}$ such that

$$\mathbb{H} := \{x \in \mathbb{R}^{n+1} : x^\top d = 0\}$$

that is,

$$v^\top d \ \geq \ 0 \text{ for all } v \in C,$$
$$v^\top d \ \leq \ 0 \text{ for all } v \in \{(r, x_f) - (\mathbf{x}_{0,*}(t_f), x_f) : r \leq \mathbf{x}_{0,*}(t_f)\} = \{(r, 0) : r \leq 0\}.$$

See Figure 6. From the last equation above, we see that the first component d_0 of d must be nonnegative. (If this first component d_0 is nonzero, we may divide d by d_0, without changing any of the above. So we may assume without loss of generality that the first component d_0 of d is equal to 1; see Remark 4.11 above.)

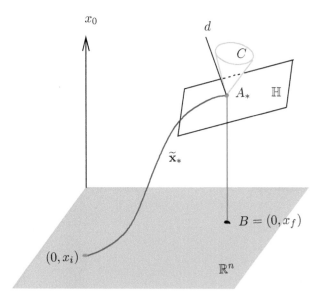

Figure 6. The hyperplane with normal d separates the cone C and the ray starting from A_* toward B.

Step 4. Setting up the co-state equation.

Let $\widetilde{\boldsymbol{\lambda}} : [t_i, t_f] \to \mathbb{R}^{n+1}$ be the solution to

$$\begin{cases} \widetilde{\boldsymbol{\lambda}}'(t) = -A(t)^\top \widetilde{\boldsymbol{\lambda}}(t), \ t \in [t_i, t_f], \\ \widetilde{\boldsymbol{\lambda}}(t_f) = d. \end{cases}$$

Then using the result in Exercise 3.13, we know that

$$\widetilde{\boldsymbol{\lambda}}(t) = (\Phi(t_f, t))^\top d, \quad t \in [t_i, t_f].$$

Define $\boldsymbol{\lambda} : [t_i, t_f] \to \mathbb{R}^n$ and λ_0 by

$$\widetilde{\boldsymbol{\lambda}}(t) = \begin{bmatrix} \lambda_0(t) \\ \boldsymbol{\lambda}(t) \end{bmatrix}, \quad t \in [t_i, t_f].$$

Then since

$$\frac{\partial \widetilde{f}}{\partial x_0} = 0,$$

we know that the first column of $A(t)$ is zero:

$$A(t) = \begin{bmatrix} 0 & \dfrac{\partial F}{\partial x}(\mathbf{x}_*(t), \mathbf{u}_*(t)) \\[2ex] 0 & \dfrac{\partial f}{\partial x}(\mathbf{x}_*(t), \mathbf{u}_*(t)) \end{bmatrix}.$$

This means that the first row of $-A(t)^\top$ is 0, and so from $\widetilde{\boldsymbol{\lambda}}'(t) = -A(t)^\top \widetilde{\boldsymbol{\lambda}}(t)$, we obtain $\lambda_0'(t) = 0$ for $t \in [t_i, t_f]$, so that λ_0 must be a constant. From the endpoint condition $\widetilde{\boldsymbol{\lambda}}(t_f) = d$, we see that λ_0 must then be the first component of d, which is nonnegative. This will serve as our λ_0. (We may assume that this λ_0 is actually 0 or 1, as mentioned at the end of Step 3.)

The equations satisfied by $\boldsymbol{\lambda}$ are exactly the ones given in (P1). Indeed,

$$\begin{aligned} \boldsymbol{\lambda}'(t) &= -\lambda_0 \left(\frac{\partial F}{\partial x}(\mathbf{x}_*(t), \mathbf{u}_*(t)) \right)^\top - \left(\frac{\partial f}{\partial x}(\mathbf{x}_*(t), \mathbf{u}_*(t)) \right)^\top \boldsymbol{\lambda}(t) \\ &= -\left(\frac{\partial H}{\partial x}(\mathbf{x}_*(t), \mathbf{u}_*(t), \lambda_0, \boldsymbol{\lambda}(t)) \right)^\top \boldsymbol{\lambda}(t), \end{aligned}$$

for all $t \in [t_i, t_f]$. So we have actually obtained (P1) for free.

Step 5. That (P2) holds.

We have

$$H(\mathbf{x}_*(\tau_\ell), v_\ell, \lambda_0, \boldsymbol{\lambda}(\tau_\ell)) - H(\mathbf{x}_*(\tau_\ell), \mathbf{u}_*(\tau_\ell), \lambda_0, \boldsymbol{\lambda}(\tau_\ell))$$

$$\begin{aligned} &= \lambda_0 \Big(F(\mathbf{x}_*(\tau_\ell), v_\ell) - F(\mathbf{x}_*(\tau_\ell), \mathbf{u}_*(\tau_\ell)) \Big) \\ &\quad + \boldsymbol{\lambda}(\tau_\ell)^\top \Big(f(\mathbf{x}_*(\tau_\ell), v_\ell) - f(\mathbf{x}_*(\tau_\ell), \mathbf{u}_*(\tau_\ell)) \Big) \\ &= (\widetilde{\boldsymbol{\lambda}}(\tau_\ell))^\top \Big(\widetilde{f}(\mathbf{x}_*(\tau_\ell), v_\ell) - \widetilde{f}(\mathbf{x}_*(\tau_\ell), \mathbf{u}_*(\tau_\ell)) \Big) \\ &= d^\top \underbrace{\Phi(t_f, \tau_\ell) \Big(\widetilde{f}(\mathbf{x}_*(\tau_\ell), v_\ell) - \widetilde{f}(\mathbf{x}_*(\tau_\ell), \mathbf{u}_*(\tau_\ell)) \Big)}_{\in C} \geq 0. \end{aligned}$$

Since $\tau_\ell \in (t_i, t_f]$ and $v_\ell \in \mathbb{U}$ were arbitrary, it follows that the first equality in (P2) holds. A proof of the constancy of $t \mapsto H(\mathbf{x}_*(t), \mathbf{u}_*(t), \lambda_0, \boldsymbol{\lambda}(t))$ on the time interval $[t_i, t_f]$ assuming that $\mathbf{u}_*$ is continuously differentiable and when $\mathbf{u}_*(t)$ belongs to the interior of $\mathbb{U}$ for each $t \in [t_i, t_f]$ can be given in the same manner as Step 5 in the proof of Theorem 4.1. We won't discuss the most general case.

Step 6. We know that

$$\begin{bmatrix} \lambda_0 \\ \boldsymbol{\lambda}(t) \end{bmatrix} = \widetilde{\boldsymbol{\lambda}}(t) = (\Phi(t_f, t))^\top d.$$

But $d \neq 0$, and by Theorem 3.12, $\Phi(t_f, t)$ (and so also its transpose $(\Phi(t_f, t))^\top$) is invertible. Consequently, $(\lambda_0, \boldsymbol{\lambda}(t)^\top) \neq 0$ for all $t \in [t_i, t_f]$, that is, (P3) holds. $\square$

Example 4.12. Consider the following optimal control problem:

$$\begin{cases} \text{maximize} & \displaystyle\int_0^{10} \mathbf{x}(t)dt \\ \text{such that} & \mathbf{x}'(t) = \mathbf{u}(t) \text{ for } t \in [0,10], \\ & \mathbf{x}(0) = 0, \\ & \mathbf{x}(10) = 2, \\ & \mathbf{u}(t) \in [0,1] \text{ for } t \in [0,10]. \end{cases}$$

Let us suppose that there exists an optimal control $\mathbf{u}_*$, and let us use Pontryagin's Minimum Principle to find $\mathbf{u}_*$.

The Hamiltonian H is given by (taking $\lambda_0 = 1$, and noting that it is a *maximization* problem)

$$H(x, u, \lambda) = -x + \lambda u.$$

Then

$$\begin{aligned} \mathbf{u}_*(t) &= \arg\min_{u \in [0,1]} H(\mathbf{x}_*(t), u, \boldsymbol{\lambda}(t)) \\[2mm] &= \arg\min_{u \in [0,1]} (-\mathbf{x}_*(t) + \boldsymbol{\lambda}(t)u) \\[2mm] &= \begin{cases} 0 & \text{if } \boldsymbol{\lambda}(t) > 0, \\ 1 & \text{if } \boldsymbol{\lambda}(t) < 0. \end{cases} \end{aligned}$$

What about when $\boldsymbol{\lambda}(t) = 0$? We only need to worry about this in case $\boldsymbol{\lambda}(t)$ is zero in an interval. Let us now show that in fact this case doesn't arise. The co-state equation is given by

$$\boldsymbol{\lambda}'(t) = -\frac{\partial H}{\partial x}(\mathbf{x}_*(t), \mathbf{u}_*(t), \boldsymbol{\lambda}(t)) = -(-1) = 1, \quad t \in [0,10],$$

and so

$$\boldsymbol{\lambda}(t) = t + C, \quad t \in [0,10],$$

for some constant C. Thus there is at most one time instant t_s (the "switching time") where $p(t_s) = 0$. Thus we have the three possible cases depicted in Figure 7. Let us find out which of these possibilities for $\mathbf{u}_*$ is admissible.

$\underline{1°}$ When $\boldsymbol{\lambda}$ is pointwise positive, $\mathbf{u}_* \equiv 0$ and so we get

$$\begin{aligned} 2 &= \mathbf{x}_*(10) = \mathbf{x}_*(10) - 0 = \mathbf{x}_*(10) - \mathbf{x}_*(0) \\[2mm] &= \int_0^{10} \mathbf{u}_*(t)dt = \int_0^{10} 0\,dt = 0, \end{aligned}$$

a contradiction. So this case (leftmost picture in Figure 7) is not possible.

$\underline{2°}$ When $\boldsymbol{\lambda}$ is pointwise negative, $\mathbf{u}_* \equiv 1$ and so we get

$$
\begin{aligned}
2 &= \mathbf{x}_*(10) = \mathbf{x}_*(10) - 0 = \mathbf{x}_*(10) - \mathbf{x}_*(0) \\
&= \int_0^{10} \mathbf{u}_*(t)dt = \int_0^{10} 1dt = 10,
\end{aligned}
$$

a contradiction. So this case (rightmost picture in Figure 7) is not possible.

$\underline{3°}$ When $\boldsymbol{\lambda}$ is first pointwise negative and then switches sign at a time instant t_s, we have

$$
\begin{aligned}
2 &= \mathbf{x}_*(10) = \mathbf{x}_*(10) - 0 = \mathbf{x}_*(10) - \mathbf{x}_*(0) \\
&= \int_0^{10} \mathbf{u}_*(t)dt = \int_0^{t_s} 1dt + \int_{t_s}^{10} 0dt = t_s,
\end{aligned}
$$

so that $t_s = 2$. Hence $\mathbf{u}_*(t) = \begin{cases} 1 & \text{if } t \in [0,2], \\ 0 & \text{if } t \in (2,10]. \end{cases}$

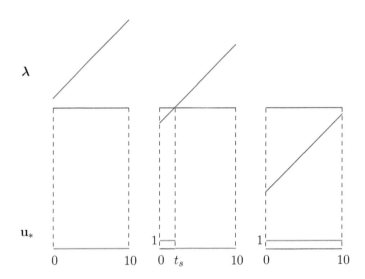

Figure 7. The possible graphs of $u \mapsto u(\lambda_0 + Cu)$ in the case when $\lambda_0 \neq 0$.

Note that the optimal control $\mathbf{u}_*$ just takes on the "extreme values" of the pointwise-constraint set $\mathbb{U} = [0,1]$. Such controls are referred to as "bang-bang controls," since they go from one extreme value (a "bang!") to another one ("bang!").   ◇

The following example shows that the case $\lambda_0 = 0$, although rare, can occur.

Example 4.13 ($\lambda_0 = 0$ can occur). Consider the optimal control problem

$$
\begin{cases}
\text{minimize} & \displaystyle\int_0^1 \mathbf{u}(t)dt \\
\text{such that} & \mathbf{x}'(t) = (\mathbf{u}(t))^2 \text{ for } t \in [0,1], \\
& \mathbf{x}(0) = 0, \\
& \mathbf{x}(1) = 0.
\end{cases}
$$

First we claim that $\mathbf{x}$ is increasing. Indeed, let $a, b \in [0,1]$ with $a \le b$. Then

$$
\mathbf{x}(b) - \mathbf{x}(a) = \int_a^b \mathbf{x}'(\tau)d\tau = \int_a^b \underbrace{(\mathbf{u}(\tau))^2}_{\ge 0} d\tau \ge 0,
$$

and so $\mathbf{x}(b) \ge \mathbf{x}(a)$. Thus $\mathbf{x}$ is increasing.

So for $0 \le t \le 1$, $0 = \mathbf{x}(0) \le \mathbf{x}(t) \le \mathbf{x}(1) = 0$, and so $\mathbf{x} \equiv 0$. But then $\mathbf{x}' \equiv 0$ on $[0,1]$, and so $(\mathbf{u}(t))^2 = \mathbf{x}'(t) = 0$ for $t \in [0,1]$. Hence $\mathbf{u} \equiv 0$. Thus the only admissible control input $\mathbf{u}$ is $\mathbf{0}$! Consequently, this is the unique optimal control. So by the Pontryagin Minimum Principle, this $\mathbf{u}_*$ must satisfy (P1), (P2), (P3). But we will now see that we must necessarily have $\lambda_0 = 0$. Let us see this. We have that the Hamiltonian is given by $H(x, u, \lambda_0, \lambda) = \lambda_0 u + \lambda u^2 = u(\lambda_0 + \lambda u)$. Thus the co-state equation is given by

$$
\boldsymbol{\lambda}'(t) = -\frac{\partial H}{\partial x}(\mathbf{x}_*(t), \mathbf{u}_*(t), \lambda_0, \boldsymbol{\lambda}(t)) = 0, \quad t \in [0,1],
$$

and so $\boldsymbol{\lambda}(t) = C$ for all $t \in [0,1]$ for some constant C. If $\lambda_0 \ne 0$, then the function $u \mapsto u(\lambda_0 + \lambda u) = u(\lambda_0 + Cu)$ has a graph of one of the forms shown in Figure 8.

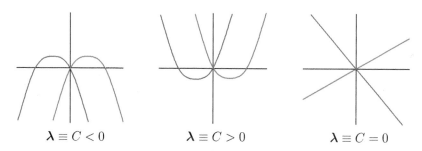

$$\lambda \equiv C < 0 \qquad\qquad \lambda \equiv C > 0 \qquad\qquad \lambda \equiv C = 0$$

Figure 8. The possible graphs of $u \mapsto u(\lambda_0 + Cu)$ in the case when $\lambda_0 \ne 0$.

From the pictures we see that C must be positive for a minimizer to exist, and when $C > 0$, the minimizer is *not* zero, a contradiction. So we can't have $\lambda_0 \ne 0$. But if $\lambda_0 = 0$, then each of (P1), (P2), (P3) *are* satisfied with $\mathbf{u}_* \equiv 0$, $\mathbf{x}_* \equiv 0$, $\boldsymbol{\lambda} \equiv C$, where $C > 0$ is arbitrary, $\lambda_0 = 0$: Indeed, for all $t \in [0,1]$, we have

(P1) $\boldsymbol{\lambda}'(t) = -\dfrac{\partial H}{\partial x}(\mathbf{x}_*(t), \mathbf{u}_*(t), \lambda_0, \boldsymbol{\lambda}(t)) = 0$,

(P2) $H(\mathbf{x}_*(t), \mathbf{u}_*(t), \lambda_0, \boldsymbol{\lambda}(t)) = 0(0 + C \cdot 0) = 0 = \min\limits_{u \in \mathbb{R}} Cu^2 = \min\limits_{u \in \mathbb{R}} H(\mathbf{x}_*(t), u, \lambda_0, \boldsymbol{\lambda}(t))$,

(P3) $(\lambda_0, \boldsymbol{\lambda}(t)) = (0, C) \neq 0$.   ◇

Exercise 4.14. Consider the following optimal control problem

$$\begin{cases} \text{minimize} & \displaystyle\int_0^1 \Big((\mathbf{x}(t))^2 - 2\mathbf{x}(t) \Big) dt \\[2mm] \text{such that} & \mathbf{x}'(t) = \mathbf{u}(t) \text{ for } t \in [0,1], \\[1mm] & \mathbf{x}(0) = 0, \\[1mm] & \mathbf{x}(1) = 0, \\[1mm] & \mathbf{u}(t) \in [-1, 1] \text{ for } t \in [0,1]. \end{cases}$$

Assuming that an optimal control $\mathbf{u}_*$ exists, we want to use the Pontryagin Minimum Principle to find $\mathbf{u}_*$. Proceed as follows:

(1) Show first that if $\mathbf{u}$ is a control such that $\mathbf{u}(t) \in [-1, 1]$ for $t \in [0, 1]$, then the corresponding state $\mathbf{x}$ satisfies $\mathbf{x}(t) < 1$ for all $t \in [0, 1)$. Conclude using the co-state equation for the co-state $\boldsymbol{\lambda}$ that $t \mapsto \boldsymbol{\lambda}(t)$ is strictly increasing in $[0, 1]$.

(2) Find the optimal control $\mathbf{u}_*$ and the corresponding state $\mathbf{x}_*$ for the optimal control problem.

4.3. Final State in a Manifold, Autonomous

In this section, we will add yet another generality, namely what happens when instead of the final state being fixed at a point, we only demand that it belongs to a specified manifold $\mathbb{X}_f$ at the final time. So we consider:

$$\begin{cases} \text{minimize} & \displaystyle\int_{t_i}^{t_f} F(\mathbf{x}(t), \mathbf{u}(t)) dt \\[2mm] \text{subject to} & \mathbf{x}'(t) = f(\mathbf{x}(t), \mathbf{u}(t)), \quad t \in [t_i, t_f], \\[1mm] & \mathbf{x}(t_i) = x_i \in \mathbb{R}^n, \\[1mm] & \mathbf{x}(t_f) \in \mathbb{X}_f \subset \mathbb{R}^n, \\[1mm] & \mathbf{u}(t) \in \mathbb{U} \subset \mathbb{R}^m, \quad t \in [t_i, t_f]. \end{cases}$$

4.3.1. What is $\mathbb{X}_f$? For us, the given "manifold" $\mathbb{X}_f$ will always be the set of common zeros of a finite set of continuously differentiable functions $g_i : \mathbb{R}^n \to \mathbb{R}$, $i \in I$, where I is the finite index set: $\mathbb{X}_f := \{x \in \mathbb{R}^n : g_i(x) = 0, \ i \in I\}$. A standing assumption will be that for all $x \in \mathbb{X}_f$, the set of vectors $\{\nabla g_i(x) : i \in I\}$ is linearly independent in $\mathbb{R}^n$. If $\#I$ is the number of elements in the finite set I, then we call $\mathbb{X}_f$ a $(n - \#I)$-*dimensional manifold*. Here are some examples.

Example 4.15.

(1) $X_f = \mathbb{R}^n$ is an n-dimensional manifold, where I is empty, so that $\#I = 0$. (This corresponds to the case in our optimal control problems when the final state is "free," that is, there are no constraints on $\mathbf{x}(t_f)$.)

(2) $X_f = \mathbb{S}^2 = \{x \in \mathbb{R}^3 : x_1^2 + x_2^2 + x_3^2 = 1\}$ is a 2-dimensional manifold. Indeed, with I being a singleton, and $g = g_1$ defined by $g(x) = x_1^2 + x_2^2 + x_3^2 - 1$, $x \in \mathbb{R}^3$, we have that $X_f = \{x \in \mathbb{R}^3 : g(x) = 0\}$. On the sphere $\mathbb{S}^2$, at least one amongst x_1, x_2, x_3 must be nonzero (otherwise we can't have $x_1^2 + x_2^2 + x_3^2 = 1!$), and so the vector $\nabla g(x) = \begin{bmatrix} 2x_1 & 2x_2 & 2x_3 \end{bmatrix} \neq 0$. Thus for each $x \in X_f$, the set $\{\nabla g(x)\}$ is linearly independent in $\mathbb{R}^3$. So $X_f = \mathbb{S}^2$ is a $(3-1) = 2$-dimensional manifold in $\mathbb{R}^3$. This is intuitively expected, since locally the surface $\mathbb{S}^2$ looks like a plane, the most vivid demonstration of which is our life on Earth!

(3) $X_f = \{x \in \mathbb{R}^3 : x_1 = 0 \text{ and } x_2 = 0\}$ is a 1-dimensional manifold. Indeed, we see that $X_f = \{x \in \mathbb{R}^3 : g_1(x) = 0, \ g_2(x) = 0\}$, where $g_1(x) := x_1$ and $g_2(x) := x_2$, $x \in \mathbb{R}^3$. We have that

$$\nabla g_1(x) = \begin{bmatrix} 1 & 0 & 0 \end{bmatrix},$$
$$\nabla g_2(x) = \begin{bmatrix} 0 & 1 & 0 \end{bmatrix},$$

and these are linearly independent in $\mathbb{R}^3$ for each $x \in X_f$. Thus X_f is a $(3-2) = 1$-dimensional manifold. From the picture below, this is clearly to be expected since X_f is a line in $\mathbb{R}^3$, which is the intersection of the two planes $x_1 = 0$ (zero set of g_1) and $x_2 = 0$ (zero set of g_2).

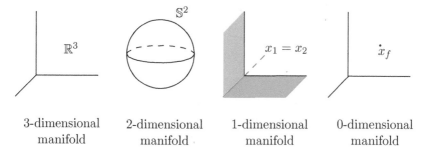

| 3-dimensional | 2-dimensional | 1-dimensional | 0-dimensional |
| manifold | manifold | manifold | manifold |

(4) If $x_f \in \mathbb{R}^3$, then $X_f := \{x_f\}$ is a 0-dimensional manifold. Indeed if $x_{f,1}, x_{f,2}, x_{f,3}$ are the three components of x_f and if we define the functions $g_1, g_2, g_3 : \mathbb{R}^3 \to \mathbb{R}$ by $g_i(x) = x_i - x_{f,i}$, $i = 1, 2, 3$, then $X_f = \{x \in \mathbb{R}^3 : g_1(x) = g_2(x) = g_3(x) = 0\}$. Moreover, $\nabla g_1(x), \nabla g_2(x), \nabla g_3(x)$ are the rows of the 3×3 identity matrix, and so they are linearly independent at each $x \in \mathbb{R}^3$, and in particular when $x = x_f$. Thus X_f is a $(3-3) = 0$-dimensional manifold. (This corresponds to the case in our optimal control problems when the final state is fixed at x_f, which is what we discussed in the previous section, §4.2.) ◇

4.3.2. How does the Pontryagin Minimum Principle change? Roughly the change is that instead of having the final co-state condition as being unspecified in (P1), this is now changed to

$$\lambda(t_f) \perp \mathbb{X}_f \text{ at } \mathbf{x}(t_f).$$

Here is the result:

Theorem 4.16. *Suppose that* $\mathbf{u}_*$ *is an optimal solution to*

$$\begin{cases} \text{minimize} & \int_{t_i}^{t_f} F(\mathbf{x}(t), \mathbf{u}(t)) dt \\ \text{subject to} & \mathbf{x}'(t) = f(\mathbf{x}(t), \mathbf{u}(t)), \quad t \in [t_i, t_f], \\ & \mathbf{x}(t_i) = x_i \in \mathbb{R}^n, \\ & \mathbf{x}(t_f) \in \mathbb{X}_f \subset \mathbb{R}^n, \\ & \mathbf{u}(t) \in \mathbb{U} \subset \mathbb{R}^m, \quad t \in [t_i, t_f]. \end{cases}$$

Define the Hamiltonian $H : \mathbb{R}^n \times \mathbb{U} \times \mathbb{R} \times \mathbb{R}^n \to \mathbb{R}$ *by*

$$H(x, u, \lambda_0, \lambda) := \lambda_0 F(x, u) + \lambda^\top f(x, u).$$

Then there exists a nonnegative number λ_0, *and a continuous and piecewise continuously differentiable function* $\lambda : [t_i, t_f] \to \mathbb{R}^n$ *such that:*

(P1a) $\lambda'(t) = -\left(\dfrac{\partial H}{\partial x}(\mathbf{x}_*(t), \mathbf{u}_*(t), \lambda_0, \lambda(t)) \right)^\top, \ t \in [t_i, t_f].$

(P1b) *For all* $v \in \bigcap_{i \in I} \ker(\nabla g_i(\mathbf{x}_*(t_f)))$, $\lambda(t_f)^\top v = 0.$

   Equivalently, there exist α_i, $i \in I$, *such that* $\lambda(t_f) = \sum_{i \in I} \alpha_i \cdot (\nabla g_i(\mathbf{x}_*(t_f)))^\top.$

(P2) *There exists a constant* C *such that for all* $t \in [t_i, t_f]$,

$$H(\mathbf{x}_*(t), \mathbf{u}_*(t), \lambda_0, \lambda(t)) = \min_{u \in \mathbb{U}} H(\mathbf{x}_*(t), u, \lambda_0, \lambda(t)) = C.$$

(P3) *For all* $t \in [t_i, t_f]$, $(\lambda_0, \lambda(t)^\top) \neq 0.$

Proof. (Sketch) We proceed with the same Steps 1 and 2 as in the proof of Theorem 4.10. However, in Step 3, we have the following difference. The optimality of $\mathbf{u}_*$ implies that the cone C cannot intersect the interior of the "surface"

$$\Omega := \left(\{x_0 \in \mathbb{R} : x_0 \leq \mathbf{x}_{0,*}(t_f)\} \times \mathbb{X}_f \right) \subset \mathbb{R}^{n+1}.$$

This gives the existence of a hyperplane $\mathbb{H}$ that has C on one side and is tangential to $\mathbf{x}_{0,*}(t_f) \times \mathbb{X}_f$. See Figure 9.

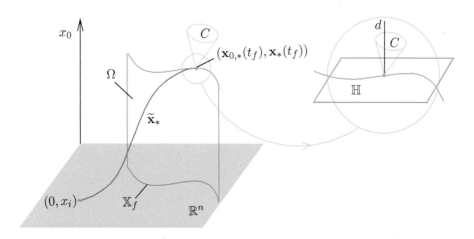

Figure 9. Hyperplane $\mathbb{H} \supset \mathbf{x}_{0,*}(t_f) \times \left(\bigcap_{i \in I} \ker(\nabla g_i(\mathbf{x}_*(t_f))) \right)$ separating C and Ω.

Given a manifold $\mathbb{X}_f$ and a curve $\gamma : \mathbb{R} \to \mathbb{X}_f$ that lives in $\mathbb{X}_f$, we have for each $i \in I$ that

$$g_i(\gamma(t)) = 0,$$
$$g_i(\gamma(0)) = 0,$$

so that for $t \neq 0$,

$$\frac{g_i(\gamma(t)) - g_i(\gamma(0))}{t} = 0,$$

and passing the limit as $t \to 0$, we obtain using the Chain Rule that

$$\nabla g_i(\gamma(0)) \cdot \gamma'(0) = 0.$$

Thus the "tangent velocity vector" $\gamma'(0)$ at the point $\gamma(0) \in \mathbb{X}_f$ lies in the kernel of each $\nabla g_i(\gamma(0))$. From our assumption that $\{\nabla g_i(x), \; i \in I\}$ is linearly independent, it can be shown that every vector that is in the kernel of each $\nabla g_i(\gamma(0))$ is the tangent velocity vector $\gamma'(0)$ at the point $\gamma(0)$ for *some* curve $\gamma : \mathbb{R} \to \mathbb{X}_f$. From these considerations, it follows that the tangent plane $\mathbb{H}$ separating the cone C and Ω contains

$$\mathbf{x}_{0,*}(t_f) \times \left(\bigcap_{i \in I} \ker(\nabla g_i(\mathbf{x}_*(t_f))) \right).$$

Thus the normal

$$d = \begin{bmatrix} \lambda_0 \\ \lambda(t_f) \end{bmatrix}$$

to $\mathbb{H}$ is perpendicular to every vector $\begin{bmatrix} 0 \\ v \end{bmatrix}$, where $v \in \bigcap_{i \in I} \ker(\nabla g_i(\mathbf{x}_*(t_f)))$.

Consequently,

$$\boldsymbol{\lambda}(t_f) \in \left(\bigcap_{i \in I} \ker(\nabla g_i(\mathbf{x}_*(t_f)))\right)^{\perp} = \mathrm{span}\{(\nabla g_i(\mathbf{x}_*(t_f)))^{\top}, \ i \in I\}.$$

So we have (P1b). The rest of the proof is completed in a manner analogous to the proof of Theorem 4.10. □

Remark 4.17 (Number of endpoint constraints on the state and co-state). Let us consider the following special cases:

$\underline{1^{\circ}}$ If $\mathbb{X}_f = \mathbb{R}^n$ (free final state), then $I = \emptyset$, and so

$$\bigcap_{i \in I} \ker(\nabla g_i(\mathbf{x}_*(t_f))) = \mathbb{R}^n.$$

Consequently, the condition (P1b) becomes $\boldsymbol{\lambda}(t_f) = 0$.

$\underline{2^{\circ}}$ On the other hand in the opposite extreme case, if $\mathbb{X}_f = \{x_f\}$ (fixed final state), then $\nabla g_i(x)$, $i \in I = \{1, \cdots, n\}$ are the rows of the $n \times n$ identity matrix. Hence

$$\bigcap_{i \in I} \ker(\nabla g_i(\mathbf{x}_*(t_f))) = \{0\}.$$

Consequently, (P1b) is superfluous, as it just states that $\boldsymbol{\lambda}(t_f) \in \mathbb{R}^n$. So we recover our older version of the Pontryagin Minimum Principle given in Theorem 4.10.

Morally, the state and co-state equations are a bunch of $2n$ scalar first order differential equations, and intuitively one expects to obtain a unique solution if the number of boundary conditions is $2n$ as well. In each of the above two special cases, and in general, this number of boundary conditions works out right:

$\underline{1^{\circ}}$ If $\mathbb{X}_f = \mathbb{R}^n$, then there are no constraints on $\mathbf{x}(t_f)$, and there are n constraints on $\boldsymbol{\lambda}(t_f)$ (all components must be 0). So together with the given initial condition $\mathbf{x}(t_i) = x_i$, the total number of boundary constraints on the state and co-state taken together is $n + 0 + n = 2n$.

$\underline{2^{\circ}}$ If $\mathbb{X}_f = \{x_f\}$, then there are n constraints on $\mathbf{x}(t_f)$ (all components are specified), and there are no constraints on $\boldsymbol{\lambda}(t_f)$ (which is free). So together with the given initial condition $\mathbf{x}(t_i) = x_i$, the total number of boundary constraints on the state and co-state taken together is equal to $n + n + 0 = 2n$.

Example 4.18. Consider the following optimal control problem

$$
\begin{cases}
\text{minimize} & \displaystyle\int_0^1 -\mathbf{x}(t)dt \\
\text{such that} & \mathbf{x}'(t) = \mathbf{x}(t) + \mathbf{u}(t) \text{ for } t \in [0,1], \\
& \mathbf{x}(0) = 0, \\
& \mathbf{u}(t) \in [-1,1] \text{ for } t \in [0,1].
\end{cases}
$$

Intuitively, it pays to have $\mathbf{x}$ large pointwise. We have

$$
\mathbf{x}(t) = e^t \cdot \underbrace{\mathbf{x}(0)}_{=0} + \int_0^t e^{t-\tau}\mathbf{u}(\tau)d\tau = \int_0^t e^{t-\tau}\mathbf{u}(\tau)d\tau. \tag{4.5}
$$

In order to make $\mathbf{x}$ large, we should have a large $\mathbf{u}$. But $\mathbf{u}(t) \leq 1$ for all t. So we guess that $\mathbf{u} \equiv 1$ is probably optimal. Let us see what the Pontryagin Minimum Principle gives us.

With $\lambda_0 = 1$, the Hamiltonian is given by

$$
H(x, u, \lambda) = -x + \lambda(x + u) = -x + \lambda x + \lambda u.
$$

Thus

$$
\mathbf{u}_*(t) = \begin{cases} +1 & \text{if } \lambda(t) < 0, \\ -1 & \text{if } \lambda(t) > 0. \end{cases}
$$

Can $\lambda(t)$ be 0? Let us look at the co-state equation. We have

$$
\lambda'(t) = -\frac{\partial H}{\partial x}(\mathbf{x}_*(t), \mathbf{u}_*(t), \lambda(t)) = 1 - \lambda(t),
$$

and so

$$
\begin{aligned}
\lambda(t) &= e^{-t}\lambda(0) + \int_0^t e^{-(t-\tau)}(+1)d\tau \\
&= e^{-t}\lambda(0) + e^{-t}\frac{e^t - e^0}{1} = e^{-t}\lambda(0) + 1 - e^{-t} = 1 + Ce^{-t},
\end{aligned}
$$

for some constant C. As $\mathbf{x}(1)$ is free, (P1b) gives $\lambda(1) = 0$. Consequently,

$$
\lambda(1) = 0 = 1 + ce^{-1},
$$

that is, $C = -e$. Hence $\lambda(t) = 1 + (-e)e^{-t} = 1 - e^{1-t} < 0$ for $t \in [0,1)$. Thus the optimal $\mathbf{u}_*$ is the function identically equal to 1 on $[0,1]$. See Figure 10.

(In fact we can check optimality of this $\mathbf{u}_*$. Using (4.5), we see that $\mathbf{u}_*$ is optimal, since the cost of any admissible $\mathbf{u}$ is

$$
\begin{aligned}
\int_0^t -\mathbf{x}(t)dt &= -\int_0^t\int_0^t e^{t-\tau}\mathbf{u}(\tau)d\tau dt = -\int_0^1\int_\tau^1 e^{t-\tau}\mathbf{u}(\tau)dt d\tau \\
&= -\int_0^1 \underbrace{\mathbf{u}(\tau)}_{\leq 1}\underbrace{(e^{1-\tau}-1)}_{\geq 0}d\tau \geq -\int_0^1 1 \cdot \underbrace{(e^{1-\tau}-1)}_{\geq 0}d\tau = \text{ cost of } \mathbf{u}_*.) \ \lozenge
\end{aligned}
$$

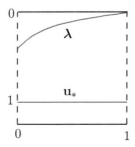

Figure 10. The co-state λ and the optimal control $\mathbf{u}_*$.

Example 4.19. Consider the problem of finding the curve of shortest length in the plane from the point $(0,0)$ to the vertical line L passing through $(1,0)$. See Figure 11.

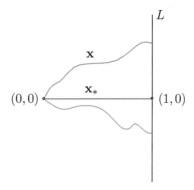

Figure 11. The curve of shortest length joining $(0,0)$ to $(1,0)$.

We know that the answer must be $\mathbf{x} \equiv 0$, that is, it is the straight line segment joining $(0,0)$ to $(1,0)$. The cost to be minimized is

$$\int_0^1 \sqrt{1 + (\mathbf{x}'(t))^2} dt,$$

where $\mathbf{x} \in C^1[0,1]$ is such that $\mathbf{x}(0) = 0$ and $\mathbf{x}(1)$ lies on L, so that $\mathbf{x}(1)$ is free. Although this problem can be solved using the Euler-Lagrange equation, let us solve it using the Pontryagin Minimum Principle by first posing this as an optimal control problem. Note that since $\mathbf{x}(0)$ is given, we can determine $\mathbf{x}$ once we know $\mathbf{x}'$ on $[0,1]$:

$$\mathbf{x}(t) = \mathbf{x}(0) + \int_0^t \mathbf{x}'(\tau)d\tau, \quad t \in [0,1].$$

Set $\mathbf{u}(t) := \mathbf{x}'(t)$. Then the problem is equivalent to the following optimal control problem:

$$\begin{cases} \text{minimize} & \int_0^1 \sqrt{1 + (\mathbf{u}(t))^2}\,dt \\ \text{such that} & \mathbf{x}'(t) = \mathbf{u}(t) \text{ for } t \in [0, 1], \\ & \mathbf{x}(0) = 0. \end{cases}$$

The Hamiltonian is $H(x, u, \lambda) = \sqrt{1 + u^2} + \lambda u$. We want to minimize

$$u \mapsto \sqrt{1 + u^2} + \lambda(t) \cdot u.$$

Let us look at the co-state equation:

$$\lambda'(t) = -\frac{\partial H}{\partial x}(\mathbf{x}_*(t), \mathbf{u}_*(t), \lambda(t)) = 0,$$

and so λ is constant on $[0, 1]$. Since $\mathbf{x}(1)$ is free, we have $\lambda(1) = 0$. Consequently, $\lambda \equiv 0$. Hence

$$\mathbf{u}_*(t) = \arg\min_{u \in \mathbb{R}} \left(\sqrt{1 + u^2} + \underbrace{\lambda(t)}_{=0} \cdot u \right) = \arg\min_{u \in \mathbb{R}} \sqrt{1 + u^2} = 0, \quad t \in [0, 1].$$

So

$$\mathbf{x}_*(t) = \mathbf{x}_*(0) + \int_0^t \mathbf{u}_*(\tau)d\tau = 0 + \int_0^t 0\,d\tau = 0, \quad t \in [0, 1],$$

as expected. For any $\mathbf{x} \in C^1[0, 1]$ such that $\mathbf{x}(0) = 0$, we have

$$(\text{cost of } \mathbf{x} =) \int_0^1 \sqrt{1 + (\mathbf{x}'(t))^2}\,dt \geq \int_0^1 \sqrt{1 + 0}\,dt = 1 \ (= \text{cost of } \mathbf{x}_*),$$

and so $\mathbf{x}_*$ is optimal for the original problem. $\diamond$

Example 4.20 (Minimum energy synchronization of two rocket cars). Consider two identical rocket cars, which can move along a line in either direction thanks to thrusters that can exert forces on them. Suppose that initially the two rocket cars are at rest, and are separated, say one is at position -1, while the other is at position 1 along the line. See Figure 12.

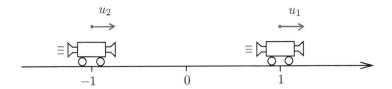

Figure 12. Rocket cars moving in a straight line.

The control task is to exert forces on the time interval $[0, 1]$ on the two rocket cars in such a manner that at the final time t, the rocket cars are "synchronized," that is, they have the same position and the same velocity. We would like to do this with the least expenditure of energy. So we arrive at an optimal control problem. Let

x_1 denote the position of the first car (initially at position 1),

$x_2 := x_1'$ be its velocity, and

u_1 be the force applied on it (one of the two control inputs).

Similarly, let

x_3 denote the position of the second car (initially at position -1),

$x_4 := x_3'$ be its velocity, and

u_2 be the force applied on it (the other control input).

Then the equations of motion are given by

$$\left\{ \begin{array}{l} x_1' = x_2, \\ x_2' = u_1 \end{array} \right\} \quad \text{and} \quad \left\{ \begin{array}{l} x_3' = x_4, \\ x_4' = u_2 \end{array} \right\}.$$

The optimal control problem is:

$$\left\{ \begin{array}{ll} \text{minimize} & \dfrac{1}{2} \displaystyle\int_0^1 \Big((u_1(t))^2 + (u_2(t))^2 \Big) \, dt \\ \text{such that} & \mathbf{x}'(t) = A\mathbf{x}(t) + B\mathbf{u}(t) \text{ for } t \in [0,1], \\ & \mathbf{x}(0) = x_i, \\ & \mathbf{x}(1) \in \mathbb{X}_f, \end{array} \right.$$

where $\mathbf{x}(t) \in \mathbb{R}^4$, $\mathbf{u}(t) \in \mathbb{R}^2$,

$$A = \begin{bmatrix} 0 & 1 & 0 & 0 \\ 0 & 0 & 0 & 0 \\ 0 & 0 & 0 & 1 \\ 0 & 0 & 0 & 0 \end{bmatrix}, \quad B = \begin{bmatrix} 0 & 0 \\ 1 & 0 \\ 0 & 0 \\ 0 & 1 \end{bmatrix}, \quad x_i = \begin{bmatrix} 1 \\ 0 \\ -1 \\ 0 \end{bmatrix},$$

and $\mathbb{X}_f \subset \mathbb{R}^4$ is given by

$$\mathbb{X}_f = \{x \in \mathbb{R}^4 : g_1(x) := x_1 - x_3 = 0 \text{ and } g_2(x) = x_2 - x_4 = 0\}.$$

We have

$$\begin{aligned} \nabla g_1(x) &= \begin{bmatrix} 1 & 0 & -1 & 0 \end{bmatrix}, \\ \nabla g_2(x) &= \begin{bmatrix} 0 & 1 & 0 & -1 \end{bmatrix}, \end{aligned}$$

and so $\mathbb{X}_f$ is a $(4-2) = 2$-dimensional manifold in $\mathbb{R}^4$. Based on the symmetry of the problem, we intuitively expect that both the cars must meet at the origin, and moreover they must have velocity 0. And so the forces must be such that they arrange this. Let us find out if the Pontryagin Minimum Principle delivers this expected answer.

The Hamiltonian is given by

$$
\begin{aligned}
H(x, u, \lambda) &= \frac{1}{2}(u_1^2 + u_2^2) + \lambda^\top (Ax + Bu) \\
&= \frac{1}{2}(u_1^2 + u_2^2) + \lambda_1 x_2 + \lambda_3 x_4 + \lambda_2 u_1 + \lambda_4 u_2 \\
&= \lambda_1 x_1 + \lambda_3 x_4 + \left(\frac{1}{2}u_1^2 + \lambda_2 u_1\right) + \left(\frac{1}{2}u_2^2 + \lambda_4 u_2\right).
\end{aligned}
$$

Thus we obtain using (P2) that

$$
\begin{aligned}
\mathbf{u}_{1,*} &= -\lambda_2(t), \\
\mathbf{u}_{2,*} &= -\lambda_4(t),
\end{aligned}
$$

$t \in [0, 1]$. The co-state equation is given by

$$
\lambda'(t) = -\left(\frac{\partial H}{\partial x}(\mathbf{x}_*(t), \mathbf{u}_*(t), \lambda(t))\right)^\top = -A^\top \lambda(t), \quad t \in [0, 1],
$$

that is,

$$
\begin{aligned}
\lambda_1'(t) &= 0 \\
\lambda_2'(t) &= -\lambda_1(t) \\
\lambda_3'(t) &= 0 \\
\lambda_4'(t) &= -\lambda_3(t).
\end{aligned}
$$

Hence

$$
\begin{aligned}
\lambda_1(t) &= c_1 \\
\lambda_2(t) &= c_2 - c_1 \cdot t \\
\lambda_3(t) &= c_3 \\
\lambda_4(t) &= c_4 - c_3 \cdot t.
\end{aligned}
$$

The first two state equations are

$$
\begin{aligned}
\mathbf{x}_{1,*}' &= \mathbf{x}_{2,*} \\
\mathbf{x}_{2,*}' &= \mathbf{u}_{1,*} = -\lambda_2 = c_1 \cdot t - c_2,
\end{aligned}
$$

and so

$$
\begin{aligned}
\mathbf{x}_{2,*} &= c_1 \cdot \frac{t^2}{2} - c_2 \cdot t + d_2 \\
\mathbf{x}_{1,*} &= c_1 \cdot \frac{t^3}{6} - c_2 \cdot \frac{t^2}{2} + d_2 \cdot t + d_1.
\end{aligned}
$$

As $\mathbf{x}_{2,*}(0) = 0$ and $\mathbf{x}_{1,*}(0) = 1$, we obtain that $d_2 = 0$ and $d_1 = 1$. Hence

$$
\begin{aligned}
\mathbf{x}_{2,*} &= c_1 \cdot \frac{t^2}{2} - c_2 \cdot t \\
\mathbf{x}_{1,*} &= c_1 \cdot \frac{t^3}{6} - c_2 \cdot \frac{t^2}{2} + 1.
\end{aligned}
$$

Similarly,

$$\mathbf{x}_{4,*} = c_3 \cdot \frac{t^2}{2} - c_4 \cdot t$$

$$\mathbf{x}_{3,*} = c_3 \cdot \frac{t^3}{6} - c_4 \cdot \frac{t^2}{2} - 1.$$

We know that $\mathbf{x}_*(1) \in \mathbb{X}_f$, and so $\mathbf{x}_{1,*}(1) = \mathbf{x}_{3,*}(1)$ and $\mathbf{x}_{2,*}(1) = \mathbf{x}_{4,*}(1)$. This yields

$$\frac{c_1}{6} - \frac{c_2}{2} + 1 = \frac{c_3}{6} - \frac{c_4}{2} - 1 \tag{4.6}$$

$$\frac{c_1}{2} - c_2 = \frac{c_3}{2} - c_4. \tag{4.7}$$

Also, (P1b) gives $(\boldsymbol{\lambda}(1))^\top = \alpha_1 \cdot \nabla g_1(\mathbf{x}_*(1)) + \alpha_2 \cdot \nabla g_2(\mathbf{x}_*(1))$, that is,

$$\begin{bmatrix} \lambda_1(1) \\ \lambda_2(1) \\ \lambda_3(1) \\ \lambda_4(1) \end{bmatrix} = \begin{bmatrix} c_1 \\ c_2 - c_1 \cdot 1 \\ c_3 \\ c_4 - c_3 \cdot 1 \end{bmatrix} = \begin{bmatrix} \alpha_1 \\ \alpha_2 \\ -\alpha_1 \\ -\alpha_2 \end{bmatrix},$$

and so,

$$c_1 = -c_3 \tag{4.8}$$

$$c_2 - c_1 = -c_4 + c_3. \tag{4.9}$$

Solving (4.6)-(4.9) gives

$$c_1 = 12 = -c_3, \quad c_2 = 6 = -c_4.$$

Consequently,

$$\mathbf{u}_{1,*} = 12 \cdot t - 6,$$
$$\mathbf{u}_{2,*} = 6 - 12 \cdot t,$$

for $t \in [0, 1]$. Then

$$\mathbf{x}_{1,*}(1) = c_1 \cdot \frac{1^3}{6} - c_2 \cdot \frac{1^2}{2} + 1 = 0,$$

$$\mathbf{x}_{2,*}(1) = c_1 \cdot \frac{1^2}{2} - c_2 \cdot 1 = 0,$$

$$\mathbf{x}_{3,*}(1) = c_3 \cdot \frac{1^3}{6} - c_4 \cdot \frac{1^2}{2} - 1 = 0,$$

$$\mathbf{x}_{4,*}(1) = c_3 \cdot \frac{1^2}{2} - c_4 \cdot 1 = 0,$$

as expected. $\diamondsuit$

Exercise 4.21. Consider the optimal control problem

$$\begin{cases} \text{maximize} & \displaystyle\int_0^2 \Big((\mathbf{x}(t))^2 - 2\mathbf{u}(t) \Big) dt \\ \text{such that} & \mathbf{x}'(t) = \mathbf{u}(t) \text{ for } t \in [0,2], \\ & \mathbf{x}(0) = 1, \\ & \mathbf{u}(t) \in [0,1] \text{ for } t \in [0,2]. \end{cases}$$

Assuming that an optimal control $\mathbf{u}_*$ exists, use the Pontryagin Minimum Principle to find $\mathbf{u}_*$.

Exercise 4.22. Consider the optimal control problem

$$\begin{cases} \text{minimize} & \displaystyle\int_0^2 \Big((\mathbf{u}_1(t))^2 + (\mathbf{u}_2(t))^2 \Big) dt \\ \text{such that} & \begin{bmatrix} \mathbf{x}_1'(t) \\ \mathbf{x}_2'(t) \end{bmatrix} = \begin{bmatrix} \mathbf{u}_1(t) \\ \mathbf{u}_2(t) \end{bmatrix}, \text{ for } t \in [0,2], \\ & \begin{bmatrix} \mathbf{x}_1(0) \\ \mathbf{x}_2(0) \end{bmatrix} = \begin{bmatrix} 0 \\ 0 \end{bmatrix}, \\ & \begin{bmatrix} \mathbf{x}_1(2) \\ \mathbf{x}_2(2) \end{bmatrix} \in \mathbb{X}_f := \left\{ \begin{bmatrix} x_1 \\ x_2 \end{bmatrix} \in \mathbb{R}^2 : x_2^2 - x_1 + 1 = 0 \right\}. \end{cases}$$

Assuming that an optimal control $\mathbf{u}_*$ exists, use the Pontryagin Minimum Principle to find $\mathbf{u}_*$.

4.4. Variable Final Time; Autonomous Case

Theorem 4.23. *Suppose that $(\mathbf{u}_*, t_{f,*})$, where $t_{f,*} > t_i$, is an optimal solution to*

$$\begin{cases} \text{minimize} & \displaystyle\int_{t_i}^{t_f} F(\mathbf{x}(t), \mathbf{u}(t)) dt \\ \text{subject to} & \mathbf{x}'(t) = f(\mathbf{x}(t), \mathbf{u}(t)), \quad t \in [t_i, t_f], \\ & \mathbf{x}(t_i) = x_i \in \mathbb{R}^n, \\ & \mathbf{x}(t_f) \in \mathbb{X}_f \subset \mathbb{R}^n, \\ & \mathbf{u}(t) \in \mathbb{U} \subset \mathbb{R}^m, \quad t \in [t_i, t_f], \\ & t_f \ (\geq t_i) \text{ is free.} \end{cases}$$

Define the Hamiltonian $H : \mathbb{R}^n \times \mathbb{U} \times \mathbb{R} \times \mathbb{R}^n \to \mathbb{R}$ by

$$H(x, u, \lambda_0, \lambda) := \lambda_0 F(x, u) + \lambda^\top f(x, u).$$

Then there exists a nonnegative number λ_0, and a continuous and piecewise continuously differentiable function $\boldsymbol{\lambda} : [t_i, t_f] \to \mathbb{R}^n$ such that:

(P1a) $\boldsymbol{\lambda}'(t) = -\left(\dfrac{\partial H}{\partial x}(\mathbf{x}_*(t), \mathbf{u}_*(t), \lambda_0, \boldsymbol{\lambda}(t))\right)^\top$, $t \in [t_i, t_f]$.

(P1b) *For all* $v \in \bigcap_{i \in I} \ker(\nabla g_i(\mathbf{x}_*(t_f)))$, $\boldsymbol{\lambda}(t_f)^\top v = 0$.

Equivalently, there exist α_i, $i \in I$, *such that* $\boldsymbol{\lambda}(t_f) = \sum_{i \in I} \alpha_i \cdot (\nabla g_i(\mathbf{x}_*(t_f)))^\top$.

(P2) *For all* $t \in [t_i, t_f]$, $H(\mathbf{x}_*(t), \mathbf{u}_*(t), \lambda_0, \boldsymbol{\lambda}(t)) = \min_{u \in \mathbb{U}} H(\mathbf{x}_*(t), u, \lambda_0, \boldsymbol{\lambda}(t)) = 0$.

(P3) *For all* $t \in [t_i, t_f]$, $(\lambda_0, \boldsymbol{\lambda}(t)^\top) \neq 0$.

Note that the only change in the Pontryagin Minimum Principle that is produced is this: in (P2), rather than the pointwise value of the Hamiltonian along the optimal trajectory being a general constant C, now we obtain that the constant C is in fact equal to 0.

Proof. (Sketch) We will recast the problem as an optimal control problem on a *fixed* time interval $[0, 1]$, by introducing a new control input, which will give "time" in the old problem as an extra state function in the new problem! To this end, we introduce the state equation

$$\begin{cases} \mathbf{t}'(\tau) &= \mathbf{U}_0, \quad \tau \in [0, 1], \\ \mathbf{t}(0) &= t_i, \\ \mathbf{t}(1) & \text{ is free}, \\ \mathbf{U}_0(\tau) &\in [0, +\infty), \end{cases}$$

where the control input $\mathbf{U}_0 : [0, 1] \to [0, \infty)$ and the state $\mathbf{t} : [0, 1] \to \mathbb{R}$. Note that if we have been given a time interval $[t_i, t_f]$, then "the running time" in this interval is produced as a state by the above equation if we take the constant input $\mathbf{U}_0(\tau) = t_f - t_i$ for all $\tau \in [0, 1]$. Indeed, the state produced by the above equation is

$$\mathbf{t}(\tau) = \mathbf{t}(t_i) + \int_0^\tau \mathbf{U}_0(\tau)d\tau = t_i + (t_f - t_i)\tau, \quad \tau \in [0, 1].$$

Define the functions $\mathbf{X} : [0, 1] \to \mathbb{R}^n$ and $\mathbf{U} : [0, 1] \to \mathbb{U}$ by

$$\mathbf{X}(\tau) = \mathbf{x}(\mathbf{t}(\tau)),$$
$$\mathbf{U}(\tau) = \mathbf{u}(\mathbf{t}(\tau)),$$

for $\tau \in [0, 1]$. By the Chain Rule, we have for $\tau \in [0, 1]$ that

$$\mathbf{X}'(\tau) = \frac{d}{d\tau}\mathbf{x}(\mathbf{t}(\tau)) = \mathbf{x}'(\mathbf{t}(\tau)) \cdot \mathbf{U}_0(\tau)$$

$$= f(\mathbf{x}(\mathbf{t}(\tau)), \mathbf{u}(\mathbf{t}(\tau))) \cdot \mathbf{U}_0(\tau) = f(\mathbf{X}(\tau), \mathbf{U}(\tau)) \cdot \mathbf{U}_0(\tau).$$

Moreover, $\mathbf{X}(0) = \mathbf{x}(\mathbf{t}(0)) = \mathbf{x}(t_i) = x_i$, and we demand that the admissible control inputs $(\mathbf{U}, \mathbf{U}_0)$ guarantee that $\mathbf{X}(1) = \mathbf{x}(\mathbf{t}(1)) \in \mathbb{X}_f$. Also, if we have chosen a $\mathbf{U}_0$, and if we set $t_f := \mathbf{U}_0(1)$, then by the change of variable $t = \mathbf{t}(\tau)$, we have

$$
\int_{t_i}^{t_f} F(\mathbf{x}(t), \mathbf{u}(t))dt = \int_0^1 F(\mathbf{x}(\mathbf{t}(\tau)), \mathbf{u}(\mathbf{t}(\tau))) \cdot \mathbf{t}'(\tau)d\tau
$$

$$
= \int_0^1 F(\mathbf{X}(\tau), \mathbf{U}(\tau)) \cdot \mathbf{U}_0(\tau)d\tau.
$$

Let us define $\mathbf{U}_{0,*} : [0,1] \to [0, \infty)$ by

$$
\mathbf{U}_{0,*}(\tau) = t_{f,*} - t_i, \quad \tau \in [0,1].
$$

Using the fact that $(\mathbf{u}_*, t_{f,*})$ is an optimal solution to the original problem, it follows that $(\mathbf{U}_*, \mathbf{U}_{0,*})$ is an optimal solution to

$$
\begin{cases}
\text{minimize} & \int_0^1 F(\mathbf{X}(\tau), \mathbf{U}(\tau)) \cdot \mathbf{U}_0(\tau)d\tau \\[2mm]
\text{subject to} & \mathbf{X}'(\tau) = f(\mathbf{X}(\tau), \mathbf{U}(\tau)) \cdot \mathbf{U}_0(\tau), \quad \tau \in [0,1], \\[2mm]
& \mathbf{t}'(\tau) = \mathbf{U}_0(\tau), \quad \tau \in [0,1], \\[2mm]
& \mathbf{X}(0) = x_i \in \mathbb{R}^n, \\[2mm]
& \mathbf{t}(0) = t_i \in \mathbb{R}, \\[2mm]
& \mathbf{X}(1) \in \mathbb{X}_f \subset \mathbb{R}^n, \\[2mm]
& \mathbf{t}(1) \text{ is free}, \\[2mm]
& \mathbf{U}(\tau) \in \mathbb{U} \subset \mathbb{R}^m, \quad \tau \in [0,1], \\[2mm]
& \mathbf{U}_0(\tau) \in [0, +\infty), \quad \tau \in [0,1].
\end{cases}
$$

But now we can use Theorem 4.16 to derive the optimality conditions to be satisfied by $(\mathbf{U}_*, \mathbf{U}_{0,*})$. The Hamiltonian is given by

$$
\widetilde{H}(X, t, U, U_0, \lambda_0, \Lambda, \Lambda_0) = \lambda_0(F(X, U) \cdot U_0) + \begin{bmatrix} \Lambda^\top & \Lambda_0 \end{bmatrix} \begin{bmatrix} f(X, U) \cdot U_0 \\ U_0 \end{bmatrix}
$$

$$
= \left(\lambda_0 F(X, U) + \Lambda^\top f(X, U) + \Lambda_0 \right) U_0.
$$

The co-state equations are

$$
\Lambda'(\tau) = -\left(\frac{\partial \widetilde{H}}{\partial X}(\mathbf{X}_*(\tau), \mathbf{U}_*(\tau), \mathbf{U}_{0,*}(\tau), \lambda_0, \Lambda(\tau), \Lambda_0(\tau)) \right)^\top \tag{4.10}
$$

$$
= -\left(\lambda_0 \frac{\partial F}{\partial x}(\mathbf{X}_*(\tau), \mathbf{U}_*(\tau)) + \Lambda(\tau)^\top \frac{\partial f}{\partial x}(\mathbf{X}_*(\tau), \mathbf{U}_*(\tau)) \right)^\top \mathbf{U}_{0,*}(\tau), \tag{4.11}
$$

$$
\Lambda'_0(\tau) = -\frac{\partial \widetilde{H}}{\partial t}(\mathbf{X}_*(\tau), \mathbf{U}_*(\tau), \mathbf{U}_{0,*}(\tau), \lambda_0, \Lambda(\tau), \Lambda_0(\tau)) = 0. \tag{4.12}
$$

The second equation shows that Λ_0 is constant on $[0,1]$. Let us show that in fact this constant must be zero. Let the manifold $\mathbb{X}_f = \{x \in \mathbb{R}^n : g_i(x) = 0, \ i \in I\}$.

Since the final state $(\mathbf{X}(1), \mathbf{t}(1)) \in \mathbb{X}_f \times \mathbb{R}$, we have from (P1b) of Theorem 4.16 that

$$\begin{bmatrix} \mathbf{\Lambda}(1)^\top & \mathbf{\Lambda}_0(1) \end{bmatrix} \in \text{span}\left\{ \begin{bmatrix} \nabla g_i(\mathbf{X}_*(1)) & 0 \end{bmatrix}, \, i \in I \right\}. \tag{4.13}$$

In particular, we see that $\mathbf{\Lambda}_0(1) = 0$. Consequently, $\mathbf{\Lambda}_0 \equiv 0$ on $[0, 1]$.

Now let us see what (P2) of Theorem 4.16 gives us. It says that there is a constant C such that for all $\tau \in [0, 1]$,

$$\begin{aligned} C &= \left(\lambda_0 F(\mathbf{X}_*(\tau), \mathbf{U}_*(\tau)) + \mathbf{\Lambda}(\tau)^\top f(\mathbf{X}_*(\tau), \mathbf{U}_*(\tau)) \right) \cdot \mathbf{U}_{0,*}(\tau) \\[2mm] &= \left(\lambda_0 F(\mathbf{X}_*(\tau), \mathbf{U}_*(\tau)) + \mathbf{\Lambda}(\tau)^\top f(\mathbf{X}_*(\tau), \mathbf{U}_*(\tau)) \right) \cdot (t_{f,*} - t_i) \\[2mm] &= \min_{\substack{U \in \mathbb{U} \\ U_0 \in [0, +\infty)}} \left(\lambda_0 F(\mathbf{X}_*(\tau), U) + \mathbf{\Lambda}(\tau)^\top f(\mathbf{X}_*(\tau), U) \right) \cdot U_0. \end{aligned}$$

But since for all $\tau \in [0, 1]$,

$$\begin{aligned} C &= \min_{\substack{U \in \mathbb{U} \\ U_0 \in [0, +\infty)}} \left(\lambda_0 F(\mathbf{X}_*(\tau), U) + \mathbf{\Lambda}(\tau)^\top f(\mathbf{X}_*(\tau), U) \right) \cdot U_0 \tag{4.14} \\[2mm] &\leq \min_{U_0 \in [0, +\infty)} \left(\lambda_0 F(\mathbf{X}_*(\tau), \mathbf{U}_*(\tau)) + \mathbf{\Lambda}(\tau)^\top f(\mathbf{X}_*(\tau), \mathbf{U}_*(\tau)) \right) \cdot U_0, \tag{4.15} \end{aligned}$$

and the rightmost expression is linear in U_0, it follows that we must have

$$\lambda_0 F(\mathbf{X}_*(\tau), \mathbf{U}_*(\tau)) + \mathbf{\Lambda}(\tau)^\top f(\mathbf{X}_*(\tau), \mathbf{U}_*(\tau)) \geq 0,$$

for all $\tau \in [0, 1]$ (for otherwise, (4.14)-(4.15) is contradicted). So the minimum in (4.15) is 0, and we obtain $C \leq 0$ from (4.14)-(4.15). But also

$$C = \Big(\underbrace{\lambda_0 F(\mathbf{X}_*(\tau), \mathbf{U}_*(\tau)) + \mathbf{\Lambda}(\tau)^\top f(\mathbf{X}_*(\tau), \mathbf{U}_*(\tau))}_{\geq 0} \Big) \cdot \underbrace{(t_{f,*} - t_i)}_{>0} \geq 0.$$

Thus $C = 0$, and so

$$\left(\lambda_0 F(\mathbf{X}_*(\tau), \mathbf{U}_*(\tau)) + \mathbf{\Lambda}(\tau)^\top f(\mathbf{X}_*(\tau), \mathbf{U}_*(\tau)) \right) \cdot (t_{f,*} - t_i) = 0.$$

Dividing by $t_{f,*} - t_i$, we obtain

$$\lambda_0 F(\mathbf{X}_*(\tau), \mathbf{U}_*(\tau)) + \mathbf{\Lambda}(\tau)^\top f(\mathbf{X}_*(\tau), \mathbf{U}_*(\tau)) = 0.$$

Now define $\boldsymbol{\lambda} : [t_i, t_{f,*}] \to \mathbb{R}^n$ by

$$\boldsymbol{\lambda}(t) = \mathbf{\Lambda}\left(\frac{t - t_i}{t_{f,*} - t_i} \right), \quad t \in [t_i, t_{f,*}].$$

Then $\boldsymbol{\lambda}(\mathbf{t}_*(\tau)) = \mathbf{\Lambda}(\tau)$, $\tau \in [0, 1]$, and

$$\mathbf{\Lambda}'(\tau) = \boldsymbol{\lambda}'(\mathbf{t}_*(\tau)) \cdot \mathbf{U}_{0,*}(\tau) = \boldsymbol{\lambda}'(\mathbf{t}_*(\tau)) \cdot (t_{f,*} - t_i), \quad \tau \in [0, 1].$$

From (4.10)-(4.11), we obtain

$$\boldsymbol{\lambda}'(\mathbf{t}_*(\tau))\cdot(t_{f,*}-t_i)=-\left(\lambda_0\frac{\partial F}{\partial x}(\mathbf{X}_*(\tau),\mathbf{U}_*(\tau))+\boldsymbol{\Lambda}(\tau)^\top\frac{\partial f}{\partial x}(\mathbf{X}_*(\tau),\mathbf{U}_*(\tau))\right)^\top(t_{f,*}-t_i),$$

and cancelling $t_{f,*}-t_i$, we get

$$\begin{aligned}\boldsymbol{\lambda}'(\mathbf{t}_*(\tau)) &= -\left(\lambda_0\frac{\partial F}{\partial x}(\mathbf{X}_*(\tau),\mathbf{U}_*(\tau))+\boldsymbol{\lambda}(\mathbf{t}_*(\tau))^\top\frac{\partial f}{\partial x}(\mathbf{X}_*(\tau),\mathbf{U}_*(\tau))\right)^\top\\ &= -\left(\lambda_0\frac{\partial F}{\partial x}(\mathbf{x}_*(\mathbf{t}_*(\tau)),\mathbf{u}_*(\mathbf{t}_*(\tau)))+\boldsymbol{\lambda}(\mathbf{t}_*(\tau))^\top\frac{\partial f}{\partial x}(\mathbf{x}_*(\mathbf{t}_*(\tau)),\mathbf{u}_*(\mathbf{t}_*(\tau)))\right)^\top\end{aligned}$$

for $\tau \in [0,1]$. Hence we obtain (P1a):

$$\begin{aligned}\boldsymbol{\lambda}'(t) &= -\left(\lambda_0\frac{\partial F}{\partial x}(\mathbf{x}_*(t),\mathbf{u}_*(t))+\boldsymbol{\lambda}(t)^\top\frac{\partial f}{\partial x}(\mathbf{x}_*(t),\mathbf{u}_*(t))\right)^\top\\ &= -\left(\frac{\partial H}{\partial x}(\mathbf{x}_*(t),\mathbf{u}_*(t),\lambda_0,\boldsymbol{\lambda}(t))\right)^\top\end{aligned}$$

for $t \in [t_i,t_f]$. (P1b) follows from (4.13). From (4.14)-(4.15), and the fact that $C = 0$, we obtain

$$\begin{aligned}0 &= \frac{C}{t_{f,*}-t_i} = \left(\lambda_0 F(\mathbf{X}_*(\tau),\mathbf{U}_*(\tau))+\boldsymbol{\lambda}(\mathbf{t}_*(\tau))^\top f(\mathbf{X}_*(\tau),\mathbf{U}_*(\tau))\right)\cdot 1\\ &\geq \min_{U\in\mathbb{U}}\left(\lambda_0 F(\mathbf{X}_*(\tau),U)+\boldsymbol{\lambda}(\mathbf{t}_*(\tau))^\top f(\mathbf{X}_*(\tau),U)\right)\cdot 1\\ &\geq \min_{\substack{U\in\mathbb{U}\\U_0\in[0,+\infty)}}\left(\lambda_0 F(\mathbf{X}_*(\tau),U)+\boldsymbol{\lambda}(\mathbf{t}_*(\tau))^\top f(\mathbf{X}_*(\tau),U)\right)\cdot U_0 = C = 0.\end{aligned}$$

Thus for all $\tau \in [0,1]$,

$$\begin{aligned}0 &= \lambda_0 F(\mathbf{X}_*(\tau),\mathbf{U}_*(\tau))+\boldsymbol{\lambda}(\mathbf{t}_*(\tau))^\top f(\mathbf{X}_*(\tau),\mathbf{U}_*(\tau))\\ &= \min_{U\in\mathbb{U}}\left(\lambda_0 F(\mathbf{X}_*(\tau),U)+\boldsymbol{\lambda}(\mathbf{t}_*(\tau))^\top f(\mathbf{X}_*(\tau),U)\right),\end{aligned}$$

that is, for all $t \in [t_i,t_{f,*}]$,

$$\begin{aligned}0 &= \lambda_0 F(\mathbf{x}_*(t),\mathbf{u}_*(t))+\boldsymbol{\lambda}(t)^\top f(\mathbf{x}_*(t),\mathbf{u}_*(t)) = H(\mathbf{x}_*(t),\mathbf{u}_*(t),\lambda_0,\boldsymbol{\lambda}(t))\\ &= \min_{u\in\mathbb{U}}\left(\lambda_0 F(\mathbf{X}_*(\tau),U)+\boldsymbol{\lambda}(\mathbf{t}_*(\tau))^\top f(\mathbf{X}_*(\tau),U)\right) = \min_{u\in\mathbb{U}} H(\mathbf{x}_*(t),u,\lambda_0,\boldsymbol{\lambda}(t)),\end{aligned}$$

that is (P2) holds. Finally, from (P3) of Theorem 4.16 we know that

$$(\lambda_0,\boldsymbol{\Lambda}(\tau),\boldsymbol{\Lambda}_0) = (\lambda_0,\boldsymbol{\Lambda}(\tau),0) \neq 0$$

for all $\tau \in [0,1]$. Thus it follows that also $(\lambda_0,\boldsymbol{\lambda}(t)) \neq 0$ for all $t \in [t_i,t_{f,*}]$. Hence (P3) holds too. This completes the proof. $\qquad\square$

As an application of the above result, let us first consider the classical problem of minimum time control of the rocket car.

Example 4.24 (Time optimal control of the rocket car). Recall Example 3.30. We want to find a pair $(\mathbf{u}_*, T_*)$, which is the solution to the optimization problem

$$
\begin{cases}
\text{minimize} \quad T = \int_0^T 1\, dt \\[2mm]
\text{subject to} \quad \mathbf{x}'(t) = A\mathbf{x}(t) + B\mathbf{u}(t), \quad t \in [0, T] \\[2mm]
\qquad\qquad \mathbf{x}(0) = x_i, \\[2mm]
\qquad\qquad \mathbf{x}(T) = 0, \\[2mm]
\qquad\qquad \mathbf{u}(t) \in [-1, 1] \text{ for all } t \in [0, T], \\[2mm]
\qquad\qquad T \text{ is free,}
\end{cases}
$$

where the matrices A, B, x_i are given by

$$
A = \begin{bmatrix} 0 & 1 \\ 0 & 0 \end{bmatrix}, \quad B = \begin{bmatrix} 0 \\ 1 \end{bmatrix}, \quad x_i = \begin{bmatrix} z_i \\ 0 \end{bmatrix},
$$

and z_i is the initial position of the rocket car. Assuming that there exists a solution $(\mathbf{u}_*, T_*)$, let us use the version of the Pontryagin Minimum Principle given in Theorem 4.23 to find $(\mathbf{u}_*, T_*)$.

The Hamiltonian is given by

$$
H(x, u, \lambda_0, \lambda) = \lambda_0 F(x, u) + \lambda^\top f(x, u) = \lambda_0 \cdot 1 + \lambda^\top (Ax + Bu) = \lambda_0 + \lambda_1 x_2 + \lambda_2 u.
$$

By (P2), we obtain

$$
\mathbf{u}_*(t) = \arg \min_{u \in [-1,1]} (\lambda_0 + \boldsymbol{\lambda}_1(t) \mathbf{x}_{2,*}(t) + \boldsymbol{\lambda}_2(t) u)
$$

and by looking at Figure 13, we see that

$$
\mathbf{u}_*(t) = \begin{cases} -1 & \text{if } \boldsymbol{\lambda}_2(t) > 0, \\ 1 & \text{if } \boldsymbol{\lambda}_2(t) < 0. \end{cases}
$$

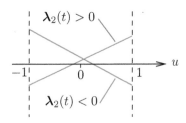

Figure 13. Possible graphs of $u \mapsto \lambda_0 + \boldsymbol{\lambda}^\top A \mathbf{x}_*(t) + \boldsymbol{\lambda}_2(t) u$.

Can $\lambda_2(t) = 0$ in an interval? Let us see the behavior of λ_2 by looking at the co-state equation:

$$\boldsymbol{\lambda}'(t) = -\left(\frac{\partial H}{\partial x}(\mathbf{x}_*(t), \mathbf{u}_*(t), \lambda_0, \boldsymbol{\lambda}(t))\right)^\top = -A^\top \boldsymbol{\lambda}(t) = \begin{bmatrix} 0 & 0 \\ -1 & 0 \end{bmatrix} \boldsymbol{\lambda}(t).$$

So for all $t \in [0, T_*]$, $\lambda_1'(t) = 0$, $\lambda_2'(t) = -\lambda_1(t)$, which gives

$$\lambda_1(t) = \alpha,$$
$$\lambda_2(t) = \beta - \alpha t,$$

for all $t \in [0, T_*]$ for some constants α, β. From here we see that if λ_2 is zero over a time interval, then we must have $\alpha = \beta = 0$, and so $\lambda_1 = \lambda_2 \equiv 0$. But then

$$H(\mathbf{x}_*(t), \mathbf{u}_*(t), \lambda_0, \boldsymbol{\lambda}(t)) = \lambda_0 = 0,$$

so that $(\lambda_0, \boldsymbol{\lambda}(t)) = 0$ for all $t \in [0, T_*]$, violating (P3)! Thus λ_2 can't be zero over a time interval. But then (since it is linear in t), it can be zero for at most one time instant, and at this time instant, $\mathbf{u}_*$ would switch from ± 1 to ∓ 1. The four possibilities for λ_2 are depicted in Figure 14 together with the corresponding $\mathbf{u}_*$:

(1) λ_2 always positive,

(2) λ_2 always negative,

(3) λ_2 first positive up to switching time t_s and then always negative after time t_s,

(4) λ_2 first negative up to switching time t_s and then always positive after time t_s.

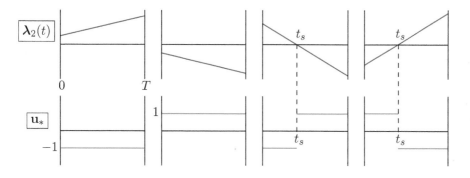

Figure 14. Possible co-state λ_2 with corresponding inputs $\mathbf{u}_*$.

Note the "bang-bang" nature of the optimal control. We can calculate the (possible) switching time t_s and the optimal time T_* by using the state equation

$$x_{1,*}' = x_{2,*},$$
$$x_{2,*}' = \mathbf{u}_*.$$

Suppose we consider an interval where $\mathbf{u}_* \equiv 1$. Then from the above two equations, we obtain

$$\mathbf{x}_{2,*} = c_1 + t, \tag{4.16}$$

$$\mathbf{x}_{1,*} = c_2 + c_1 t + \frac{t^2}{2}. \tag{4.17}$$

It will turn out that one can understand the optimal control action better by drawing a "phase plane" picture. What do we mean by this? If we fix a time instant t, then $(\mathbf{x}_{1,*}(t), \mathbf{x}_{2,*}(t)) \in \mathbb{R}^2$, that is, $(\mathbf{x}_{1,*}(t), \mathbf{x}_{2,*}(t))$ is a point in the plane (we refer to this as the "phase plane"). If we now vary t, then the point $(\mathbf{x}_{1,*}(t), \mathbf{x}_{2,*}(t))$ moves in the phase plane $\mathbb{R}^2$, and describes a curve. Such a curve is called a "phase plane trajectory." Let us draw the phase plane trajectories corresponding to the formulae we obtained in (4.16) and (4.17). To this end, we "eliminate t":

$$\mathbf{x}_{1,*} = c_2 + c_1(\mathbf{x}_{2,*} - c_1) + \frac{(\mathbf{x}_{2,*} - c_1)^2}{2} = c_2 + c_1 \mathbf{x}_{2,*} - c_1^2 + \frac{\mathbf{x}_{2,*}^2}{2} - c_1 \mathbf{x}_{2,*} + \frac{c_1^2}{2} = C + \frac{\mathbf{x}_{2,*}^2}{2}, \tag{4.18}$$

for some constant C. Thus the phase plane trajectories are parabolas, as shown in Figure 15. Since $\mathbf{x}_{2,*}(t) = c_1 + t \xrightarrow{t \to \infty} \infty$, the motion is "upward," and this explains the arrows on the phase plane trajectories depicted in Figure 15.

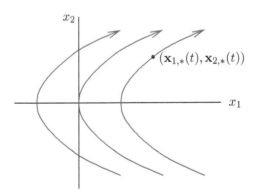

Figure 15. Phase plane trajectories when $\mathbf{u}_* \equiv 1$.

Similarly, if $\mathbf{u}_* \equiv -1$ in an interval, then we have

$$\mathbf{x}_{2,*} = c_1 - t, \tag{4.19}$$

$$\mathbf{x}_{1,*} = c_2 + c_2 t - \frac{t^2}{2}. \tag{4.20}$$

Again eliminating t, we have

$$\mathbf{x}_{1,*} = c_2 + c_1(c_1 - \mathbf{x}_{2,*}) - \frac{(c_1 - \mathbf{x}_{2,*})^2}{2} = c_2 + c_1^2 - c_1\mathbf{x}_{2,*} - \frac{c_1^2}{2} + c_1\mathbf{x}_{2,*} - \frac{\mathbf{x}_{2,*}^2}{2} = C - \frac{\mathbf{x}_{2,*}^2}{2}.$$

See Figure 16, where besides the old phase plane trajectories (when $\mathbf{u}_*$ was 1), we have now also shown the new possible phase plane trajectories (when $\mathbf{u}_*$ is -1).

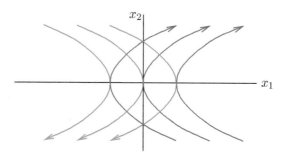

Figure 16. Phase plane trajectories when $\mathbf{u}_* \equiv -1$ are the ones with downwards motion, since $\mathbf{x}_{2,*}(t) = c_1 - t \overset{t \to \infty}{\longrightarrow} -\infty$.

So for the intervals when $\mathbf{u}_* \equiv 1$, the state evolves along the upward motion curves, while for the intervals when $\mathbf{u}_* \equiv -1$, the state evolves along the downward motion curves. To bring the state from the initial condition $(z_i, 0)$ to $(0, 0)$, we proceed like this:

  $\underline{1}°$ $z_i = 0$: Do nothing! In this case $T_* = 0$.

  $\underline{2}°$ $z_i > 0$. In this case, we first take $\mathbf{u}_* \equiv -1$ until time t_s and for times $t > t_s$, $\mathbf{u}_*(t) = +1$. See the picture on the left in Figure 17.

  $\underline{3}°$ $z_i < 0$. In this case, we first take $\mathbf{u}_* \equiv +1$ until time t_s and for times $t > t_s$, $\mathbf{u}_*(t) = -1$. See the picture on the right in Figure 17.

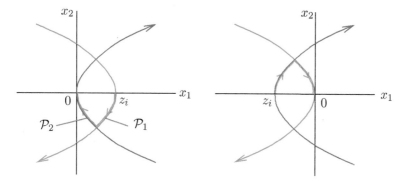

Figure 17. The optimal state $\mathbf{x}_*$ in the two cases when $z_i > 0$ (left picture) and when $z_i < 0$ (right picture).

Let us calculate the switching time t_s and the optimal time T_* for the case when $z_i > 0$. Look at the picture on the left of Figure 17. The optimal state trajectory comprises portions $\mathcal{P}_1$, $\mathcal{P}_2$ of two parabolas:

(1) Points on $\mathcal{P}_1$ are of the form $\left(c_2 + c_1 t - \dfrac{t^2}{2},\ c_1 - t \right)$.

(2) Points on $\mathcal{P}_2$ are of the form $\left(\tilde{c}_2 + \tilde{c}_1 t + \dfrac{t^2}{2},\ \tilde{c}_1 - t \right)$.

Since $\mathbf{x}_{2,*}(0) = 0$, $c_1 - 0 = 0$, and so $c_1 = 0$. Thus $\mathbf{x}_{2,*} = -t$ on $\mathcal{P}_1$. Also, $\mathbf{x}_{1,*}(0) = z_i$, which gives

$$c_2 + c_1 \cdot 0 - \frac{0^2}{2} = z_i,$$

and so $c_2 = z_i$. Consequently, $\mathbf{x}_{1,*} = z_i - \dfrac{t^2}{2}$ on $\mathcal{P}_1$.

As $(\mathbf{x}_{1,*}(t_s), \mathbf{x}_{2,*}(t_s))$ lies on $\mathcal{P}_2$, we have from (4.18) that

$$\mathbf{x}_{1,*}(t_s) - \frac{(\mathbf{x}_{2,*}(t_s))^2}{2} = \mathbf{x}_{1,*}(T_*) - \frac{(\mathbf{x}_{2,*}(T_*))^2}{2} = 0 - \frac{0^2}{2} = 0.$$

Thus $z_i - \dfrac{t_s^2}{2} = \dfrac{(-t_s)^2}{2}$, and so $z_i = t_s^2$.

Hence $t_s = \sqrt{z_i}$. What is the optimal time T_*? We have

$$\tilde{c}_1 + t \Big|_{t_s} = c_1 - t \Big|_{t_s},$$

which gives $\tilde{c}_1 = -2\sqrt{z_i}$. Finally, $0 = \mathbf{x}_{2,*}(T_*) = \tilde{c}_1 + T_*$ yields $T_* = 2\sqrt{z_i}$. So the pair $(\mathbf{u}_*, T_*)$ is given by

$$\mathbf{u}_*(t) = \begin{cases} -1 & \text{for } t \in [0, \sqrt{z_i}], \\ 1 & \text{for } t \in (\sqrt{z_i}, 2\sqrt{z_i}], \end{cases}$$

and $T_* = 2\sqrt{z_i}$.   ◇

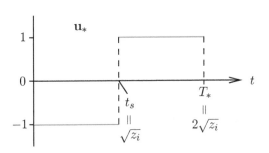

Let us now see an example of an optimization problem, where despite the input being constrained and the Hamiltonian being linear in the control input, the optimal control is not "bang-bang."

Example 4.25 (Non-bang-bang control; singular control). Consider the optimization problem of determining $(\mathbf{u}_*, T_*)$ that is the solution to

$$
\left\{
\begin{array}{ll}
\text{minimize} & T \\
\text{subject to} & \mathbf{x}_1'(t) = 1 - (\mathbf{x}_2(t))^2, \ t \in [0, T], \\
& \mathbf{x}_2'(t) = \mathbf{u}(t), \ t \in [0, T], \\
& \mathbf{x}_1(0) = 0, \\
& \mathbf{x}_2(0) = 0, \\
& \mathbf{x}_1(T) = 1, \\
& \mathbf{x}_2(T) = 0, \\
& \mathbf{u}(t) \in [-1, 1], \ t \in [0, T], \\
& T \, (\geq 0) \text{ is free.}
\end{array}
\right.
$$

Let us first check that $(\mathbf{u}_*, T_*) = (\mathbf{0}, 1)$ is optimal.

Note that $(\mathbf{u}_*, T_*) = (\mathbf{0}, 1)$ constitutes an admissible pair. The second state equation $\mathbf{x}_{2,*}'(t) = \mathbf{u}_*(t) = 0$ gives $\mathbf{x}_{2,*} \equiv$ a constant, and since $\mathbf{x}_{2,*}(0) = 0$, it follows that $\mathbf{x}_{2,*} \equiv 0$ on $[0, 1]$. In particular, $\mathbf{x}_{2,*}(1) = 0$, as required. Furthermore, the first state equation, namely $\mathbf{x}_{1,*}'(t) = 1 - (\mathbf{x}_{2,*}(t))^2 = 1 - 0^2 = 1$, together with the initial condition $\mathbf{x}_{1,*}(0) = 0$ yields $\mathbf{x}_{1,*}(t) = t$ for $t \in [0, 1]$. In particular, $\mathbf{x}_{1,*}(1) = 1$, as wanted. Finally, $\mathbf{u}_*(t) = 0 \in [-1, 1]$ for all $t \in [0, 1]$ and $T_* = 1 \geq 0$. Thus $(\mathbf{u}_*, T_*) = (\mathbf{0}, 1)$ is indeed admissible.

The cost associated with $(\mathbf{u}_*, T_*) = (\mathbf{0}, 1)$ is equal to $T_* = 1$. We show that the cost of any other admissible pair $(\mathbf{u}, T)$ is at least equal to 1. We have

$$
1 = 1 - 0 = \mathbf{x}_1(T) - \mathbf{x}_1(0) = \int_0^T \mathbf{x}_1'(t) dt = \int_0^T (1 - (\mathbf{x}_2(t))^2) dt = T - \int_0^T (\mathbf{x}_2(t))^2 dt,
$$

and so

$$
T = 1 + \int_0^T (\mathbf{x}_2(t))^2 dt \geq 1 + \int_0^T 0 \, dt = 1 + 0 = 1 = T_*.
$$

Thus $(\mathbf{u}_*, T_*) = (\mathbf{0}, 1)$ is optimal. In fact we also note that if $T = 1$, then we must have $\mathbf{x}_2 \equiv 0$, and then $\mathbf{u} = \mathbf{x}_2' = \mathbf{0}$. So in fact the *unique* optimal solution is $(\mathbf{u}_*, T_*) = (\mathbf{0}, 1)$.

From the above, we know that $(\mathbf{u}_*, T_*) = (\mathbf{0}, 1)$ must satisfy (P1)-(P3) in the Pontryagin Minimum Principle (Theorem 4.23). The Hamiltonian is given by

$$
\begin{aligned}
H(x, u, \lambda_0, \lambda) &= \lambda_0 \cdot 1 + \lambda^\top \begin{bmatrix} 1 - x_2^2 \\ u \end{bmatrix} \\
&= \lambda_0 \cdot 1 + \begin{bmatrix} \lambda_1 & \lambda_2 \end{bmatrix} \begin{bmatrix} 1 - x_2^2 \\ u \end{bmatrix} \\
&= \lambda_0 + \lambda_1(1 - x_2^2) + \lambda_2 u.
\end{aligned}
$$

Thus (P2) becomes

$$
\mathbf{u}_*(t) = \min_{u \in [-1,1]} \left(\lambda_0 + \boldsymbol{\lambda}_1(t)(1 - (\mathbf{x}_{2,*}(t))^2) + \boldsymbol{\lambda}_2(t) u \right).
$$

So we know that

$$
\mathbf{u}_*(t) = \begin{cases} 1 & \text{if } \boldsymbol{\lambda}_2(t) < 0, \\ \boxed{?} & \text{if } \boldsymbol{\lambda}_2(t) = 0, \\ -1 & \text{if } \boldsymbol{\lambda}_2(t) > 0. \end{cases}
$$

We note that the $\mathbf{u}_*$ is determined at the time instances where $\boldsymbol{\lambda}_2$ is nonzero, and is ± 1 there (bang-bang!). But from our earlier discussion, we know that the optimal control is in fact identically zero, and so it must be the case that $\boldsymbol{\lambda}_2$ is identically 0. It is for this reason, that such optimal controls are referred to as "singular."

We also know that $\mathbf{x}_{1,*}(t) = t$ and $\mathbf{x}_{2,*}(t) = 0$, for all $t \in [0,1]$. And from (P2), it follows that for all $t \in [0,1]$,

$$
\begin{aligned}
0 &= H(\mathbf{x}_*(t), \mathbf{u}_*(t), \lambda_0, \boldsymbol{\lambda}(t)) \\
&= \lambda_0 + \boldsymbol{\lambda}_1(t)(1 - (\mathbf{x}_{2,*}(t))^2) + \boldsymbol{\lambda}_2(t)\mathbf{u}_*(t) \\
&= \lambda_0 + \boldsymbol{\lambda}_1(t)(1 - 0^2) + 0 \cdot 0 \\
&= \lambda_0 + \boldsymbol{\lambda}_1(t).
\end{aligned}
$$

So $\boldsymbol{\lambda}_1 \equiv -\lambda_0$. Hence the conditions (P1)-(P3) are all satisfied with

$$
\begin{aligned}
T_* &= 1, \\
\mathbf{u}_* &= \mathbf{0}, \\
\mathbf{x}_{1,*} &= t, \\
\mathbf{x}_{2,*} &= \mathbf{0}, \\
\lambda_0 &\neq 0, \\
\boldsymbol{\lambda}_1 &\equiv -\lambda_0, \\
\boldsymbol{\lambda}_2 &= \mathbf{0}.
\end{aligned}
$$

So what we have illustrated by this example is that a non-bang-bang optimal solution can appear in optimization problems.   ◇

Exercise 4.26. A firm has an order of B units of a commodity to be delivered at the final time T, and T is a variable. Let $\mathbf{x}(t)$ be the stock at time t. We assume that the cost per unit time of storing $\mathbf{x}(t)$ is $a \cdot \mathbf{x}(t)$, where a is a positive constant. The production per unit time is $\mathbf{u}(t) = \mathbf{x}'(t)$. We assume that the total cost of production per unit time is $b \cdot (\mathbf{u}(t))^2$, where b is a positive constant. Thus the firm's natural problem is

$$
\left\{
\begin{aligned}
&\text{minimize} && \int_0^T \Big(a \cdot \mathbf{x}(t) + b \cdot (\mathbf{u}(t))^2 \Big) dt \\
&\text{subject to} && \mathbf{x}'(t) = \mathbf{u}(t), \ \ t \in [0, T], \\
& && \mathbf{x}(0) = 0, \\
& && \mathbf{x}(T) = B, \\
& && \mathbf{u}(t) \geq 0, \ \ t \in [0, T], \\
& && T \ (\geq 0) \text{ is free.}
\end{aligned}
\right.
$$

Assuming $(\mathbf{u}_*, T_*)$ is optimal, find *all* possibilities for $(\mathbf{u}_*, T_*)$ using Pontryagin's Minimum Principle.

Exercise 4.27. Consider the optimal control problem

$$
\left\{
\begin{aligned}
&\text{minimize} && \int_0^T \Big(9 + \frac{(\mathbf{u}(t))^2}{4} \Big) dt \\
&\text{such that} && \mathbf{x}'(t) = \mathbf{u}(t) \text{ for } t \in [0, T], \\
& && \mathbf{x}(0) = 0, \\
& && \mathbf{x}(T) = 16, \\
& && T \ (\geq 0) \text{ is free.}
\end{aligned}
\right.
$$

Assuming $(\mathbf{u}_*, T_*)$ is optimal, find $(\mathbf{u}_*, T_*)$ using Pontryagin's Minimum Principle.

4.5. Cost Involving the Final State; Autonomous Case

Theorem 4.28. *Suppose that $\mathbf{u}_*$ is an optimal solution to*

$$
\left\{
\begin{aligned}
&\text{minimize} && \varphi(\mathbf{x}(t_f)) + \int_{t_i}^{t_f} F(\mathbf{x}(t), \mathbf{u}(t)) dt \\
&\text{subject to} && \mathbf{x}'(t) = f(\mathbf{x}(t), \mathbf{u}(t)), \ \ t \in [t_i, t_f], \\
& && \mathbf{x}(t_i) = x_i \in \mathbb{R}^n, \\
& && \mathbf{x}(t_f) \in \mathbb{X}_f \subset \mathbb{R}^n, \\
& && \mathbf{u}(t) \in \mathbb{U} \subset \mathbb{R}^m, \ \ t \in [t_i, t_f].
\end{aligned}
\right.
$$

Define the Hamiltonian $H : \mathbb{R}^n \times \mathbb{U} \times \mathbb{R} \times \mathbb{R}^n \to \mathbb{R}$ by

$$
H(x, u, \lambda_0, \lambda) := \lambda_0 F(x, u) + \lambda^\top f(x, u).
$$

Then there exists a nonnegative number λ_0, and a continuous and piecewise con-tinuously differentiable function $\boldsymbol{\lambda} : [t_i, t_f] \to \mathbb{R}^n$ such that:

(P1a) $\boldsymbol{\lambda}'(t) = -\left(\dfrac{\partial H}{\partial x}(\mathbf{x}_*(t), \mathbf{u}_*(t), \lambda_0, \boldsymbol{\lambda}(t))\right)^{\top}, \ t \in [t_i, t_f].$

(P1b) *For all* $v \in \bigcap_{i \in I} \ker(\nabla g_i(\mathbf{x}_*(t_f)))$, $\left(\boldsymbol{\lambda}(t_f)^{\top} - \lambda_0 \nabla \varphi(\mathbf{x}_*(t_f))\right) \cdot v = 0.$

Equivalently, there exist scalars α_i, $i \in I$, *such that*

$$\boldsymbol{\lambda}(t_f) = \lambda_0 \cdot (\nabla \varphi(\mathbf{x}_*(t_f)))^{\top} + \sum_{i \in I} \alpha_i \cdot (\nabla g_i(\mathbf{x}_*(t_f)))^{\top}.$$

(P2) *There exists a constant* C *such that for all* $t \in [t_i, t_f]$,

$$H(\mathbf{x}_*(t), \mathbf{u}_*(t), \lambda_0, \boldsymbol{\lambda}(t)) = \min_{u \in \mathbb{U}} H(\mathbf{x}_*(t), u, \lambda_0, \boldsymbol{\lambda}(t)) = C.$$

(P3) *For all* $t \in [t_i, t_f]$, $(\lambda_0, \boldsymbol{\lambda}(t)^{\top}) \neq 0.$

If $t_f \ (\geq t_i)$ is free, then the same conclusions (P1a), (P1b), (P3) hold in the above theorem, and in (P2), the constant C is 0. Thus one has:

(P2) There exists a constant C such that for all $t \in [t_i, t_{f,*}]$,

$$H(\mathbf{x}_*(t), \mathbf{u}_*(t), \lambda_0, \boldsymbol{\lambda}(t)) = \min_{u \in \mathbb{U}} H(\mathbf{x}_*(t), u, \lambda_0, \boldsymbol{\lambda}(t)) = 0.$$

Proof. (Sketch.) Let $\mathbf{u}$ be any admissible input and let the corresponding state be denoted by $\mathbf{x}$. Let $\mathbf{x}_{n+1} : [t_i, t_f] \to \mathbb{R}$ be any continuously differentiable function such that

$$\begin{aligned} \mathbf{x}_{n+1}(t_i) &:= 0, \\ \mathbf{x}_{n+1}(t_f) &:= \varphi(\mathbf{x}(t_f)). \end{aligned}$$

Define $\mathbf{u}_{m+1} : [t_i, t_f] \to \mathbb{R}$ by

$$\mathbf{u}_{m+1}(t) := \mathbf{x}'_{n+1}(t), \quad t \in [t_i, t_f].$$

Then we obtain

$$\varphi(\mathbf{x}(t_f)) = \varphi(\mathbf{x}(t_f)) - 0 = \mathbf{x}_{n+1}(t_f) - \mathbf{x}_{n+1}(t_i) = \int_{t_i}^{t_f} \mathbf{x}'_{n+1}(t)dt = \int_{t_i}^{t_f} \mathbf{u}_{m+1}(t)dt.$$

So the cost corresponding to $\mathbf{u}$ is given by

$$\begin{aligned} \varphi(\mathbf{x}(t_f)) + \int_{t_i}^{t_f} F(\mathbf{x}(t), \mathbf{u}(t))dt &= \int_{t_i}^{t_f} \mathbf{u}_{m+1}(t)dt + \int_{t_i}^{t_f} F(\mathbf{x}(t), \mathbf{u}(t))dt \\ &= \int_{t_i}^{t_f} \Big(F(\mathbf{x}(t), \mathbf{u}(t)) + \mathbf{u}_{m+1}(t)\Big)dt. \end{aligned}$$

Suppose that $\mathbf{u}_*$ is optimal, with corresponding state $\mathbf{x}_*$. Construct any $\mathbf{x}_{n+1,*}$ as above, and set $\mathbf{u}_{m+1,*} = \mathbf{x}'_{n+1,*}$. Then $(\mathbf{u}_*, \mathbf{u}_{m+1,*})$ is optimal for the following problem:

$$
\begin{cases}
\text{minimize} & \displaystyle\int_{t_i}^{t_f} \Big(F(\mathbf{x}(t), \mathbf{u}(t)) + \mathbf{u}_{m+1}(t) \Big)\, dt \\[2mm]
\text{subject to} & \mathbf{x}'(t) = f(\mathbf{x}(t), \mathbf{u}(t)), \quad t \in [t_i, t_f], \\[1mm]
& \mathbf{x}'_{n+1}(t) = \mathbf{u}_{m+1}(t), \quad t \in [t_i, t_f], \\[1mm]
& \mathbf{x}(t_i) = x_i \in \mathbb{R}^n, \\[1mm]
& \mathbf{x}_{n+1}(t_i) = 0, \\[1mm]
& \mathbf{x}(t_f) \in \mathbb{X}_f \subset \mathbb{R}^n, \\[1mm]
& \mathbf{x}_{n+1}(t_f) = \varphi(\mathbf{x}(t_f)), \\[1mm]
& \mathbf{u}(t) \in \mathbb{U} \subset \mathbb{R}^m, \quad t \in [t_i, t_f] \\[1mm]
& \mathbf{u}_{m+1}(t) \in \mathbb{R}, \quad t \in [t_i, t_f].
\end{cases}
$$

Now we use our old Pontryagin Minimum Principle for this new problem, and see what it tells us about $\mathbf{u}_*$. The Hamiltonian is

$$
\begin{aligned}
\widetilde{H}\left(\begin{bmatrix} x \\ x_{n+1} \end{bmatrix}, \begin{bmatrix} u \\ u_{m+1} \end{bmatrix}, \lambda_0, \begin{bmatrix} \lambda \\ \lambda_{n+1} \end{bmatrix} \right) & \\
&\hspace{-5cm}= \lambda_0\Big(F(x, u) + u_{m+1} \Big) + \lambda^\top f(x, u) + \lambda_{n+1} u_{m+1} \\
&\hspace{-5cm}= H(x, u, \lambda_0, \lambda) + (\lambda_0 + \lambda_{n+1}) u_{m+1}.
\end{aligned}
$$

From (P2) in the old version of the Pontryagin Minimum Principle, we obtain the existence of a constant C such that for all $t \in [t_i, t_f]$,

$$
\begin{aligned}
C &= \widetilde{H}\left(\begin{bmatrix} \mathbf{x}_*(t) \\ \mathbf{x}_{n+1,*}(t) \end{bmatrix}, \begin{bmatrix} \mathbf{u}_*(t) \\ \mathbf{u}_{m+1,*}(t) \end{bmatrix}, \lambda_0, \begin{bmatrix} \boldsymbol{\lambda}(t) \\ \lambda_{n+1}(t) \end{bmatrix} \right) \\
&= \min_{\substack{u \in \mathbb{U} \\ u_{m+1} \in \mathbb{R}}} \widetilde{H}\left(\begin{bmatrix} \mathbf{x}_*(t) \\ \mathbf{x}_{n+1,*}(t) \end{bmatrix}, \begin{bmatrix} u \\ u_{m+1} \end{bmatrix}, \lambda_0, \begin{bmatrix} \boldsymbol{\lambda}(t) \\ \lambda_{n+1}(t) \end{bmatrix} \right) \\
&= \min_{\substack{u \in \mathbb{U} \\ u_{m+1} \in \mathbb{R}}} \Big(H(\mathbf{x}_*(t), u, \lambda_0, \boldsymbol{\lambda}(t)) + (\lambda_0 + \lambda_{n+1}(t)) u_{m+1} \Big).
\end{aligned}
$$

From here it follows that for each time instant $t \in [t_i, t_f]$, $\lambda_{n+1}(t) = -\lambda_0$, and

$$
C = H(\mathbf{x}_*(t), \mathbf{u}_*(t), \lambda_0, \boldsymbol{\lambda}(t)) = \min_{u \in \mathbb{U}} H(\mathbf{x}_*(t), u, \lambda_0, \boldsymbol{\lambda}(t)).
$$

So we obtain (P2) in Theorem 4.28.

Also from (P3) in the old version of the Pontryagin Minimum Principle, we know that for each $t \in [t_i, t_f]$, $(\lambda_0, \boldsymbol{\lambda}(t), \lambda_{n+1}(t)) \neq 0$, that is, $(\lambda_0, \boldsymbol{\lambda}(t), -\lambda_0) \neq 0$, and from here we obtain that $(\lambda_0, \boldsymbol{\lambda}(t)) \neq 0$ for each $t \in [t_i, t_f]$. Hence (P3) holds in Theorem 4.28.

From (P1a) in the old version of the Pontryagin Minimum Principle, we have

$$
\boldsymbol{\lambda}'(t) = -\left(\frac{\partial \widetilde{H}}{\partial x}\left(\begin{bmatrix} \mathbf{x}_*(t) \\ \mathbf{x}_{n+1,*}(t) \end{bmatrix}, \begin{bmatrix} \mathbf{u}_*(t) \\ \mathbf{u}_{m+1,*}(t) \end{bmatrix}, \lambda_0, \begin{bmatrix} \boldsymbol{\lambda}(t) \\ \boldsymbol{\lambda}_{n+1}(t) \end{bmatrix}\right)\right)^{\top}
$$

$$
= -\left(\frac{\partial H}{\partial x}(\mathbf{x}_*(t), \mathbf{u}_*(t), \lambda_0, \boldsymbol{\lambda}(t))\right)^{\top},
$$

$$
\boldsymbol{\lambda}'_{n+1}(t) = -\left(\frac{\partial \widetilde{H}}{\partial x_{n+1}}\left(\begin{bmatrix} \mathbf{x}_*(t) \\ \mathbf{x}_{n+1,*}(t) \end{bmatrix}, \begin{bmatrix} \mathbf{u}_*(t) \\ \mathbf{u}_{m+1,*}(t) \end{bmatrix}, \lambda_0, \begin{bmatrix} \boldsymbol{\lambda}(t) \\ \boldsymbol{\lambda}_{n+1}(t) \end{bmatrix}\right)\right)^{\top} = 0.
$$

The first of these is precisely (P1a) in Theorem 4.28. (Note that the last equation says that $\boldsymbol{\lambda}_{n+1}$ is constant, but we already knew this since we had shown above that $\boldsymbol{\lambda}_{n+1} \equiv -\lambda_0$ on $[t_i, t_f]$.)

It remains to show (P1b). To this end, let us define the manifold $\widetilde{\mathbb{X}}_f$ in $\mathbb{R}^{n+1}$ as follows:

$$
\widetilde{\mathbb{X}}_f = \left\{ \begin{bmatrix} x \\ x_{n+1} \end{bmatrix} : x \in \mathbb{X}_f,\ x_{n+1} = \varphi(x) \right\} = \left\{ \begin{bmatrix} x \\ x_{n+1} \end{bmatrix} : \begin{array}{l} g_i(x) = 0,\ i \in I, \\ x_{n+1} - \varphi(x) = 0. \end{array} \right\}.
$$

From (P1b) in the old version of the Pontryagin Minimum Principle, there exist scalars α_i, $i \in I$ and α_{n+1} such that

$$
\begin{bmatrix} \boldsymbol{\lambda}(t_f) \\ \boldsymbol{\lambda}_{n+1}(t_f) \end{bmatrix} = \sum_{i \in I} \alpha_i \cdot \begin{bmatrix} (\nabla g_i(\mathbf{x}_*(t_f)))^{\top} \\ 0 \end{bmatrix} + \alpha_{n+1} \begin{bmatrix} -(\nabla\varphi(\mathbf{x}_*(t_f)))^{\top} \\ 1 \end{bmatrix}.
$$

From the equality for the last component, we see that $\alpha_{n+1} = \boldsymbol{\lambda}_{n+1}(t_f)$, and we already knew that $\boldsymbol{\lambda}_{n+1}(t_f) = -\lambda_0$. So $\alpha_{n+1} = -\lambda_0$. The equalities for the first n components now yield

$$
\boldsymbol{\lambda}(t_f) = \sum_{i \in I} \alpha_i \cdot (\nabla g_i(\mathbf{x}_*(t_f)))^{\top} + \alpha_{n+1} \cdot (-(\nabla\varphi(\mathbf{x}_*(t_f)))^{\top})
$$

$$
= \sum_{i \in I} \alpha_i \cdot (\nabla g_i(\mathbf{x}_*(t_f)))^{\top} + \lambda_0 \cdot (\nabla\varphi(\mathbf{x}_*(t_f)))^{\top},
$$

that is, (P1b) in Theorem 4.28 holds. $\qquad\square$

In the case when t_f ($\geq t_i$) is free, the proof of the analogous result, with $C = 0$ in (P2), is the same, mutatis mutandis, as the above.

Example 4.29 (Optimal fish farming). Let

$\mathbf{x}(t)$ = weight of the fish at time t,

$p(x)$ = price per kilogram of fish whose weight is x
     = $a_0 + a_1 x$ (a_0, a_1 are positive constants),

$\mathbf{u}(t)$ = amount of fish food per unit of time measured as a proportion of the weight of fish,

c = constant cost of one kilogram of fish food.

The profit from feeding the fish over a time interval $[0, T]$ and then harvesting at the fixed final time T is

$$\mathbf{x}(T)p(\mathbf{x}(T)) - \int_0^T c \cdot \mathbf{x}(t) \cdot \mathbf{u}(t)dt.$$

Suppose that the known law for the growth in weight of the fish in terms of $\mathbf{u}$ and $\mathbf{x}$ is given by

$$\mathbf{x}'(t) = \mathbf{x}(t)\left(a - \frac{b}{\mathbf{u}(t)}\right),$$

where a, b are constants. The initial condition is known: $\mathbf{x}(0) = x_i$. For simplicity, we will assume that all the constants $a_0, a_1, c, a, b, x_i, T$ are equal to 1.

The question the fish farming company is faced with is: How should it feed the fish so that the profit is maximized? In other words, it would like to find $\mathbf{u}_*$ which is the solution to the following optimization problem:

$$\begin{cases} \text{minimize} & -\mathbf{x}(1)(1 + \mathbf{x}(1)) + \displaystyle\int_0^1 \mathbf{x}(t) \cdot \mathbf{u}(t)dt \\[2mm] \text{subject to} & \mathbf{x}'(t) = \mathbf{x}(t)\left(1 - \dfrac{1}{\mathbf{u}(t)}\right), \quad t \in [0, 1], \\[2mm] & \mathbf{x}(0) = 1, \\[2mm] & \mathbf{u}(t) > 0, \quad t \in [0, 1]. \end{cases}$$

The Hamiltonian is given by

$$H(x, u, \lambda_0, \lambda) = \lambda_0 xu + \lambda x\left(1 - \frac{1}{u}\right).$$

From the state equation and the fact that $x_i = 1 > 0$, we have that $\mathbf{x}_*(t)$ is always positive, since

$$\mathbf{x}_*(t) = \underbrace{\exp\left(\int_0^t \left(1 - \frac{1}{\mathbf{u}_*(\tau)}\right)d\tau\right) x_i}_{> 0}, \quad t \geq 0.$$

(P2) gives for all $t \in [0, 1]$,

$$\begin{aligned} C &= \lambda(t)\mathbf{x}_*(t) + \mathbf{x}_*(t)\left(\lambda_0 \mathbf{u}_*(t) - \frac{\lambda(t)}{\mathbf{u}_*(t)}\right) \\ &= \min_{u>0}\left(\lambda(t)\mathbf{x}_*(t) + \mathbf{x}_*(t)\left(\lambda_0 u - \frac{\lambda(t)}{u}\right)\right). \end{aligned}$$

So we see that if $\lambda_0 = 0$, we must have $\boldsymbol\lambda \equiv 0$, violating (P3). Hence we take $\lambda_0 = 1$. Also (P2) shows that $\boldsymbol\lambda$ must be nonpositive, and it can't be zero (for otherwise $\mathbf{u}_*$ is zero, and not admissible). So $\lambda(t) < 0$, and

$$\mathbf{u}_*(t) = \sqrt{-\lambda(t)}, \quad t \in [0, 1].$$

The co-state equation is given by

$$\boldsymbol{\lambda}'(t) = -\frac{\partial H}{\partial x} H(\mathbf{x}_*(t), \mathbf{u}_*(t), \lambda_0, \boldsymbol{\lambda}(t))$$

$$= -\mathbf{u}_*(t) - \boldsymbol{\lambda}(t)\left(1 - \frac{1}{\mathbf{u}_*(t)}\right) = -\boldsymbol{\lambda}(t) - 2\sqrt{-\boldsymbol{\lambda}(t)}.$$

Since $\boldsymbol{\lambda}(t) = -(\mathbf{u}_*(t))^2$, we obtain upon differentiating with respect to t that

$$\boldsymbol{\lambda}'(t) = -2\mathbf{u}_*(t) \cdot \mathbf{u}_*'(t),$$

and so, using the co-state equation, we arrive at

$$-2\mathbf{u}_*(t) \cdot \mathbf{u}_*'(t) = (\mathbf{u}_*(t))^2 - 2 \cdot \mathbf{u}_*(t),$$

that is,

$$\mathbf{u}_*'(t) = -\frac{1}{2}\mathbf{u}_*(t) + 1.$$

Hence

$$\mathbf{u}_*(t) = e^{-t/2}\alpha + \int_0^t e^{-(t-\tau)/2} \cdot 1 d\tau = e^{-t/2}\alpha + 2,$$

for some constant α. We can use (P1b) to determine α:

$$-(\mathbf{u}_*(1))^2 = \boldsymbol{\lambda}(1) = -1 - 2\mathbf{x}_*(1).$$

Let us find an expression for $\mathbf{x}_*$ using the state equation. We have

$$\mathbf{x}_*(1) = \exp\left(-\int_0^1 \left(1 - \frac{1}{\mathbf{u}_*(t)}\right) dt\right) \cdot x_i$$

$$= \exp\left(-\int_0^1 \left(1 - \frac{1}{e^{-t/2}\alpha + 2}\right) dt\right) \cdot 1$$

$$= \frac{e^{-1/2}\alpha + 2}{e^{1/2}(\alpha + 2)}.$$

Thus (P1b) gives the equation

$$(e^{-1/2}\alpha + 2)^2 = 1 + 2 \cdot \frac{e^{-1/2}\alpha + 2}{e^{1/2}(\alpha + 2)},$$

and it can be shown that $\alpha \approx -0.705$ is a solution. (There are others, but they make $\mathbf{u}_*$ negative on the time interval $[0, 1]$, and so we discard them.) So if $\mathbf{u}_*$ is optimal, then it is given by $\mathbf{u}_*(t) \approx 2 - 0.705 \cdot e^{-t/2}$, $t \in [0, 1]$.   ◇

Exercise 4.30. Consider the optimal control problem

$$\begin{cases} \text{minimize} & \int_0^1 \frac{(\mathbf{u}(t))^2}{2} dt - \sqrt{\mathbf{x}(1)} \\ \text{such that} & \mathbf{x}'(t) = \mathbf{x}(t) + \mathbf{u}(t) \text{ for } t \in [0, 1], \\ & \mathbf{x}(0) = 0. \end{cases}$$

Assuming $\mathbf{u}_*$ is optimal, find $\mathbf{u}_*$ using Pontryagin's Minimum Principle.

4.6. Nonautonomous Case

Theorem 4.31. *Suppose that $\mathbf{u}_*$ is an optimal solution to*

$$\begin{cases} \text{minimize} & \varphi(\mathbf{x}(t_f)) + \int_{t_i}^{t_f} F(t, \mathbf{x}(t), \mathbf{u}(t))dt \\ \text{subject to} & \mathbf{x}'(t) = f(t, \mathbf{x}(t), \mathbf{u}(t)), \;\; t \in [t_i, t_f], \\ & \mathbf{x}(t_i) = x_i \in \mathbb{R}^n, \\ & \mathbf{x}(t_f) \in \mathbb{X}_f \subset \mathbb{R}^n, \\ & \mathbf{u}(t) \in \mathbb{U} \subset \mathbb{R}^m, \;\; t \in [t_i, t_f]. \end{cases}$$

Define the Hamiltonian $H : [t_i, t_f] \times \mathbb{R}^n \times \mathbb{U} \times \mathbb{R} \times \mathbb{R}^n \to \mathbb{R}$ by

$$H(t, x, u, \lambda_0, \lambda) := \lambda_0 F(t, x, u) + \lambda^\top f(t, x, u).$$

Then there exists a nonnegative number λ_0, and a continuous and piecewise continuously differentiable function $\boldsymbol{\lambda} : [t_i, t_f] \to \mathbb{R}^n$ such that:

(P1a) $\boldsymbol{\lambda}'(t) = -\left(\dfrac{\partial H}{\partial x}(t, \mathbf{x}_*(t), \mathbf{u}_*(t), \lambda_0, \boldsymbol{\lambda}(t)) \right)^\top, \; t \in [t_i, t_f].$

(P1b) *For all $v \in \bigcap_{i \in I} \ker(\nabla g_i(\mathbf{x}_*(t_f)))$, $\left(\boldsymbol{\lambda}(t_f)^\top - \lambda_0 \nabla \varphi(\mathbf{x}_*(t_f)) \right) \cdot v = 0.$*

    Equivalently, there exist scalars α_i, $i \in I$, such that

$$\boldsymbol{\lambda}(t_f) = \lambda_0 \cdot (\nabla \varphi(\mathbf{x}_*(t_f)))^\top + \sum_{i \in I} \alpha_i \cdot (\nabla g_i(\mathbf{x}_*(t_f)))^\top.$$

(P2a) *For all $t \in [t_i, t_f]$,*

$$H(t, \mathbf{x}_*(t), \mathbf{u}_*(t), \lambda_0, \boldsymbol{\lambda}(t)) = \min_{u \in \mathbb{U}} H(t, \mathbf{x}_*(t), u, \lambda_0, \boldsymbol{\lambda}(t)).$$

(P2b) *Let $H_*(t) := H(t, \mathbf{x}_*(t), \mathbf{u}_*(t), \lambda_0, \boldsymbol{\lambda}(t))$, $t \in [t_i, t_f]$. Then*

$$\text{for all } \tau \in [t_i, t_f], \;\; H_*(\tau) = H_*(t_f) - \int_\tau^{t_f} \frac{\partial H}{\partial t}(t, \mathbf{x}_*(t), \mathbf{u}_*(t), \lambda_0, \boldsymbol{\lambda}(t))dt.$$

(P3) *For all $t \in [t_i, t_f]$, $(\lambda_0, \boldsymbol{\lambda}(t)^\top, H_*(t_f) - H_*(t)) \neq 0$.*

If $t_f \; (\geq t_i)$ is free, then the same conclusions (P1a), (P1b), (P2a), (P3) hold in the above theorem, and in (P2b), $H_*(t_f) = 0$, so that we have:

(P2b) *Let $H_*(t) := H(t, \mathbf{x}_*(t), \mathbf{u}_*(t), \lambda_0, \boldsymbol{\lambda}(t))$, $t \in [t_i, t_f]$. Then for all $\tau \in [t_i, t_f]$,*

$$H_*(\tau) = - \int_\tau^{t_f} \frac{\partial H}{\partial t}(t, \mathbf{x}_*(t), \mathbf{u}_*(t), \lambda_0, \boldsymbol{\lambda}(t))dt.$$

Proof. (Sketch) We eliminate the dependence of f and F on time by introducing an extra state variable $\mathbf{x}_{n+1}$ that satisfies the differential equation

$$\mathbf{x}'_{n+1}(t) = 1, \quad t \in [t_i, t_f],$$

with the boundary conditions

$$\mathbf{x}_{n+1}(t_i) = t_i,$$
$$\mathbf{x}_{n+1}(t_f) = t_f.$$

Thus $\mathbf{x}_{n+1}$ "is" time. Then we have for all $t \in [t_i, t_f]$ that

$$f(t, \mathbf{x}(t), \mathbf{u}(t)) = f(\mathbf{x}_{n+1}(t), \mathbf{x}(t), \mathbf{u}(t)),$$
$$F(t, \mathbf{x}(t), \mathbf{u}(t)) = F(\mathbf{x}_{n+1}(t), \mathbf{x}(t), \mathbf{u}(t)).$$

So if $\mathbf{u}_*$ is optimal for the original problem, then $\mathbf{u}_*$ is also a solution to the following *autonomous* optimization problem:

$$\left\{ \begin{array}{ll} \text{minimize} & \varphi(\mathbf{x}(t_f)) + \displaystyle\int_{t_i}^{t_f} F(\mathbf{x}_{n+1}(t), \mathbf{x}(t), \mathbf{u}(t)) dt \\ \text{subject to} & \mathbf{x}'(t) = f(\mathbf{x}_{n+1}(t), \mathbf{x}(t), \mathbf{u}(t)), \ t \in [t_i, t_f], \\ & \mathbf{x}'_{n+1}(t) = 1, \ t \in [t_i, t_f], \\ & \mathbf{x}(t_i) = x_i \in \mathbb{R}^n, \\ & \mathbf{x}_{n+1}(t_i) = t_i, \\ & \mathbf{x}(t_f) \in \mathbb{X}_f \subset \mathbb{R}^n, \\ & \mathbf{x}_{n+1}(t_f) = t_f, \\ & \mathbf{u}(t) \in \mathbb{U} \subset \mathbb{R}^m, \ t \in [t_i, t_f]. \end{array} \right.$$

So we can use our old Pontryagin Minimum Principle given in Theorem 4.28 in order to derive the new conditions. Let us define the Hamiltonian $\widetilde{H}$ by

$$\widetilde{H}(x, x_{n+1}, u, \lambda_0, \lambda, \lambda_{n+1}) = \lambda_0 F(x_{n+1}, x, u) + \lambda^\top f(x_{n+1}, x, u) + \lambda_{n+1} \cdot 1.$$

From (P2) in the old Pontryagin Minimum Principle, there exists a constant C such that for all $t \in [t_i, t_f]$, we have

$$\begin{aligned} C &= \widetilde{H}(\mathbf{x}_*(t), \mathbf{x}_{n+1,*}(t), \mathbf{u}_*(t), \lambda_0, \boldsymbol{\lambda}(t), \boldsymbol{\lambda}_{n+1}(t)) \\ &= \lambda_0 F(\mathbf{x}_{n+1,*}(t), \mathbf{x}_*(t), \mathbf{u}_*(t)) + \boldsymbol{\lambda}(t)^\top f(\mathbf{x}_{n+1,*}(t), \mathbf{x}_*(t), \mathbf{u}_*(t)) + \boldsymbol{\lambda}_{n+1}(t) \\ &= \lambda_0 F(t, \mathbf{x}_*(t), \mathbf{u}_*(t)) + \boldsymbol{\lambda}(t)^\top f(t, \mathbf{x}_*(t), \mathbf{u}_*(t)) + \boldsymbol{\lambda}_{n+1}(t) \\ &= H(t, \mathbf{x}_*(t), \mathbf{u}_*(t), \lambda_0, \boldsymbol{\lambda}(t)) + \boldsymbol{\lambda}_{n+1}(t) \\ &= \min_{u \in \mathbb{U}} \left(\lambda_0 F(\mathbf{x}_{n+1,*}(t), \mathbf{x}_*(t), u) + \boldsymbol{\lambda}(t)^\top f(\mathbf{x}_{n+1,*}(t), \mathbf{x}_*(t), u) + \boldsymbol{\lambda}_{n+1}(t) \right) \\ &= \left(\min_{u \in \mathbb{U}} H(t, \mathbf{x}_*(t), u, \lambda_0, \boldsymbol{\lambda}(t)) \right) + \boldsymbol{\lambda}_{n+1}(t). \end{aligned}$$

From here, we obtain

$$H(t, \mathbf{x}_*(t), \mathbf{u}_*(t), \lambda_0, \boldsymbol{\lambda}(t)) = \min_{u \in \mathbb{U}} H(t, \mathbf{x}_*(t), u, \lambda_0, \boldsymbol{\lambda}(t)),$$

that is, (P2a) in Theorem 4.31 holds. Also from the above equation array, we obtain for all $t \in [t_i, t_f]$ that

$$C = H_*(t) + \lambda_{n+1}(t). \tag{4.21}$$

From (P1a) in the old Pontryagin Minimum Principle, we have

$$\lambda'(t)^\top$$

$$= -\frac{\partial \widetilde{H}}{\partial x}(\mathbf{x}_*(t), \mathbf{x}_{n+1,*}(t), \mathbf{u}_*(t), \lambda_0, \boldsymbol{\lambda}(t), \boldsymbol{\lambda}_{n+1}(t))$$

$$= -\frac{\partial}{\partial x}(\lambda_0 F(x_{n+1}, x, u) + \lambda^\top f(x_{n+1}, x, u) + \lambda_{n+1})\Big|_{(\mathbf{x}_*(t), \mathbf{x}_{n+1,*}(t), \mathbf{u}_*(t), \lambda_0, \boldsymbol{\lambda}(t), \boldsymbol{\lambda}_{n+1}(t))}$$

$$= -\lambda_0 \frac{\partial F}{\partial x}(\mathbf{x}_{n+1,*}(t), \mathbf{x}_*(t), \mathbf{u}_*(t)) - \boldsymbol{\lambda}(t)^\top \frac{\partial f}{\partial x}(\mathbf{x}_{n+1,*}(t), \mathbf{x}_*(t), \mathbf{u}_*(t))$$

$$= -\lambda_0 \frac{\partial F}{\partial x}(t, \mathbf{x}_*(t), \mathbf{u}_*(t)) - \boldsymbol{\lambda}(t)^\top \frac{\partial f}{\partial x}(t, \mathbf{x}_*(t), \mathbf{u}_*(t))$$

$$= -\frac{\partial H}{\partial x}(t, \mathbf{x}_*(t), \mathbf{u}_*(t), \lambda_0, \boldsymbol{\lambda}(t)).$$

Thus (P1a) in Theorem 4.31 holds. Also, from (P1a) in the old Pontryagin Minimum Principle, we have

$$\lambda'_{n+1}(t)$$

$$= -\frac{\partial \widetilde{H}}{\partial x_{n+1}}(\mathbf{x}_*(t), \mathbf{x}_{n+1,*}(t), \mathbf{u}_*(t), \lambda_0, \boldsymbol{\lambda}(t), \boldsymbol{\lambda}_{n+1}(t))$$

$$= -\frac{\partial(\lambda_0 F(x_{n+1}, x, u) + \lambda^\top f(x_{n+1}, x, u) + \lambda_{n+1})}{\partial x_{n+1}}\Big|_{(\mathbf{x}_*(t), \mathbf{x}_{n+1,*}(t), \mathbf{u}_*(t), \lambda_0, \boldsymbol{\lambda}(t), \boldsymbol{\lambda}_{n+1}(t))}$$

$$= -\lambda_0 \frac{\partial F}{\partial t}(\mathbf{x}_{n+1,*}(t), \mathbf{x}_*(t), \mathbf{u}_*(t)) - \boldsymbol{\lambda}(t)^\top \frac{\partial f}{\partial t}(\mathbf{x}_{n+1,*}(t), \mathbf{x}_*(t), \mathbf{u}_*(t))$$

$$= -\lambda_0 \frac{\partial F}{\partial t}(t, \mathbf{x}_*(t), \mathbf{u}_*(t)) - \boldsymbol{\lambda}(t)^\top \frac{\partial f}{\partial t}(t, \mathbf{x}_*(t), \mathbf{u}_*(t))$$

$$= -\frac{\partial H}{\partial t}(t, \mathbf{x}_*(t), \mathbf{u}_*(t), \lambda_0, \boldsymbol{\lambda}(t)).$$

From here we obtain by integrating from τ to t_f that

$$\lambda_{n+1}(t_f) - \lambda_{n+1}(\tau) = \int_\tau^{t_f} -\frac{\partial H}{\partial t}(t, \mathbf{x}_*(t), \mathbf{u}_*(t), \lambda_0, \boldsymbol{\lambda}(t))dt. \tag{4.22}$$

From (P1b) in the old Pontryagin Minimum Principle, there exist scalars α_i, $i \in I$, such that

$$\lambda(t_f) = \lambda_0(\nabla\varphi(\mathbf{x}_*(t_f)))^\top + \sum_{i \in I} \alpha_i \cdot \nabla g_i(\mathbf{x}_*(t_f)))^\top,$$

$$\lambda_{n+1}(t_f) = 0.$$

From the first of the above equations, we see that (P1b) in Theorem 4.31 holds. Moreover using $\lambda_{n+1}(t_f) = 0$, (4.22) and (4.21), we obtain for all $\tau \in [t_i, t_f]$ that

$$H_*(t_f) - H_*(\tau) = \lambda_{n+1}(\tau) - \underbrace{\lambda_{n+1}(t_f)}_{=0} = \int_\tau^{t_f} -\frac{\partial H}{\partial t}(t, \mathbf{x}_*(t), \mathbf{u}_*(t), \lambda_0, \boldsymbol{\lambda}(t))dt,$$

that is, (P2a) in Theorem 4.31 holds.

Finally, from (P3) in the old Pontryagin Minimum Principle, we know that for all $t \in [t_i, t_f]$, $(\lambda_0, \boldsymbol{\lambda}(t), \lambda_{n+1}(t)) \neq 0$. Using $H_*(t_f) - H_*(t) = \lambda_{n+1}(t)$, it follows that (P3) in Theorem 4.31 holds.   □

In the case when t_f $(\geq t_i)$ is free, the proof of the analogous result is the same, mutatis mutandis, as the above. Indeed, the fact that C is 0 in (4.21) implies, when $t = t_f$, that

$$H_*(t_f) = -\lambda_{n+1}(t_f) = 0.$$

Example 4.32. Consider the optimal control problem

$$\begin{cases} \text{minimize} & \displaystyle\int_0^1 \left((\mathbf{u}(t))^2 + t \cdot \mathbf{u}(t) - 1\right)dt - 2\mathbf{x}(1) - 3 \\ \text{subject to} & \mathbf{x}'(t) = \mathbf{u}(t), \quad t \in [0, 1], \\ & \mathbf{x}(0) = 1. \end{cases}$$

The Hamiltonian is given by

$$H(t, x, u, \lambda_0, \lambda) = \lambda_0(u^2 + tu - 1) + \lambda u.$$

From (P2a), we have

$$\lambda_0((\mathbf{u}_*(t))^2 + t \cdot \mathbf{u}_*(t) - 1) + \boldsymbol{\lambda}(t) \cdot \mathbf{u}_*(t) = \min_{u \in \mathbb{R}} \left(\lambda_0(u^2 + t \cdot u - 1) + \boldsymbol{\lambda}(t) \cdot u\right) \quad (4.23)$$

for all $t \in [0, 1]$. The co-state equation is given by

$$\boldsymbol{\lambda}'(t) = -\frac{\partial H}{\partial x}(t, \mathbf{x}_*(t), \mathbf{u}_*(t), \lambda_0, \boldsymbol{\lambda}(t)) = 0, \quad t \in [0, 1],$$

and so $\boldsymbol{\lambda} \equiv C$ on $[0, 1]$ for some constant C. From (P1b), we have

$$\boldsymbol{\lambda}(1) = \varphi'(\mathbf{x}_*(1)) = -2.$$

It follows from (4.23) that $\lambda_0 \neq 0$. So we take $\lambda_0 = 1$, and then from (4.23) we obtain that

$$\mathbf{u}_*(t) = -\frac{t + \boldsymbol{\lambda}(t)}{2} = -\frac{t - 2}{2} = 1 - \frac{t}{2},$$

for $t \in [0, 1]$.   ◇

Exercise 4.33. Consider the optimal control problem

$$
\begin{cases}
\text{minimize} & \displaystyle\int_0^1 -e^{-t}\sqrt{\mathbf{u}(t)}\,dt \\
\text{such that} & \mathbf{x}'(t) = \mathbf{x}(t) - \mathbf{u}(t) \text{ for } t \in [0,1], \\
& \mathbf{x}(0) = 1, \\
& \mathbf{x}(1) = 0, \\
& \mathbf{u}(t) \geq 0 \text{ for } t \in [0,1].
\end{cases}
$$

Assuming $\mathbf{u}_*$ is optimal, find $\mathbf{u}_*$ using Pontryagin's Minimum Principle.

4.7. Do Optimal Solutions Always Exist?

The Pontryagin Minimum Principle gives a *necessary* condition for optimal controls. This means that *if* an optimal control *exists*, then it must be something that satisfies these conditions (P1)-(P3). But we end this chapter by looking at a sample of optimal control problems that lack a solution, highlighting some of the reasons why an optimal solution may not exist. This will help us to appreciate the main result in the next chapter, which gives a *sufficient* condition for optimality!

Example 4.34 (Set of admissible inputs may be empty!). Consider the optimization problem

$$
\begin{cases}
\text{minimize} & T \\
\text{such that} & \mathbf{x}'(t) = \mathbf{x}(t) + \mathbf{u}(t) \text{ for } t \in [0,T], \\
& \mathbf{x}(0) = -1, \\
& \mathbf{x}(T) = 0, \\
& \mathbf{u}(t) \in [-1,1], \quad t \in [0,T], \\
& T \ (\geq 0) \text{ is free.}
\end{cases}
$$

Note that we should have $\mathbf{x}(T) = 0 = e^T \cdot (-1) + \displaystyle\int_0^T e^{T-t}\mathbf{u}(t)\,dt$, and so

$$
1 = \int_0^T \underbrace{e^{-t}}_{>0}\underbrace{\mathbf{u}(t)}_{\leq 1}\,dt \leq \int_0^T e^{-t}\cdot 1\,dt = 1 - e^{-T} < 1 \quad \text{for all } T \geq 0,
$$

a contradiction. So there is no $\mathbf{u}$ that satisfies the constraints. Hence there is no optimal pair $(\mathbf{u}_*, T_*)$ either. This situation is analogous to the following one in ordinary calculus:

$$
\begin{cases}
\text{minimize} & f(x) \\
\text{such that} & x \in \mathbb{R} \text{ and } x^2 + 1 = 0,
\end{cases}
$$

where again the feasible set $\{x \in \mathbb{R} : x^2 + 1 = 0\} = \emptyset$! $\diamond$

Example 4.35 (The cost may not be bounded below!). Consider

$$
\begin{cases}
\text{minimize} & \displaystyle\int_0^1 -(\mathbf{u}(t))^2 dt \\
\text{such that} & \mathbf{x}'(t) = \mathbf{u}(t) \text{ for } t \in [0,1], \\
& \mathbf{x}(0) = 0, \\
& \mathbf{x}(1) = 1.
\end{cases}
$$

We must have $1 = 1 - 0 = \mathbf{x}(1) - \mathbf{x}(0) = \displaystyle\int_0^1 \mathbf{u}(t)dt.$

Take $\mathbf{u}_n : [0,1] \to \mathbb{R}$, $n \in \mathbb{N}$, to be $\mathbf{u}_n(t) := \begin{cases} 2n & \text{for } 0 \le t \le 1/2, \\ -2(n-1) & \text{for } 1/2 < t \le 1. \end{cases}$

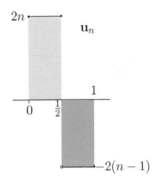

Then we have

$$
\int_0^1 \mathbf{u}_n(t)dt = 2n \cdot \frac{1}{2} + (-2(n-1)) \cdot \frac{1}{2} = (2n - 2n + 2) \cdot \frac{1}{2} = 1,
$$

and so these control inputs $\mathbf{u}_n$ are all admissible. What is their cost? We have

$$
\int_0^1 -(\mathbf{u}_n(t))^2 dt = -\left(\frac{1}{2} \cdot 4n^2 + \frac{1}{2} \cdot 4(n-1)^2 \right) = -n^2 \cdot \left(4 - \frac{4}{n} + \frac{1}{2n^2} \right) \overset{n\to\infty}{\longrightarrow} -\infty.
$$

So there is no minimizing input! This situation is analogous to the following one in ordinary calculus:

$$
\begin{cases}
\text{minimize} & -x^2 \\
\text{such that} & x \in \mathbb{R},
\end{cases}
$$

where again the cost is not bounded below. $\diamond$

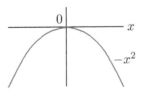

Example 4.36 (Cost bounded below, but infimum not achieved by an admissible control!). Consider the optimization problem

$$
\begin{cases}
\text{minimize} & \displaystyle\int_0^1 (\mathbf{u}(t))^2 dt \\[2mm]
\text{such that} & \mathbf{x}'(t) = (\mathbf{u}(t))^4 \text{ for } t \in [0,1], \\[2mm]
& \mathbf{x}(0) = 0, \\[2mm]
& \mathbf{x}(1) = 1.
\end{cases}
$$

We must have $1 = 1 - 0 = \mathbf{x}(1) - \mathbf{x}(0) = \displaystyle\int_0^1 (u(t))^4 dt$.

Take $\mathbf{u}_n : [0,1] \to \mathbb{R}$, $n \in \mathbb{N}$, to be $\mathbf{u}_n(t) := \begin{cases} \sqrt[4]{n} & \text{for } 0 \le t \le 1/n, \\ 0 & \text{for } 1/n < t \le 1. \end{cases}$

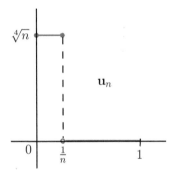

Then we have

$$
\int_0^1 (\mathbf{u}_n(t))^4 dt = \int_0^{1/n} (\sqrt[4]{n})^4 dt = \frac{1}{n} \cdot n = 1,
$$

and so these control inputs $\mathbf{u}_n$ are all admissible. What is their cost? We have

$$
\int_0^1 (\mathbf{u}_n(t))^2 dt = \int_0^{1/n} (\sqrt[4]{n})^2 dt = \frac{1}{n} \cdot \sqrt{n} = \frac{1}{\sqrt{n}} \overset{n \to \infty}{\longrightarrow} 0.
$$

So if an optimal $\mathbf{u}_*$ exists, then its cost must be 0. Thus $\mathbf{u}_* \equiv 0$ on $[0,1]$ in the class of piecewise continuous control inputs. But then the corresponding state must be $\mathbf{0}$ too, and so we can't accomplish the control task of having $\mathbf{x}_*(1) = 1$. Hence there is no optimal control, although there are admissible controls and the cost is bounded below (by 0). This situation is analogous to the following one in ordinary calculus:

$$
\begin{cases}
\text{minimize} & x^2 \\
\text{such that} & x \in (0, \infty),
\end{cases}
$$

where again the cost is bounded below, but there is no positive minimizer; see the following picture. $\Diamond$

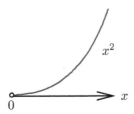

Chapter 5

Bellman's Equation

So far we have only studied *necessary* conditions for an optimal control given by the Pontryagin Minimum Principle. We now study a *sufficient* condition, given by Bellman's equation.

Consider the optimal control problem

$$\begin{cases} \text{minimize} & \varphi(\mathbf{x}(t_f)) + \int_{t_i}^{t_f} F(t, \mathbf{x}(t), \mathbf{u}(t)) dt \\ \text{subject to} & \mathbf{x}'(t) = f(t, \mathbf{x}(t), \mathbf{u}(t)), \ \ t \in [t_i, t_f], \\ & \mathbf{x}(t_i) = x_i \in \mathbb{R}^n, \\ & \mathbf{u}(t) \in \mathbb{U} \subset \mathbb{R}^m, \ \ t \in [t_i, t_f]. \end{cases}$$

We call the partial differential equation

$$\frac{\partial V}{\partial t}(x, t) + \min_{u \in \mathbb{U}} \left(\frac{\partial V}{\partial x}(x, t) \cdot f(t, x, u) + F(t, x, u) \right) = 0, \quad x \in \mathbb{R}^n, \ t \in [t_i, t_f]$$

with the boundary condition

$$V(x, t_f) = \varphi(x) \text{ for all } x \in \mathbb{R}^n$$

the *Bellman equation* associated with the optimal control problem above. The unknown in this equation is

$$V : \mathbb{R}^n \times [t_i, t_f] \to \mathbb{R}.$$

We will learn in this chapter that, roughly speaking, if Bellman's equation has a solution, then there exists an optimal control, and it can be built from V.

First, we will show that the existence of a solution to Bellman's equation gives a lower bound on the cost.

Lemma 5.1. *Let* $V : \mathbb{R}^n \times [t_i, t_f] \to \mathbb{R}$ *be a solution to*

$$\frac{\partial V}{\partial t}(x, t) + \min_{u \in \mathbb{U}} \left(\frac{\partial V}{\partial x}(x, t) \cdot f(t, x, u) + F(t, x, u) \right) = 0, \quad x \in \mathbb{R}^n, \, t \in [t_i, t_f]$$

with $V(x, t_f) = \varphi(x)$ *for all* $x \in \mathbb{R}^n$. *Suppose that* $\mathbf{u} : [t_i, t_f] \to \mathbb{U}$ *is piecewise continuous and let* $\mathbf{x}$ *be the solution to*

$$\begin{aligned} \mathbf{x}'(t) &= f(t, \mathbf{x}(t), \mathbf{u}(t)), \, t \in [t_i, t_f], \\ \mathbf{x}(t_i) &= x_i. \end{aligned}$$

Then

$$\varphi(\mathbf{x}(t_f)) + \int_{t_i}^{t_f} F(t, \mathbf{x}(t), \mathbf{u}(t)) dt \geq V(x_i, t_i).$$

Thus for an admissible control input, the cost is bounded below by the number $V(x_i, t_i)$.

Proof. We have

$$\begin{aligned} \int_{t_i}^{t_f} F(t, \mathbf{x}(t), \mathbf{u}(t)) dt &= \int_{t_i}^{t_f} \left(\frac{\partial V}{\partial x}(\mathbf{x}(t), t) \cdot f(t, \mathbf{x}(t), \mathbf{u}(t)) + F(t, \mathbf{x}(t), \mathbf{u}(t)) \right) dt \\ &\quad - \int_{t_i}^{t_f} \frac{\partial V}{\partial x}(\mathbf{x}(t), t) \cdot f(t, \mathbf{x}(t), \mathbf{u}(t)) dt \\ &\geq \int_{t_i}^{t_f} \min_{u \in \mathbb{U}} \left(\frac{\partial V}{\partial x}(\mathbf{x}(t), t) \cdot f(t, \mathbf{x}(t), u) + F(t, \mathbf{x}(t), u) \right) dt \\ &\quad - \int_{t_i}^{t_f} \frac{\partial V}{\partial x}(\mathbf{x}(t), t) \cdot \mathbf{x}'(t) dt \\ &= \int_{t_i}^{t_f} -\frac{\partial V}{\partial t}(\mathbf{x}(t), t) dt - \int_{t_i}^{t_f} \frac{\partial V}{\partial x}(\mathbf{x}(t), t) \cdot \mathbf{x}'(t) dt \\ &= -\int_{t_i}^{t_f} \left(\frac{\partial V}{\partial t}(\mathbf{x}(t), t) + \frac{\partial V}{\partial x}(\mathbf{x}(t), t) \cdot \mathbf{x}'(t) \right) dt \\ &= -\int_{t_i}^{t_f} \frac{d}{dt} V(\mathbf{x}(t), t) dt = V(\mathbf{x}(t_i), t_i) - V(\mathbf{x}(t_f), t_f) \\ &= V(x_i, t_i) - \varphi(\mathbf{x}(t_f)). \end{aligned}$$

This completes the proof.   □

We will now give the main result of this chapter.

Theorem 5.2. *Consider the optimal control problem*

$$
\begin{cases}
\text{minimize} & \varphi(\mathbf{x}(t_f)) + \displaystyle\int_{t_i}^{t_f} F(t, \mathbf{x}(t), \mathbf{u}(t))dt \\
\text{subject to} & \mathbf{x}'(t) = f(t, \mathbf{x}(t), \mathbf{u}(t)), \quad t \in [t_i, t_f], \\
& \mathbf{x}(t_i) = x_i \in \mathbb{R}^n, \\
& \mathbf{u}(t) \in \mathbb{U} \subset \mathbb{R}^m, \quad t \in [t_i, t_f].
\end{cases}
$$

Let $V : \mathbb{R}^n \times [t_i, t_f] \to \mathbb{R}$ be a solution to

$$
\frac{\partial V}{\partial t}(x, t) + \min_{u \in \mathbb{U}} \left(\frac{\partial V}{\partial x}(x, t) \cdot f(t, x, u) + F(t, x, u) \right) = 0, \quad x \in \mathbb{R}^n, \ t \in [t_i, t_f]
$$

with $V(x, t_f) = \varphi(x)$ for all $x \in \mathbb{R}^n$.

(1) *Let $U : \mathbb{R}^n \times [t_i, t_f] \to \mathbb{U}$ be given by*

$$
U(x, t) := \arg\min_{u \in \mathbb{U}} \left(\frac{\partial V}{\partial x}(x, t) \cdot f(t, x, u) + F(t, x, u) \right), \quad x \in \mathbb{R}^n, \ t \in [t_i, t_f].
$$

(2) *Let $\mathbf{x}_*$ be the solution to*

$$
\begin{aligned}
\mathbf{x}'(t) &= f(t, \mathbf{x}(t), U(\mathbf{x}(t), t)), \quad t \in [t_i, t_f], \\
\mathbf{x}(t_i) &= x_i.
\end{aligned}
$$

(3) *Define $\mathbf{u}_* : [t_i, t_f] \to \mathbb{U}$ by $\mathbf{u}_*(t) := U(\mathbf{x}_*(t), t), \ t \in [t_i, t_f].$*

Then $\mathbf{u}_*$ is an optimal control, and its cost is given by $V(x_i, t_i)$.

Before we go on to prove this, let us see what this result says. Roughly, if there exists a solution V to Bellman's equation, then there exists an optimal control $\mathbf{u}_*$, which can be constructed from V using the steps (1),(2),(3).

In step (1), we look at the function

$$
u \mapsto \frac{\partial V}{\partial x}(x, t) \cdot f(t, x, u) + F(t, x, u) : \mathbb{U} \to \mathbb{R},
$$

where we have imagined that we have fixed the "parameters" $x \in \mathbb{R}^n$ and $t \in [t_i, t_f]$. We *assume* that there is a minimizer for this function in $\mathbb{U}$, and we call it $U(x, t)$. (It depends on x and t, since the function being minimized changes if we change x and t.) In this manner, we get a map $U : \mathbb{R}^n \times [t_i, t_f] \to \mathbb{U}$. This finishes step (1).

Now that we have the function U from step (1), we look at the differential equation

$$
\mathbf{x}'(t) = f(t, \mathbf{x}(t), U(\mathbf{x}(t), t)), \quad t \in [t_i, t_f],
$$

with the initial condition $\mathbf{x}(t_i) = x_i$. *Assuming* that this has a solution, we denote this solution by $\mathbf{x}_*$.

Finally, in step (3), we simply define $\mathbf{u}_*$ by $\mathbf{u}_*(t) := U(\mathbf{x}_*(t), t), \ t \in [t_i, t_f].$

This is the construction procedure for $\mathbf{u}_*$. The result then says that this $\mathbf{u}_*$ is an optimal control, and that its cost is equal to $V(x_i, t_i)$. Recall that $V(x_i, t_i)$ was the lower bound on the costs of all admissible control inputs; see Lemma 5.1.

Let us now see the proof of Theorem 5.2.

Proof. We have

$$\frac{\partial V}{\partial x}(\mathbf{x}_*(t), t) \cdot f(t, \mathbf{x}_*(t), \mathbf{u}_*(t)) + F(t, \mathbf{x}_*(t), \mathbf{u}_*(t))$$

$$= \frac{\partial V}{\partial x}(\mathbf{x}_*(t), t) \cdot f(t, \mathbf{x}_*(t), U(\mathbf{x}_*(t), t)) + F(t, \mathbf{x}_*(t), U(\mathbf{x}_*(t), t))$$

$$= \min_{u \in \mathbb{U}} \left(\frac{\partial V}{\partial x}(\mathbf{x}_*(t), t) \cdot f(t, \mathbf{x}_*(t), u) + F(t, \mathbf{x}_*(t), u) \right).$$

From (2) and (3), we have

$$\mathbf{x}'_*(t) = f(t, \mathbf{x}_*(t), U(\mathbf{x}_*(t), t)) = f(t, \mathbf{x}_*(t), \mathbf{u}_*(t)), \quad t \in [t_i, t_f],$$
$$\mathbf{x}_*(t_i) = x_i,$$

and so $\mathbf{x}_*$ is the state corresponding to the input $\mathbf{u}_*$. Let us calculate its cost. There holds

$$\int_{t_i}^{t_f} F(t, \mathbf{x}_*(t), \mathbf{u}_*(t)) dt = \int_{t_i}^{t_f} \left(\frac{\partial V}{\partial x}(\mathbf{x}_*(t), t) \, f(t, \mathbf{x}_*(t), \mathbf{u}_*(t)) + F(t, \mathbf{x}_*(t), \mathbf{u}_*(t)) \right) dt$$

$$- \int_{t_i}^{t_f} \frac{\partial V}{\partial x}(\mathbf{x}_*(t), t) \, f(t, \mathbf{x}_*(t), \mathbf{u}_*(t)) dt$$

$$= \int_{t_i}^{t_f} \min_{u \in \mathbb{U}} \left(\frac{\partial V}{\partial x}(\mathbf{x}_*(t), t) \, f(t, \mathbf{x}_*(t), u) + F(t, \mathbf{x}_*(t), u) \right) dt$$

$$- \int_{t_i}^{t_f} \frac{\partial V}{\partial x}(\mathbf{x}_*(t), t) \mathbf{x}'_*(t) dt$$

$$= \int_{t_i}^{t_f} -\frac{\partial V}{\partial t}(\mathbf{x}_*(t), t) dt - \int_{t_i}^{t_f} \frac{\partial V}{\partial x}(\mathbf{x}_*(t), t) \, \mathbf{x}'_*(t) dt$$

$$= - \int_{t_i}^{t_f} \left(\frac{\partial V}{\partial t}(\mathbf{x}_*(t), t) + \frac{\partial V}{\partial x}(\mathbf{x}_*(t), t) \, \mathbf{x}'_*(t) \right) dt$$

$$= - \int_{t_i}^{t_f} \frac{d}{dt} V(\mathbf{x}_*(t), t) dt = V(\mathbf{x}_*(t_i), t_i) - V(\mathbf{x}(t_f), t_f)$$

$$= V(x_i, t_i) - \varphi(\mathbf{x}_*(t_f)).$$

Thus

$$\varphi(\mathbf{x}_*(t_f)) + \int_{t_i}^{t_f} F(t, \mathbf{x}_*(t), \mathbf{u}_*(t)) dt = V(x_i, t_i).$$

But from Lemma 5.1 we know that the cost of any admissible input is at least $V(x_i, t_i)$. Consequently, $\mathbf{u}_*$ is optimal. □

Example 5.3. Consider the optimal control problem

$$
\begin{cases}
\text{minimize} & \displaystyle\int_0^1 \left((\mathbf{x}(t))^2 + (\mathbf{u}(t)^2 \right) dt \\
\text{subject to} & \mathbf{x}'(t) = \mathbf{u}(t), \quad t \in [0,1], \\
& \mathbf{x}(0) = x_i.
\end{cases}
$$

Bellman's equation for this problem is

$$
\frac{\partial V}{\partial t}(x,t) + \min_{u \in U} \left(\frac{\partial V}{\partial x}(x,t) \cdot u + x^2 + u^2 \right) = 0, \quad x \in \mathbb{R},\ t \in [0,1],
$$

with the boundary condition $V(x,1) = 0$ for all $x \in \mathbb{R}$. We have

$$
U(x,t) = \arg\min_{u \in \mathbb{R}} \left(\frac{\partial V}{\partial x}(x,t) \cdot u + x^2 + u^2 \right) = -\frac{1}{2}\frac{\partial V}{\partial x}(x,t).
$$

So Bellman's equation becomes

$$
\frac{\partial V}{\partial t}(x,t) + x^2 - \frac{1}{4}\left(\frac{\partial V}{\partial x}(x,t) \right)^2 = 0, \quad x \in \mathbb{R},\ t \in [0,1],
$$

with the boundary condition $V(x,1) = 0$ for all $x \in \mathbb{R}$. This is a nonlinear partial differential equation!

We make an inspired guess: $V(x,t) = x^2 \cdot \mathbf{p}(t)$ for some function $\mathbf{p}$. (What is the rationale behind this guess? Well, we know that if $\mathbf{u}_*$ is an optimal control when the initial condition is x_i, then it is not hard to see that $\alpha \cdot \mathbf{u}_*$ is the optimal control when the initial condition is scaled to $\alpha \cdot x_i$, and that the cost of $\alpha \cdot \mathbf{u}_*$ in the new problem with the scaled initial condition is α^2 times the cost of $\mathbf{u}_*$ in the original problem. So in light of Theorem 5.2, we have $V(\alpha \cdot x_i, 0) = \alpha^2 \cdot V(x_i, 0)$. Thus $V(x,0) = V(x \cdot 1, 0) = x^2 \cdot V(1,0)$. But there is nothing special about the initial time $t = 0$ in this autonomous case. And so we expect that $V(x,t) = x^2 \cdot V(1,t)$. With $\mathbf{p}(t) := V(1,t)$, the above guess is reasonable.)

With our guess, the Bellman equation gives

$$
x^2 \cdot \mathbf{p}'(t) + x^2 - \frac{1}{4}(2x \cdot \mathbf{p}(t))^2 = 0, \quad x \in \mathbb{R},\ t \in [0,1],
$$

with the boundary condition $x^2 \cdot \mathbf{p}(1) = 0$ for all $x \in \mathbb{R}$.

So we arrive at the following *Riccati Equation*:

$$
\mathbf{p}'(t) = (\mathbf{p}(t))^2 - 1, \quad t \in [0,1],
$$

with $\mathbf{p}(1) = 0$. How do we solve this? From Exercise 3.29, we know a procedure, and following it, we obtain

$$
\mathbf{p}(t) = \frac{1 - e^{2(t-1)}}{1 + e^{2(t-1)}}, \quad t \in [0,1].
$$

Thus

$$V(x,t) = x^2 \cdot \frac{1 - e^{2(t-1)}}{1 + e^{2(t-1)}}, \quad x \in \mathbb{R}, \ t \in [0,1].$$

Now that we have found a solution V to Bellman's equation, we follow the algorithm in the steps (1),(2),(3) of Theorem 5.2 in order to construct the optimal $\mathbf{u}_*$. We have

$$U(x,t) = -\frac{1}{2}\frac{\partial V}{\partial x}(x,t) = -\frac{1}{2} \cdot 2x \cdot \mathbf{p}(t) = -x \cdot \mathbf{p}(t).$$

We need to find the solution $\mathbf{x}_*$ to

$$\begin{aligned}
\mathbf{x}_*'(t) &= f(\mathbf{x}_*(t), U(\mathbf{x}_*(t),t)) = U(\mathbf{x}_*(t),t) = -\mathbf{x}_*(t) \cdot \mathbf{p}(t), \quad t \in [0,1],\\
\mathbf{x}_*(0) &= x_i.
\end{aligned}$$

Thus

$$\begin{aligned}
\mathbf{x}_*(t) &= \exp\left(-\int_0^t \mathbf{p}(\tau)d\tau\right) \cdot x_i = \exp\left(-\int_0^t \frac{1 - e^{2(\tau-1)}}{1 + e^{2(\tau-1)}}d\tau\right) \cdot x_i\\
&= \exp\left(\int_0^t \frac{e^{\tau-1} - e^{-(\tau-1)}}{e^{\tau-1} + e^{-(\tau-1)}}d\tau\right) \cdot x_i = \exp\left(\int_0^t \frac{\sinh(\tau-1)}{\cosh(\tau-1)}d\tau\right) \cdot x_i\\
&= \exp\left(\int_{\cosh(-1)}^{\cosh(t-1)} \frac{1}{v}dv\right) \cdot x_i \quad \text{(with } v := \cosh(\tau-1))\\
&= \exp\left(\log\frac{\cosh(t-1)}{\cosh(-1)}\right) \cdot x_i = \frac{\cosh(t-1)}{\cosh(-1)} \cdot x_i\\
&= \frac{e^{t-1} + e^{-(t-1)}}{e^{-1} + e^{1}} \cdot x_i.
\end{aligned}$$

Finally,

$$\begin{aligned}
\mathbf{u}_*(t) &= U(\mathbf{x}_*(t),t) = -\mathbf{x}_*(t) \cdot \mathbf{p}(t) = \frac{e^{t-1} + e^{-(t-1)}}{e^{-1} + e^{1}} \cdot x_i \cdot \frac{1 - e^{2(t-1)}}{1 + e^{2(t-1)}}\\
&= \frac{e^{-(t-1)} - e^{t-1}}{e^{-1} + e^{1}} \cdot x_i
\end{aligned}$$

for $t \in [0,1]$. By Theorem 5.2, this $\mathbf{u}_*$ is optimal. $\Diamond$

Theorem 5.2 gives a sufficient condition for an optimal control. So we know that the moment we can solve Bellman's equation, the steps (1),(2),(3) of Theorem 5.2 deliver a $\mathbf{u}_*$ that we know surely to be optimal (and not just a candidate for an optimal control provided by the necessary conditions given in Pontryagin Minimum Principle). But as illustrated by the above simple example, we see that Bellman's equation is hard to solve in general. So we face a problem. But here is where the Pontryagin Minimum Principle can be used to make a guess for a solution to Bellman's equation! We explain this below.

Suppose we are given an optimal control problem as in Theorem 5.2 on a time interval $[t_i, t_f]$.

(1) We first fix $t_0 \in [t_i, t_f]$ and take an $x_0 \in \mathbb{R}$.

(2) We consider an auxiliary problem on the time interval $[t_0, t_f]$, with initial condition x_0, and with the same other data as the original problem. We write down the Hamiltonian equations for this auxiliary problem given by the Pontryagin Minimum Principle, and we find a candidate for the optimal control $\mathbf{u}_*^{[t_0]}$ for this auxiliary problem. We also calculate the corresponding cost. We know that the optimal cost should be $V(x_0, t_0)$ (since if the Bellman's equation is satisfied for all $t \in [t_i, t_f]$, then in particular it is also satisfied on $[t_0, t_f]$!). So we simply check if the candidate expression for $V(x_0, t_0)$ obtained in this manner provides a solution to Bellman's equation. If it does, then subsequently by following the steps in Theorem 5.2, we can find an optimal control. Done!

The best way to see this procedure is to carry it out in a simple example, as done below.

Example 5.4. Consider the optimal control problem

$$\begin{cases} \text{minimize} & \int_0^1 -\mathbf{x}(t)dt \\ \text{subject to} & \mathbf{x}'(t) = \mathbf{x}(t) + \mathbf{u}(t), \quad t \in [0,1], \\ & \mathbf{x}(0) = 0, \\ & \mathbf{u}(t) \in [-1,1], \quad t \in [0,1]. \end{cases}$$

Bellman's equation is given by

$$\frac{\partial V}{\partial t}(x,t) + \min_{u \in [-1,1]} \left(\frac{\partial V}{\partial x}(x,t) \cdot (x+u) - x \right) = 0, \quad t \in [0,1], \ x \in \mathbb{R}. \qquad (5.1)$$

Let $x_0 \in \mathbb{R}$ and $t_0 \in [0,1]$, and consider the auxiliary problem

$$\begin{cases} \text{minimize} & \int_{t_0}^1 -\mathbf{x}(t)dt \\ \text{subject to} & \mathbf{x}'(t) = \mathbf{x}(t) + \mathbf{u}(t), \quad t \in [t_0, 1], \\ & \mathbf{x}(t_0) = x_0, \\ & \mathbf{u}(t) \in [-1,1], \quad t \in [t_0, 1]. \end{cases}$$

Our expectation is that the optimal cost for the auxiliary problem will be $V(x_0, t_0)$, and this should give us a candidate for solving Bellman's equation (5.1) above.

The Hamiltonian for the auxiliary problem is given by

$$H(x, u, \lambda_0, \lambda) = -\lambda_0 x + \lambda(x+u).$$

So

$$\mathbf{u}_*(t) = \arg \min_{u \in [-1,1]} (-\lambda_0 \mathbf{x}_*(t) + \lambda(t)\mathbf{x}_*(t) + \lambda(t)u) = \begin{cases} 1 & \text{if } \lambda(t) < 0, \\ -1 & \text{if } \lambda(t) > 0. \end{cases}$$

Can $\lambda(t)$ be 0 in an interval? Let us look at the co-state equation. We have

$$\lambda'(t) = -\frac{\partial H}{\partial x}(\mathbf{x}_*(t), \mathbf{u}_*(t), \lambda_0, \lambda(t)) = \lambda_0 - \lambda(t), \quad t \in [t_0, 1].$$

Thus

$$\lambda(t) = e^{(t-t_0)(-1)}\lambda(t_0) + \int_{t_0}^t e^{(t-\tau)(-1)}\lambda_0 d\tau = \lambda_0 - Ce^{t_0-t},$$

for some constant C. As $\mathbf{x}(1)$ is free, we have $\lambda(1) = 0$, and so $C = \lambda_0 e^{1-t_0}$. Hence

$$\lambda(t) = \lambda_0(1 - e^{1-t}), \quad t \in [t_0, 1].$$

By (P3), $(\lambda_0, \lambda(t)) = \lambda_0(1, 1 - e^{1-t}) \neq 0$ for all $t \in [t_0, 1]$. So $\lambda_0 \neq 0$. Hence we can take $\lambda_0 = 1$. Thus $\lambda(t) < 0$ for all $t \in [t_0, 1)$, and it follows that $\mathbf{u}_* \equiv 1$ on $[t_0, 1]$. Let us now calculate the corresponding cost. We have

$$\begin{aligned} \mathbf{x}'_*(t) &= \mathbf{x}_*(t) + 1, \ t \in [t_0, 1], \\ \mathbf{x}_*(t_0) &= x_0, \end{aligned}$$

and so

$$\mathbf{x}_*(t) = e^{t-t_0}x_0 + \int_{t_0}^t e^{t-\tau} \cdot 1 d\tau = e^t e^{-t_0}(x_0 + 1) - 1.$$

So the cost of $\mathbf{u}_*$ is

$$\int_{t_0}^1 -\mathbf{x}_*(t)dt = -\int_{t_0}^1 \left(e^t e^{-t_0}(x_0 + 1) - 1\right)dt = (1 - e^{1-t_0})(x_0 + 1) + 1 - t_0.$$

Hence our candidate solution V to the Bellman equation (5.1) is given by

$$V(x, t) = (1 - e^{1-t})(x + 1) + 1 - t, \quad x \in \mathbb{R}, \ t \in [0, 1].$$

Let us first find out if this works. Since

$$V(x, 1) = (1 - e^{1-1})(x + 1) + 1 - 1 = 0, \quad x \in \mathbb{R},$$

at least the boundary condition is satisfied. Moreover,

$$\begin{aligned} \frac{\partial V}{\partial x}(x, t) &= 1 - e^{1-t}, \\ \frac{\partial V}{\partial t}(x, t) &= e^{1-t}(x + 1) - 1. \end{aligned}$$

Thus

$$\frac{\partial V}{\partial t}(x,t) + \min_{u\in[-1,1]} \left(\frac{\partial V}{\partial x}(x,t)\cdot(x+u) - x \right)$$

$$= e^{1-t}(x+1) - 1 + \min_{u\in[-1,1]} \left((1-e^{1-t})x + (1-e^{1-t})u - x \right)$$

$$= e^{1-t}(x+1) - 1 + (1-e^{1-t})x + (1-e^{1-t})\cdot 1 - x \quad (\text{since } 1 - e^{1-t} < 0)$$

$$= 0.$$

So our candidate V is indeed a solution for the Bellman equation (5.1) associated with the original problem. Let us now use Theorem 5.2 to determine the optimal control. We have

$$U(x,t) = \arg\min_{u\in[-1,1]} \left(\frac{\partial V}{\partial x}(x,t)\cdot(x+u) - x \right)$$

$$= \arg\min_{u\in[-1,1]} \left((1-e^{1-t})x + (1-e^{1-t})u - x \right) = 1,$$

since $1 - e^{1-t} < 0$ for all $t \in [0,1]$. Next we find the solution $\mathbf{x}_*$ to

$$\mathbf{x}_*(t) = \mathbf{x}_*(t) + U(\mathbf{x}_*(t),t) = \mathbf{x}_*(t) + 1, \ t \in [t_0,1],$$

$$\mathbf{x}_*(t_i) = x_i,$$

and so

$$\mathbf{x}_*(t) = e^t\cdot 0 + \int_0^t e^{t-\tau}\cdot 1 d\tau = e^t - 1, \quad t \in [0,1].$$

Finally, $\mathbf{u}_*(t) = U(\mathbf{x}_*(t),t) = 1$ on $[0,1]$. By Theorem 5.2, it follows that $\mathbf{u}_*$ is an optimal control. ◇

Exercise 5.5. Consider the optimal control problem

$$\begin{cases} \text{minimize} & \int_0^1 \left((\mathbf{x}(t))^4 + (\mathbf{x}(t)\cdot\mathbf{u}(t))^2 \right) dt \\ \text{subject to} & \mathbf{x}'(t) = \mathbf{u}(t), \ t \in [0,1], \\ & \mathbf{x}(0) = 1. \end{cases}$$

(1) Write Bellman's equation associated with this problem.

(2) Let $\mathbf{p}$ denote the unique solution to the Riccati equation $\mathbf{p}'(t) = 4(\mathbf{p}(t))^2 - 1, \ t \in [0,1]$, $\mathbf{p}(1) = 0$. Verify that V given by $V(x,t) = x^4\mathbf{p}(t)$ for $(x,t) \in \mathbb{R} \times [0,1]$ satisfies Bellman's equation.

(3) Using Bellman's theorem, conclude that $\mathbf{u}_*$ given by $\mathbf{u}_*(t) = -2\mathbf{x}_*(t)\mathbf{p}(t), \ t \in [0,1]$, where $\mathbf{p}$ is the unique solution to the Riccati equation from part (2), and $\mathbf{x}_*$ is the unique solution to $\mathbf{x}_*'(t) = -2\mathbf{p}(t)\mathbf{x}_*(t), \ t \in [0,1]$, with the initial condition $\mathbf{x}_*(0) = 1$.

(4) It can be shown that the Riccati equation from part (2) has the solution given by

$$\mathbf{p}(t) = \frac{e^{-4t} - e^{-4}}{2(e^{-4t} + e^{-4})}, \quad t \in [0,1].$$

Without calculating the optimal control $\mathbf{u}_*$, determine the corresponding cost.

Exercise 5.6. Consider the optimal control problem

$$
\begin{cases}
\text{minimize} & \displaystyle\int_0^1 \mathbf{x}(t)dt \\[2mm]
\text{such that} & \mathbf{x}'(t) = \mathbf{u}(t) \text{ for } t \in [0,1], \\[1mm]
& \mathbf{x}(0) = x_0, \\[1mm]
& -1 \le \mathbf{u}(t) \le 1 \text{ for } t \in [0,1].
\end{cases}
$$

The Bellman equation for this problem is given by

$$
\frac{\partial V}{\partial t}(x,t) + \min_{u \in [-1,1]} \left(\frac{\partial V}{\partial x}(x,t) \cdot u + x \right) = 0, \quad x \in \mathbb{R}, \ t \in [0,1],
$$

with $V(x,1) = 0$ for all $x \in \mathbb{R}$.

(1) Show that $V : \mathbb{R} \times [0,1] \to \mathbb{R}$ defined by

$$
V(x,t) = -\frac{1}{2}(t-1)^2 - x \cdot (t-1), \quad x \in \mathbb{R}, \ t \in [0,1]
$$

satisfies the Bellman equation.

(2) Find an optimal control $\mathbf{u}_*$ and the corresponding state $\mathbf{x}_*$. Also find the cost corresponding to $\mathbf{u}_*$.

(3) For any control input $\mathbf{u}$, the control system

$$
\begin{aligned}
\mathbf{x}'(t) &= \mathbf{u}(t), \quad t \in [0,1], \\
\mathbf{x}(0) &= x_0,
\end{aligned}
$$

has the solution given by

$$
\mathbf{x}(t) = x_0 + \int_0^t \mathbf{u}(\tau)d\tau, \quad t \in [0,1].
$$

So if we substitute this in the cost function for the optimal control problem, we obtain

$$
\int_0^1 \mathbf{x}(t)dt = \int_0^1 \left(x_0 + \int_0^t \mathbf{u}(\tau)d\tau \right) dt = x_0 + \int_0^1 \int_0^t \mathbf{u}(\tau)d\tau dt.
$$

Perform an exchange of the order of integration in the last double integral above, and prove directly with a suitable inequality that the control input found in part (2) is optimal.

Exercise 5.7. Consider the optimization problem from Exercise 4.9 again:

$$
\begin{cases}
\text{maximize} & \displaystyle\int_0^1 \left(1 - (\mathbf{u}(t))^2 \right) \cdot (\mathbf{x}(t))^2 dt \\[2mm]
\text{subject to} & \mathbf{x}'(t) = \mathbf{u}(t) \cdot \mathbf{x}(t), \ t \in [0,1], \\[1mm]
& \mathbf{x}(0) = 1.
\end{cases}
$$

Show, using Bellman's equation, that the unique $\mathbf{u}_*$ obtained in the solution to Exercise 4.9 (using the Pontryagin Minimum Principle) is indeed a maximizer.

Solutions

Solutions to the Exercises from the Introduction

Solution to Exercise 0.1. Suppose that x_* is a minimizer of f. Then for every $x \in S$, $f(x) \geq f(x_*)$, that is, $(-f)(x_*) = -f(x_*) \geq -f(x) = (-f)(x)$. Hence x_* is a maximizer of $-f$.

Now suppose that x_* is a maximizer of $-f$. Then for every $x \in S$, $(-f)(x_*) \geq (-f)(x)$, that is, $-f(x_*) \geq -f(x)$, and so, upon rearranging, $f(x) \geq f(x_*)$. Hence x_* is a minimizer of f.

Solution to Exercise 0.3. We have

$$f(\mathbf{x}_1) = \int_0^T \left(aQ \frac{t}{T} + bQ \frac{1}{T} \right) Q \frac{1}{T} dt = \frac{aQ^2}{2} + \frac{bQ^2}{T}.$$

On the other hand,

$$f(\mathbf{x}_2) = \int_0^T \left(aQ \frac{t^2}{T^2} + bQ \frac{2t}{T^2} \right) Q \frac{2t}{T^2} dt = \frac{aQ^2}{2} + \frac{4}{3} \frac{bQ^2}{T}.$$

Clearly $f(\mathbf{x}_2) > f(\mathbf{x}_1)$, and so the mining operation $\mathbf{x}_1$ is preferred to $\mathbf{x}_2$ because it incurs a lower cost.

Solutions to the Exercises from Chapter 1

Solution to Exercise 1.3.

("If" part.) Suppose that $y_a = y_b = 0$. Then we have:

(S1) If $\mathbf{x}_1, \mathbf{x}_2 \in S$, then $\mathbf{x}_1 + \mathbf{x}_2 \in S$. Indeed, as $\mathbf{x}_1, \mathbf{x}_2 \in C^1[a, b]$, it follows that $\mathbf{x}_1 + \mathbf{x}_2 \in C^1[a, b]$, and

$$\mathbf{x}_1(a) + \mathbf{x}_2(a) = 0 + 0 = 0 = y_a,$$
$$\mathbf{x}_1(b) + \mathbf{x}_2(b) = 0 + 0 = 0 = y_b.$$

(S2) If $\mathbf{x} \in S$ and $\alpha \in \mathbb{R}$, then $\alpha \cdot \mathbf{x} \in S$. Indeed, as $\mathbf{x} \in C^1[a, b]$, and $\alpha \in \mathbb{R}$, it follows that $\alpha \cdot \mathbf{x} \in C^1[a, b]$, and $(\alpha \cdot \mathbf{x})(a) = \alpha 0 = 0 = y_a$, $(\alpha \cdot \mathbf{x})(b) = \alpha 0 = 0 = y_b$.

(S3) $\mathbf{0} \in S$, since $\mathbf{0} \in C^1[a, b]$ and $\mathbf{0}(a) = 0 = y_a = y_b = \mathbf{0}(b)$.

Hence, S is a subspace of a vector space $C^1[a, b]$.

("Only if" part.) Suppose that S is a subspace of $C^1[a, b]$. Let $\mathbf{x} \in S$. Then $2 \cdot \mathbf{x} \in S$. Therefore, $(2 \cdot \mathbf{x})(a) = y_a$, and so $y_a = (2 \cdot \mathbf{x})(a) = 2\mathbf{x}(a) = 2y_a$. Thus $y_a = 0$. Moreover, $(2 \cdot \mathbf{x})(b) = y_b$, and so $y_b = (2 \cdot \mathbf{x})(b) = 2\mathbf{x}(b) = 2y_b$. Hence also $y_b = 0$.

Solution to Exercise 1.4. We prove this by contradiction. Suppose that $C[0, 1]$ has dimension d. Consider functions

$$\mathbf{x}_n(t) = t^n, \quad t \in [0, 1], \quad n = 1, \cdots, d.$$

Since polynomials are continuous, we have $\mathbf{x}_n \in C[0, 1]$ for all $n = 1, \cdots, d$.

First we prove that $\mathbf{x}_n$, $n = 1, \cdots, d$ are linearly independent in $C[0, 1]$. Suppose not. Then there exist $\alpha_n \in \mathbb{R}$, $n = 1, \cdots, d$, not all zeroes, such that

$$\alpha_1 \cdot \mathbf{x}_1 + \cdots + \alpha_d \cdot \mathbf{x}_d = \mathbf{0}.$$

Let $m \in \{1, \cdots, d\}$ be the smallest index such that $\alpha_m \neq 0$. Then for all $t \in [0, 1]$,

$$\alpha_m t^m + \cdots + \alpha_d t^d = 0.$$

In particular, for all $t \in (0, 1]$, we have

$$\alpha_m + \alpha_{m+1} t + \cdots + \alpha_d t^{d-m} = 0.$$

Thus for all $n \in \mathbb{N}$ we have

$$\alpha_m + \frac{\alpha_{m+1}}{n} + \cdots + \frac{\alpha_d}{n^{d-m}} = 0.$$

Passing the limit as $n \to \infty$, we obtain $\alpha_m = 0$, a contradiction. So $\mathbf{x}_n$, $n = 1, \cdots, d$, are linearly independent in $C[0, 1]$.

Next, we demonstrate the contradiction to our assumption that $C[0, 1]$ has dimension d. Indeed, since any independent set of cardinality d in a d-dimensional vector space is a basis for this vector space, we obtain that $\mathbf{x}_n$, $n = 1, \cdots, d$ is a basis in $C[0, 1]$. Since the constant function $\mathbf{1}$ (taking value 1 everywhere on $[0, 1]$) belongs to $C[0, 1]$, there exist $\beta_n \in \mathbb{R}$, $n = 1, \cdots, d$, such that

$$\mathbf{1} = \beta_1 \cdot \mathbf{x}_1 + \cdots + \beta_d \cdot \mathbf{x}_d.$$

In particular, by looking at the value at $t = 0$, we obtain the contradiction that $1 = 0$:

$$1 = \mathbf{1}(0) = (\beta_1 \cdot \mathbf{x}_1 + \cdots + \beta_d \cdot \mathbf{x}_d)(0) = 0.$$

Solution to Exercise 1.10. We have

$$\|t\|_\infty = \max_{t\in[0,1]} |t| = \max_{t\in[0,1]} t = 1,$$

$$\|-t\|_\infty = \max_{t\in[0,1]} |-t| = \max_{t\in[0,1]} t = 1,$$

$$\|t^n\|_\infty = \max_{t\in[0,1]} |t^n| = \max_{t\in[0,1]} t^n = 1,$$

$$\|\sin(2\pi nt)\|_\infty = \max_{t\in[0,1]} |\sin(2\pi nt)| = 1.$$

Solution to Exercise 1.12. From the triangle inequality, it follows that for all $x, y \in X$, we have $\|x\| = \|y + x - y\| \leq \|y\| + \|x - y\|$. Therefore,

$$\text{for all } x, y \in X, \ \|x\| - \|y\| \leq \|x - y\|. \tag{5.2}$$

Interchanging x and y, we also obtain

$$\begin{aligned}
\text{for all } x, y \in X, \ -(\|x\| - \|y\|) = \|y\| - \|x\| &\leq \|y - x\| \\
&= \|(-1)\cdot(x - y)\| = |-1|\|x - y\| \\
&= \|x - y\|. \tag{5.3}
\end{aligned}$$

Combining (5.2) and (5.3), we conclude that $|\|x\| - \|y\|| \leq \|x - y\|$ for all $x, y \in X$.

Solution to Exercise 1.13. No. Indeed, although (N1) is satisfied, neither (N2) nor (N3) are satisfied:

(N1) For all $x \in \mathbb{R}$, $\|x\| = x^2 \geq 0$. If $\|x\| = 0$ for some $x \in \mathbb{R}$, then $x^2 = 0$, and so $x = 0$. Thus (N1) holds.

(N2) We take $x = 1$ and $\alpha = 2$. Then $\|2\cdot 1\| = \|2\| = 2^2 = 4$ whereas $|2|\|1\| = 2\cdot 1^2 = 2$. Since $4 \neq 2$, (N2) does not hold.

(N3) We take $x = y = 1$. Clearly, $\|1 + 1\| = \|2\| = 2^2 = 4 \nleq 2 = 1^2 + 1^2 = \|1\| + \|1\|$, and so (N3) does not hold.

Solution to Exercise 1.14. We verify that (N1), (N2), (N3) are satisfied by $\|\cdot\|_Y$:

(N1) For all $y \in Y$, $\|y\|_Y = \|y\|_X \geq 0$. If $y \in Y$ and $\|y\|_Y = 0$, then $\|y\|_X = 0$, and so $y = \mathbf{0} \in X$. But $\mathbf{0} \in Y$, and so $y = \mathbf{0} \in Y$.

(N2) If $y \in Y$ and $\alpha \in \mathbb{R}$, then $\alpha \cdot y \in Y$ and

$$\|\alpha \cdot y\|_Y = \|\alpha \cdot y\|_X = |\alpha|\|y\|_X = |\alpha|\|y\|_Y.$$

(N3) If $y_1, y_2 \in Y$, then $y_1 + y_2 \in Y$ and

$$\|y_1 + y_2\|_Y = \|y_1 + y_2\|_X \leq \|y_1\|_X + \|y_2\|_X = \|y_1\|_Y + \|y_2\|_Y.$$

Solution to Exercise 1.15. We first consider $p = 1, 2$ and subsequently the case of $\|\cdot\|_\infty$. We verify that (N1), (N2), (N3) hold for $\|\cdot\|_p$ when $p = 1$ or $p = 2$:

(N1) Clearly $\|x\|_p = \sqrt[p]{|x_1|^p + \cdots + |x_n|^p} \geq 0$. Also if $\|x\|_p = 0$ for some $x \in \mathbb{R}^n$, then

$$\sqrt[p]{|x_1|^p + \cdots + |x_n|^p} = 0,$$

and so $|x_1|^p + \cdots + |x_n|^p = 0$, which in turn implies that $x_1 = \cdots = x_n = 0$. Hence $x = \mathbf{0}$.

(N2) For $x \in \mathbb{R}^n$ and $\alpha \in \mathbb{R}$, we have
$$\|\alpha \cdot x\|_p = \sqrt[p]{|\alpha \cdot x_1|^p + \cdots + |\alpha \cdot x_n|^p} = \sqrt[p]{|\alpha|^p(|x_1|^p + \cdots + |x_n|^p)} = |\alpha| \|x\|_p.$$

(N3) Let $x, y \in \mathbb{R}^n$. If $p = 1$, then
$$
\begin{aligned}
\|x + y\|_1 &= |x_1 + y_1| + \cdots + |x_n + y_n| \\
&\leq |x_1| + |y_1| + \cdots + |x_n| + |y_n| \\
&= \|x\|_1 + \|y\|_1.
\end{aligned}
$$

On the other hand, if $p = 2$, then
$$
\begin{aligned}
\|x + y\|_2^2 &= (x_1 + y_1)^2 + \cdots + (x_n + y_n)^2 \\
&= x_1^2 + 2x_1y_1 + y_1^2 + \cdots + x_n^2 + 2x_ny_n + y_n^2 \\
&= (x_1^2 + \cdots + x_n^2) + (y_1^2 + \cdots + y_n^2) + 2(x_1y_1 + \cdots + x_ny_n) \\
&= \|x\|_2^2 + \|y\|_2^2 + 2(x_1y_1 + \cdots + x_ny_n) \\
&\leq \|x\|_2^2 + \|y\|_2^2 + 2\|x\|_2 \cdot \|y\|_2 = (\|x\|_2 + \|y\|_2)^2.
\end{aligned}
$$

As $\|x\|_2, \|y\|_2, \|x+y\|_2$ are all nonnegative, it follows that $\|x+y\|_2 \leq \|x\|_2 + \|y\|_2$.

Now we consider the case of $\| \cdot \|_\infty$.

(N1) Clearly $\|x\|_\infty = \max\{|x_1|, \cdots, |x_n|\} \geq 0$. Also if $\|x\|_\infty = 0$ for some $x \in \mathbb{R}^n$, then
$$\max\{|x_1|, \cdots, |x_n|\} = 0,$$
and so $|x_1| = \cdots = |x_n| = 0$, which in turn implies that $x_1 = \cdots = x_n = 0$. Hence $x = \mathbf{0}$.

(N2) For $x \in \mathbb{R}^n$ and $\alpha \in \mathbb{R}$, we have
$$\|\alpha \cdot x\|_\infty = \max\{|\alpha \cdot x_1|, \cdots, |\alpha \cdot x_n|\} = \max\{|\alpha||x_1|, \cdots, |\alpha||x_n|\} = |\alpha| \|x\|_\infty.$$

(N3) Let $x, y \in \mathbb{R}^n$. Then for each $k = 1, \cdots, n$,
$$
\begin{aligned}
|x_k + y_k| &\leq |x_k| + |y_k| \leq \max\{|x_1|, \cdots, |x_n|\} + \max\{|y_1|, \cdots, |y_n|\} \\
&= \|x\|_\infty + \|y\|_\infty
\end{aligned}
$$
Thus $\|x + y\|_\infty = \max\{|x_1 + y_1|, \cdots, |x_n + y_n|\} \leq \|x\|_\infty + \|y\|_\infty$.

For pictures of the balls $B_1(\mathbf{0}, 1), B_2(\mathbf{0}, 1), B_\infty(\mathbf{0}, 1)$, see Figure 1.

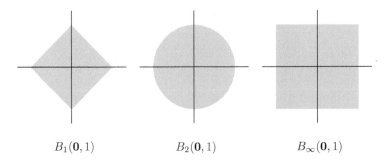

$B_1(\mathbf{0}, 1)$ $B_2(\mathbf{0}, 1)$ $B_\infty(\mathbf{0}, 1)$

Figure 1. Open balls in $\mathbb{R}^2$ in the $\| \cdot \|_1$, $\| \cdot \|_2$ and $\| \cdot \|_\infty$-norms.

Solution to Exercise 1.16.

(1) If $x, y \in B(\mathbf{0}, 1)$, then for $\alpha \in (0, 1)$, we have

$$\|(1-\alpha)x + \alpha y\| \leq \|(1-\alpha)x\| + \|\alpha y\| = |1-\alpha| \|x\| + |\alpha| \|y\|$$
$$< |1-\alpha| \cdot 1 + |\alpha| \cdot 1 = 1 - \alpha + \alpha = 1.$$

Hence if $x, y \in B(\mathbf{0}, 1)$ and $\alpha \in (0, 1)$, then we have that $(1-\alpha)x + \alpha y \in B(\mathbf{0}, 1)$. Therefore $B(\mathbf{0}, 1)$ is convex.

(2) See Figure 2.

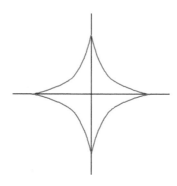

Figure 2. Points $(x_1, x_2) \in \mathbb{R}^2$ satisfying $\sqrt{|x_1|} + \sqrt{|x_2|} = 1$.

(3) From the previous exercise, we visually see that $B_{1/2}(\mathbf{0}, 1)$ is not convex, and we will check this below $B_{1/2}(\mathbf{0}, 1)$. It will then follow from the first part of the exercise that $\|\cdot\|_{1/2}$ is not a norm on $\mathbb{R}^2$. Consider $x = (0, 9/10)$ and $y = (9/10, 0)$. Then since

$$\left(\sqrt{0} + \sqrt{\frac{9}{10}}\right)^2 = \frac{9}{10} < 1,$$

it follows that $x, y \in B_{1/2}(\mathbf{0}, 1)$. On the other hand, with $\alpha := 1/2$, we have

$$(1-\alpha)x + \alpha y = \frac{x+y}{2} = \left(\frac{9}{20}, \frac{9}{20}\right) \notin B_{1/2}(\mathbf{0}, 1)$$

because

$$\left(\sqrt{\frac{9}{20}} + \sqrt{\frac{9}{20}}\right)^2 = 4 \cdot \frac{9}{20} = \frac{9}{5} > 1.$$

Solution to Exercise 1.17. We will verify that (N1), (N2), (N3) hold.

(N1) If $\mathbf{x} \in C[a, b]$, then $|\mathbf{x}(t)| \geq 0$ for all $t \in [a, b]$ and so

$$\|\mathbf{x}\|_1 = \int_a^b |\mathbf{x}(t)| dt \geq 0.$$

Now suppose that $\mathbf{x} \in C[a, b]$ is such that $\|\mathbf{x}\|_1 = 0$. If $\mathbf{x} \neq \mathbf{0}$, then there exists a $t_0 \in [a, b]$ such that $\mathbf{x}(t_0) \neq 0$. Since $\mathbf{x}$ is continuous, there exists a $\delta > 0$ such that for all $t \in I$, where

$$I := \{t \in [a, b] : |t - t_0| < \delta\},$$

there holds that $|\mathbf{x}(t) - \mathbf{x}(t_0)| < |\mathbf{x}(t_0)|/2$, and so $|\mathbf{x}(t)| \geq |\mathbf{x}(t_0)|/2$ for in $t \in I$, by the triangle inequality. Consequently,

$$\|\mathbf{x}\|_1 = \int_a^b |\mathbf{x}(t)|dt = \int_{[a,b]\setminus I} |\mathbf{x}(t)|dt + \int_I |\mathbf{x}(t)|dt \geq 0 + \frac{|\mathbf{x}(t_0)|}{2} 2\delta > 0,$$

a contradiction. Consequently, $\mathbf{x} = \mathbf{0}$.

(N2) For $\mathbf{x} \in C[a,b]$ and $\alpha \in \mathbb{R}$, we have

$$\|\alpha \cdot \mathbf{x}\|_1 = \int_a^b |(\alpha \cdot \mathbf{x})(t)|dt = \int_a^b |\alpha||\mathbf{x}(t)|dt = |\alpha| \int_a^b |\mathbf{x}(t)|dt = |\alpha|\|\mathbf{x}\|_1.$$

(N3) Suppose that $\mathbf{x}, \mathbf{y} \in C[a,b]$. For each $t \in [a,b]$, we have

$$|(\mathbf{x} + \mathbf{y})(t)| = |\mathbf{x}(t) + \mathbf{y}(t)| \leq |\mathbf{x}(t)| + |\mathbf{y}(t)|,$$

and so

$$\begin{aligned}
\|\mathbf{x} + \mathbf{y}\|_1 = \int_a^b |(\mathbf{x} + \mathbf{y})(t)|dt &\leq \int_a^b (|\mathbf{x}(t)| + |\mathbf{y}(t)|)dt \\
&= \int_a^b |\mathbf{x}(t)|dt + \int_a^b |\mathbf{y}(t)|dt = \|\mathbf{x}\|_1 + \|\mathbf{y}\|_1.
\end{aligned}$$

Solution to Exercise 1.18. We will check (N1), (N2), (N3) below.

(N1) For $\mathbf{x} \in C^n[a,b]$, clearly $\|\mathbf{x}\|_{n,\infty} = \|\mathbf{x}\|_\infty + \cdots + \|\mathbf{x}^{(n)}\|_\infty \geq 0 + \cdots + 0 = 0$. Also, if $\mathbf{x} \in C^n[a,b]$ is such that $\|\mathbf{x}\|_{n,\infty} = 0$, then we have $\|\mathbf{x}\|_\infty + \cdots + \|\mathbf{x}^{(n)}\|_\infty = 0$, and since each term in this sum is nonnegative, it follows from here that $\|\mathbf{x}\|_\infty = 0$. Hence $\mathbf{x} = \mathbf{0}$.

(N2) Let $\mathbf{x} \in C^n[a,b]$ and $\alpha \in \mathbb{R}$. Then

$$\begin{aligned}
\|\alpha \cdot \mathbf{x}\|_{n\infty} &= \|\alpha \cdot \mathbf{x}\|_\infty + \|(\alpha \cdot \mathbf{x})'\|_\infty + \cdots + \|(\alpha \cdot \mathbf{x})^{(n)}\|_\infty \\
&= \|\alpha \cdot \mathbf{x}\|_\infty + \|\alpha \cdot (\mathbf{x}')\|_\infty + \cdots + \|\alpha \cdot (\mathbf{x}^{(n)})\|_\infty \\
&= |\alpha|\|\mathbf{x}\|_\infty + |\alpha|\|\mathbf{x}'\|_\infty + \cdots + |\alpha|\|\mathbf{x}^{(n)}\|_\infty \\
&= |\alpha|\Big(\|\mathbf{x}\|_\infty + \|\mathbf{x}'\|_\infty + \cdots + \|\mathbf{x}^{(n)}\|_\infty\Big) = |\alpha|\|\mathbf{x}\|_{n,\infty}.
\end{aligned}$$

(N3) Let $\mathbf{x}, \mathbf{y} \in C^n[a,b]$. By the triangle inequality for $\|\cdot\|_\infty$, we have

$$\|\mathbf{x}^{(k)} + \mathbf{y}^{(k)}\|_\infty \leq \|\mathbf{x}^{(k)}\|_\infty + \|\mathbf{y}^{(k)}\|_\infty, \quad k = 0, 1, \cdots, n.$$

Consequently,

$$\begin{aligned}
\|\mathbf{x} + \mathbf{y}\|_{n,\infty} &= \|\mathbf{x} + \mathbf{y}\|_\infty + \cdots + \|(\mathbf{x} + \mathbf{y})^{(n)}\|_\infty \\
&= \|\mathbf{x} + \mathbf{y}\|_\infty + \cdots + \|\mathbf{x}^{(n)} + \mathbf{y}^{(n)}\|_\infty \\
&\leq \|\mathbf{x}\|_\infty + \|\mathbf{y}\|_\infty + \cdots + \|\mathbf{x}^{(n)}\|_\infty + \|\mathbf{y}^{(n)}\|_\infty \\
&= \|\mathbf{x}\|_{n,\infty} + \|\mathbf{y}\|_{n,\infty}.
\end{aligned}$$

Solution to Exercise 1.22. Let $x_0 \in X$. Given $\epsilon > 0$, set $\delta := \epsilon$. Then for all $x \in X$ satisfying $\|x - x_0\| < \delta$, we have

$$\big| \|x\| - \|x_0\| \big| \le \|x - x_0\| < \delta = \epsilon,$$

(where the first inequality was established in Exercise 1.12).

Solution to Exercise 1.23.

(1) We have

$$f(\mathbf{0}) = \int_0^1 \sqrt{1 + (\mathbf{0}'(t))^2} dt = \int_0^1 \sqrt{1 + 0^2} dt = \int_0^1 1 dt = 1.$$

(As expected, the arc length is simply the length of the line segment $[0,1]$.)

(2) We have $\mathbf{x}'_n(t) = -2\pi\sqrt{n}\sin(2\pi nt)$, $t \in [0,1]$, and so

$$
\begin{aligned}
f(\mathbf{x}_n) &= \int_0^1 \sqrt{1 + (\mathbf{x}'_n(t))^2} dt = \int_0^1 \sqrt{1 + 4\pi^2 n(\sin(2\pi nt))^2} dt \\
&\ge n \int_{\frac{1}{8n}}^{\frac{3}{8n}} \sqrt{1 + 4\pi^2 n(\sin(2\pi nt))^2} dt \\
&\ge n \int_{\frac{1}{8n}}^{\frac{3}{8n}} \sqrt{1 + 2\pi^2 n} \, dt = n \cdot \left(\frac{3}{8n} - \frac{1}{8n} \right) \sqrt{1 + 2\pi^2 n} = \frac{\sqrt{1 + 2\pi^2 n}}{4}.
\end{aligned}
$$

(3) Suppose that f is continuous at $\mathbf{0}$. Then with $\epsilon := 1 > 0$, there is a $\delta > 0$ such that whenever $\mathbf{x} \in C^1[0,1]$ and $\|\mathbf{x} - \mathbf{0}\|_\infty < \delta$, we must have $|f(\mathbf{x}) - f(\mathbf{0})| < 1$. We have

$$\|\mathbf{x}_n - \mathbf{0}\|_\infty = \max_{x \in [0,1]} \frac{|\cos(2\pi nt)|}{\sqrt{n}} = \frac{1}{\sqrt{n}} < \delta$$

for all $n > 1/\delta^2$, and so for such n there must hold that $|f(\mathbf{x}_n) - f(\mathbf{0})| = |f(\mathbf{x}_n) - 1| < 1$. Hence for all $n > n > 1/\delta^2$, there holds

$$\frac{\sqrt{1 + 2\pi^2 n}}{4} \le |f(\mathbf{x}_n)| \le |f(\mathbf{x}_n) - 1| + 1 < 1 + 1 = 2,$$

which is a contradiction. So f is not continuous at $\mathbf{0}$.

Let $\mathbf{x}_0, \mathbf{x} \in C^1[a,b]$. Using the triangle inequality in $(\mathbb{R}^2, \|\cdot\|_2)$, we obtain

$$\left| \sqrt{1^2 + (\mathbf{x}'(t))^2} - \sqrt{1^2 + (\mathbf{x}'_0(t))^2} \right| \le \sqrt{(1-1)^2 + (\mathbf{x}'(t) - \mathbf{x}'_0(t))^2} = |\mathbf{x}'(t) - \mathbf{x}'_0(t)|,$$

and so

$$
\begin{aligned}
|f(\mathbf{x}) - f(\mathbf{x}_0)| &= \left| \int_0^1 \left(\sqrt{1^2 + (\mathbf{x}'(t))^2} - \sqrt{1^2 + (\mathbf{x}'_0(t))^2} \right) dt \right| \\
&\le \int_0^1 \left| \sqrt{1^2 + (\mathbf{x}'(t))^2} - \sqrt{1^2 + (\mathbf{x}'_0(t))^2} \right| dt \\
&\le \int_0^1 |\mathbf{x}'(t) - \mathbf{x}'_0(t)| dt \le \int_0^1 \|\mathbf{x}' - \mathbf{x}'_0\|_\infty dt = \|\mathbf{x}' - \mathbf{x}'_0\|_\infty \le \|\mathbf{x} - \mathbf{x}_0\|_{1,\infty}.
\end{aligned}
$$

Thus given $\epsilon > 0$, if we set $\delta := \epsilon$, then we have for all $\mathbf{x} \in C^1[0,1]$ satisfying $\|\mathbf{x} - \mathbf{x}_0\|_{1,\infty} < \delta$ that $|f(\mathbf{x}) - f(\mathbf{x}_0)| \le \|\mathbf{x} - \mathbf{x}_0\|_{1,\infty} < \delta = \epsilon$. So f is continuous at $\mathbf{x}_0$. As the choice of $\mathbf{x}_0$ was arbitrary, f is continuous.

Solution to Exercise 1.28. f is a linear transformation:

(L1) For all $\mathbf{x}_1, \mathbf{x}_2 \in C[0,1]$, we have

$$
\begin{aligned}
f(\mathbf{x}_1 + \mathbf{x}_2) &= \int_0^1 (\mathbf{x}_1 + \mathbf{x}_2)(t^2)dt = \int_0^1 \Big(\mathbf{x}_1(t^2) + \mathbf{x}_2(t^2)\Big)dt \\
&= \int_0^1 \mathbf{x}_1(t^2)dt + \int_0^1 \mathbf{x}_2(t^2)dt = f(\mathbf{x}_1) + f(\mathbf{x}_2).
\end{aligned}
$$

(L2) For all $\mathbf{x} \in C[0,1]$ and all $\alpha \in \mathbb{R}$,

$$
f(\alpha \cdot \mathbf{x}) = \int_0^1 (\alpha \cdot \mathbf{x})(t^2)dt = \int_0^1 (\alpha \mathbf{x}(t^2))dt = \alpha \int_0^1 \mathbf{x}(t^2)dt = \alpha f(\mathbf{x}).
$$

Solution to Exercise 1.32. We have for $\mathbf{x} \in C[0,1]$ that

$$
|f(\mathbf{x})| = \left| \int_0^1 \mathbf{x}(t^2)dt \right| \le \int_0^1 |\mathbf{x}(t^2)|dt \le \int_0^1 \|\mathbf{x}\|_\infty dt = \|\mathbf{x}\|_\infty \cdot \underbrace{1}_{=:M},
$$

and so it follows from Theorem 1.29 that f is continuous.

Solution to Exercise 1.37. We had seen earlier in Exercises 1.28 and 1.32 that f is a continuous linear transformation. Thus it follows that $f'(\mathbf{x}_0) = f$ for all $\mathbf{x}_0$, and in particular also for $\mathbf{x}_0 = \mathbf{0}$.

Solution to Exercise 1.39. Suppose that $f'(x_0) = L$ and, since L is a continuous linear transformation, let $M > 0$ be such that

$$
\|Lh\| \le M\|h\|, \quad h \in X.
$$

Let $\epsilon > 0$. Then there exists a $\delta_1 > 0$ such that whenever $x \in X$ satisfies $0 < \|x - x_0\| < \delta_1$, we have

$$
\frac{\|f(x) - f(x_0) - L(x - x_0)\|}{\|x - x_0\|} < \epsilon.
$$

From here it follows that for all $x \in X$ satisfying $\|x - x_0\| < \delta_1$, we have

$$
\|f(x) - f(x_0) - L(x - x_0)\| \le \epsilon \|x - x_0\|.
$$

Now let $\delta := \min\left\{ \delta_1, \dfrac{\epsilon}{\epsilon + M} \right\}$.

Then for all $x \in X$ satisfying $\|x - x_0\| < \delta$, we have, using the triangle inequality, that

$$
\begin{aligned}
\|f(x) - f(x_0)\| &\le \|f(x) - f(x_0) - L(x - x_0)\| + \|L(x - x_0)\| \\
&\le \epsilon\|x - x_0\| + M\|x - x_0\| = (\epsilon + M)\|x - x_0\| \\
&< (\epsilon + M) \cdot \frac{\epsilon}{\epsilon + M} = \epsilon.
\end{aligned}
$$

Hence f is continuous at x_0.

Solution to Exercise 1.40. We have for $\mathbf{x} \in C^1[0,1]$ that

$$f(\mathbf{x}) - f(\mathbf{x}_0) = (\mathbf{x}'(1))^2 - (\mathbf{x}_0'(1))^2 = (\mathbf{x}'(1) + \mathbf{x}_0'(1))(\mathbf{x} - \mathbf{x}_0)'(1)$$
$$\approx 2\mathbf{x}_0'(1)(\mathbf{x} - \mathbf{x}_0)'(1) = L(\mathbf{x} - \mathbf{x}_0),$$

where $L : C^1[0,1] \to \mathbb{R}$ is the map given by

$$L\mathbf{h} = 2\mathbf{x}_0'(1)\mathbf{h}'(1), \quad \mathbf{h} \in C^1[0,1].$$

So we make the guess that $f'(\mathbf{x}_0) = L$. Let us first check that L is a continuous linear transformation. L is linear because:

(L1) For all $\mathbf{h}_1, \mathbf{h}_2 \in C^1[0,1]$, we have

$$L(\mathbf{h}_1 + \mathbf{h}_2) = 2\mathbf{x}_0'(1)(\mathbf{h}_1 + \mathbf{h}_2)'(1) = 2\mathbf{x}_0'(1)\mathbf{h}_1'(1) + 2\mathbf{x}_0'(1)\mathbf{h}_2'(1)$$
$$= L(\mathbf{h}_1) + L(\mathbf{h}_2).$$

(L2) For all $\mathbf{h} \in C^1[0,1]$ and $\alpha \in \mathbb{R}$, we have

$$L(\alpha \cdot \mathbf{h}) = 2\mathbf{x}_0'(1)(\alpha \cdot \mathbf{h})'(1) = 2\mathbf{x}_0'(1)\alpha\mathbf{h}'(1) = \alpha L(\mathbf{h}).$$

Also, L is continuous since for all $\mathbf{h} \in C^1[0,1]$, we have

$$|L(\mathbf{h})| = |2\mathbf{x}_0'(1)\mathbf{h}'(1)| = 2|\mathbf{x}_0'(1)||\mathbf{h}'(1)| \leq 2|\mathbf{x}_0'(1)|\|\mathbf{h}'\|_\infty \leq \underbrace{2|\mathbf{x}_0'(1)|}_{=:M} \|\mathbf{h}\|_{1,\infty}.$$

So L is a continuous linear transformation. Moreover, we have for all $\mathbf{x} \in C^1[0,1]$ that

$$f(\mathbf{x}) - f(\mathbf{x}_0) - L(\mathbf{x} - \mathbf{x}_0) = (\mathbf{x}'(1))^2 - (\mathbf{x}_0'(1))^2 - 2\mathbf{x}_0'(1)(\mathbf{x} - \mathbf{x}_0)'(1) = ((\mathbf{x} - \mathbf{x}_0)'(1))^2,$$

so that $|f(\mathbf{x}) - f(\mathbf{x}_0) - L(\mathbf{x} - \mathbf{x}_0)| = |(\mathbf{x} - \mathbf{x}_0)'(1)|^2 \leq \|(\mathbf{x} - \mathbf{x}_0)'\|_\infty^2 \leq \|\mathbf{x} - \mathbf{x}_0\|_{1,\infty}^2$. Given $\epsilon > 0$, set $\delta = \epsilon$. Then for all $\mathbf{x} \in C^1[0,1]$ satisfying $0 < \|\mathbf{x} - \mathbf{x}_0\|_{1,\infty} < \delta$, we have

$$\frac{|f(\mathbf{x}) - f(\mathbf{x}_0) - L(\mathbf{x} - \mathbf{x}_0)|}{\|\mathbf{x} - \mathbf{x}_0\|_{1,\infty}} \leq \frac{\|\mathbf{x} - \mathbf{x}_0\|_{1,\infty}^2}{\|\mathbf{x} - \mathbf{x}_0\|_{1,\infty}} = \|\mathbf{x} - \mathbf{x}_0\|_{1,\infty} < \delta = \epsilon.$$

Solution to Exercise 1.41. Given $\epsilon > 0$, let $\epsilon' > 0$ be such that $\epsilon'\|x_2 - x_1\| < \epsilon$. Let $\delta' > 0$ such that whenever $0 < \|x - \gamma(t_0)\| < \delta'$, we have

$$\frac{|f(x) - f(\gamma(t_0)) - f'(\gamma(t_0))(x - \gamma(t_0))|}{\|x - \gamma(t_0)\|} < \epsilon'.$$

Let $\delta > 0$ be such that $\delta\|x_2 - x_1\| < \delta'$. Then for all $t \in \mathbb{R}$ satisfying $0 < |t - t_0| < \delta$,

$$\gamma(t) - \gamma(t_0) = (1-t)x_1 + tx_2 - (1-t_0)x_1 - t_0x_2 = (t_0 - t)x_1 + (t - t_0)x_2 = (t - t_0)(x_2 - x_1).$$

and so $\|\gamma(t) - \gamma(t_0)\| = |t - t_0|\|x_2 - x_1\| \leq \delta\|x_2 - x_1\| < \delta'$. Thus for all $t \in \mathbb{R}$ satisfying $0 < |t - t_0| < \delta$, we have

$$\left| \frac{f(\gamma(t)) - f(\gamma(t_0))}{t - t_0} - f'(\gamma(t_0))(x_2 - x_1) \right|$$
$$= \frac{|f(\gamma(t)) - f(\gamma(t_0)) - f'(\gamma(t_0))((t - t_0)(x_2 - x_1))|}{|t - t_0|\|x_2 - x_1\|}\|x_2 - x_1\|$$
$$= \frac{|f(\gamma(t)) - f(\gamma(t_0)) - f'(\gamma(t_0))(\gamma(t) - \gamma(t_0))|}{\|\gamma(t) - \gamma(t_0)\|}\|x_2 - x_1\|$$
$$< \epsilon'\|x_2 - x_1\| < \epsilon.$$

Thus $f \circ \gamma$ is differentiable at t_0 and $\dfrac{d}{dt}(f \circ \gamma)(t_0) = f'(\gamma(t_0))(x_2 - x_1)$.

Let $x_1, x_2 \in X$ be such that $g(x_1) \neq g(x_2)$. With γ the same as above, we have for *all* $t \in \mathbb{R}$ that

$$\frac{d}{dt}(g \circ \gamma)(t) = g'(\gamma(t))(x_2 - x_1) = \mathbf{0}(x_2 - x_1) = 0.$$

By the Mean Value Theorem, it follows that $g \circ \gamma$ is constant. In particular,

$$(g \circ \gamma)(1) = g(x_2) = g(x_1) = (g \circ \gamma)(0),$$

a contradiction. So our assumption that there exist $x_1, x_2 \in X$ satisfying $g(x_1) \neq g(x_2)$ is false. Consequently, g is constant.

Solution to Exercise 1.44. Suppose that $f'(\mathbf{x}_0) = \mathbf{0}$. Then for every $\mathbf{h} \in C[a,b]$,

$$2 \int_a^b \mathbf{x}_0(t)\mathbf{h}(t)dt = 0.$$

In particular, if we set $\mathbf{h} = \mathbf{x}_0$, we obtain

$$\int_a^b (\mathbf{x}_0(t))^2 dt = 0,$$

which implies that $\mathbf{x}_0 = \mathbf{0} \in C[a,b]$. Vice versa, if $\mathbf{x}_0 = \mathbf{0}$, then

$$(f'(\mathbf{x}_0))(\mathbf{h}) = 2 \int_a^b \mathbf{x}_0(t)\mathbf{h}(t)dt = 2 \int_a^b \mathbf{0}(t)\mathbf{h}(t)dt = 2 \int_a^b \mathbf{0} \cdot \mathbf{h}(t)dt = 2 \int_a^b 0 dt = 0$$

for all $\mathbf{h} \in C[a,b]$, that is, $f'(\mathbf{0}) = \mathbf{0}$. Consequently,

$$f'(\mathbf{x}_0) = \mathbf{0} \quad \Leftrightarrow \quad \mathbf{x}_0 = \mathbf{0}.$$

So we see that if $\mathbf{x}_*$ is a minimizer, then $f'(\mathbf{x}_*) = \mathbf{0}$, and so from the above $\mathbf{x}_* = \mathbf{0}$. We remark that $\mathbf{0}$ is easily seen to be the minimizer because

$$f(\mathbf{x}) = \int_a^b \underbrace{(\mathbf{x}(t))^2}_{\geq 0} dt \geq 0 = f(\mathbf{0}) \text{ for all } \mathbf{x} \in C[a,b].$$

Solution to Exercise 1.46. If $\mathbf{x}_1, \mathbf{x}_2 \in S$, and $\alpha \in (0,1)$, then $\mathbf{x}_1, \mathbf{x}_2 \in C^1[a,b]$, and so clearly $(1-\alpha)\mathbf{x}_1 + \alpha\mathbf{x}_2 \in C^1[a,b]$. Moreover, because we have $\mathbf{x}_1(a) = \mathbf{x}_2(a) = y_a$ and $\mathbf{x}_1(b) = \mathbf{x}_2(b) = y_b$, we also have that

$$((1-\alpha)\mathbf{x}_1 + \alpha\mathbf{x}_2)(a) = (1-\alpha)\mathbf{x}_1(a) + \alpha\mathbf{x}_2(a) = (1-\alpha)y_a + \alpha y_a = y_a,$$
$$((1-\alpha)\mathbf{x}_1 + \alpha\mathbf{x}_2)(b) = (1-\alpha)\mathbf{x}_1(b) + \alpha\mathbf{x}_2(b) = (1-\alpha)y_b + \alpha y_b = y_b.$$

Thus $(1-\alpha)\mathbf{x}_1 + \alpha\mathbf{x}_2 \in S$. Consequently, S is convex.

Solution to Exercise 1.47. For $\mathbf{x}_1, \mathbf{x}_2 \in X$ and $\alpha \in (0,1)$ we have by the triangle inequality that

$$\|(1-\alpha)\mathbf{x}_1 + \alpha\mathbf{x}_2\| \leq \|(1-\alpha)\mathbf{x}_1\| + \|\alpha\mathbf{x}_2\| = |1-\alpha|\|\mathbf{x}_1\| + |\alpha|\|\mathbf{x}_2\| = (1-\alpha)\|\mathbf{x}_1\| + \alpha\|\mathbf{x}_2\|.$$

Thus $\|\cdot\|$ is convex.

Solution to Exercise 1.48.

("If" part) Let $x_1, x_2 \in C$ and $\alpha \in (0,1)$. Then we have that $(x_1, f(x_1)) \in U(f)$ and $(x_2, f(x_2)) \in U(f)$. Since $U(f)$ is convex,

$$(1-\alpha) \cdot (x_1, f(x_1)) + \alpha \cdot (x_2, f(x_2)) = (\underbrace{(1-\alpha) \cdot x_1 + \alpha \cdot x_2}_{=:x \in C}, \underbrace{(1-\alpha)f(x_1) + \alpha f(x_2)}_{=:y}) \in U(f).$$

Consequently, $(1-\alpha)f(x_1) + \alpha f(x_2) = y \geq f(x) = f((1-\alpha) \cdot x_1 + \alpha \cdot x_2)$. Hence f is convex.

("Only if" part) Let $(x_1, y_1), (x_2, y_2) \in U(f)$ and $\alpha \in (0,1)$. Then we know that $y_1 \geq f(x_1)$ and $y_2 \geq f(x_2)$ and so we also have that

$$(1-\alpha)y_1 + \alpha y_2 \geq (1-\alpha)f(x_1) + \alpha f(x_2) \geq f((1-\alpha) \cdot x_1 + \alpha \cdot x_2).$$

Consequently,

$$((1-\alpha) \cdot x_1 + \alpha \cdot x_2, (1-\alpha)y_1 + \alpha y_2) \in U(f),$$

that is, $(1-\alpha) \cdot (x_1, f(x_1)) + \alpha \cdot (x_2, f(x_2)) \in U(f)$. So $U(f)$ is convex.

Solution to Exercise 1.49. We prove this using induction on n. The result is trivially true when $n = 1$, and in fact we have equality in this case. Suppose the inequality has been established for some $n \in \mathbb{N}$. If $x_1, \cdots, x_n, x_{n+1}$ are $n+1$ vectors, then we have with $\alpha := \frac{1}{n+1} \in (0,1)$ that

$$
\begin{aligned}
f\left(\frac{1}{n+1}(x_1 + \cdots + x_n + x_{n+1})\right) &= f\left(\frac{n}{n+1} \cdot \frac{1}{n}(x_1 + \cdots + x_n) + \frac{1}{n+1} \cdot x_{n+1}\right) \\
&= f\left(\left(1 - \frac{1}{n+1}\right) \cdot \frac{1}{n}(x_1 + \cdots + x_n) + \frac{1}{n+1}x_{n+1}\right) \\
&= f\left((1-\alpha) \cdot \frac{1}{n}(x_1 + \cdots + x_n) + \alpha \cdot x_{n+1}\right) \\
&\leq (1-\alpha) \cdot f\left(\frac{1}{n}(x_1 + \cdots + x_n)\right) + \alpha \cdot f(x_{n+1}) \\
&\leq (1-\alpha) \cdot \frac{f(x_1) + \cdots + f(x_n)}{n} + \alpha \cdot f(x_{n+1}) \\
&= \frac{n}{n+1} \cdot \frac{f(x_1) + \cdots + f(x_n)}{n} + \frac{1}{n+1} \cdot f(x_{n+1}) \\
&= \frac{f(x_1) + \cdots + f(x_n) + f(x_{n+1})}{n+1},
\end{aligned}
$$

and so the claim follows for all n.

Solution to Exercise 1.52. We have for all $x \in \mathbb{R}$

$$
\begin{aligned}
f'(x) &= \frac{2x}{2\sqrt{1+x^2}} = \frac{x}{\sqrt{1+x^2}}, \\
f''(x) &= \frac{1}{\sqrt{1+x^2}} + x \cdot \left(-\frac{1}{2}\right) \cdot \frac{2x}{\sqrt{(1+x^2)^3}} = \frac{1+x^2-x^2}{\sqrt{(1+x^2)^3}} = \frac{1}{\sqrt{(1+x^2)^3}} > 0.
\end{aligned}
$$

Thus f is convex.

(Alternately, one could note that $(x,y) \mapsto \|(x,y)\|_2 = \sqrt{x^2 + y^2}$ is a norm on $\mathbb{R}^2$, and so it is convex. Now fixing $y = 1$, and keeping x variable, we get convexity of $x \mapsto \sqrt{1+x^2}$.)

Solution to Exercise 1.54. For $\mathbf{x}_1, \mathbf{x}_2 \in C^1[0,1]$ and $\alpha \in (0,1)$, we have, using the convexity of $\xi \mapsto \sqrt{1+\xi^2} : \mathbb{R} \to \mathbb{R}$, that

$$
\begin{aligned}
f((1-\alpha)\mathbf{x}_1 + \alpha\mathbf{x}_2) &= \int_0^1 \sqrt{1 + ((1-\alpha)\mathbf{x}_1'(t) + \alpha\mathbf{x}_2'(t))^2}\, dt \\
&\leq \int_0^1 \left((1-\alpha)\sqrt{1+(\mathbf{x}_1'(t))^2} + \alpha\sqrt{1+(\mathbf{x}_2'(t))^2} \right) dt \\
&= (1-\alpha)\int_0^1 \sqrt{1+(\mathbf{x}_1'(t))^2}\,dt + \alpha\int_0^1 \sqrt{1+(\mathbf{x}_2'(t))^2}\,dt \\
&= (1-\alpha)f(\mathbf{x}_1) + \alpha f(\mathbf{x}_2).
\end{aligned}
$$

So f is convex.

Solution to Exercise 1.59. That C is convex is obvious, since the points in C lie on the straight line described by $\xi = \eta$ in $\mathbb{R}^2$, and so clearly the line segment joining two points on this line is contained inside the line. (If $\xi_1 = \eta_1$ and $\xi_2 = \eta_2$ and $\alpha \in (0,1)$, then also $(1-\alpha)\xi_1 + \alpha\xi_2 = (1-\alpha)\eta_1 + \alpha\eta_2$.)

For $u := (a, a), v := (b, b) \in C$ and $\alpha \in (0,1)$, we have

$$
f((1-\alpha)u + \alpha v) = ((1-\alpha)a + \alpha b)^2 \leq (1-\alpha)a^2 + \alpha b^2 = (1-\alpha)f(u) + \alpha f(v).
$$

So f is convex.

The Hessian of f at a point is given by

$$
H_f(\xi, \eta) = \begin{bmatrix} 0 & 1 \\ 1 & 0 \end{bmatrix},
$$

and with $v := (1, -1)$, we have $v^\top H_f(\xi, \eta)v = -2 < 0$, showing that for each $(\xi, \eta) \in C$, $H_f(\xi, \eta)$ is not positive semidefinite.

Solution to Exercise 1.61.

(If) Suppose that $\mathbf{x}_0(t) = 0$ for all $t \in [0, 1]$. Then we have that for all $\mathbf{h} \in C[0, 1]$,

$$
(f'(\mathbf{x}_0))(\mathbf{h}) = 2\int_0^1 \mathbf{x}_0(t)\mathbf{h}(t)dt = 2\int_0^1 0 \cdot \mathbf{h}(t)dt = 2\int_0^1 0\,dt = 2 \cdot 0 = 0,
$$

and so $f'(\mathbf{x}_0) = \mathbf{0}$.

(Only if) Now suppose that $f'(\mathbf{x}_0) = \mathbf{0}$. Thus for every $\mathbf{h} \in C[0, 1]$, we have

$$
(f'(\mathbf{x}_0))(\mathbf{h}) = 2\int_0^1 \mathbf{x}_0(t)\mathbf{h}(t)dt = 0.
$$

In particular, taking $\mathbf{h} := \mathbf{x}_0 \in C[0, 1]$, we obtain from here that

$$
2\int_0^1 \mathbf{x}_0(t)\mathbf{x}_0(t)dt = 0,
$$

and so

$$
\int_0^1 (\mathbf{x}_0(t))^2 dt = 0.
$$

As $\mathbf{x}_0$ is continuous on $[0, 1]$, it follows that for all $t \in [0, 1]$, $\mathbf{x}_0(t) = 0$.

By the necessary condition for $\mathbf{x}_0$ to be a minimizer, we have that $f'(\mathbf{x}_0) = \mathbf{0}$ and so $\mathbf{x}_0$ must be the zero function $\mathbf{0}$ on $[0, 1]$. Furthermore, as f is convex and $f'(\mathbf{0}) = \mathbf{0}$, it follows that the zero function is a minimizer. Consequently, there exists a unique solution to the optimization problem, namely the zero function $\mathbf{0} \in C[0, 1]$. This conclusion is of course also obvious from the inequality that for all $\mathbf{x} \in C[0, 1]$,

$$f(\mathbf{x}) = \int_0^1 \underbrace{(\mathbf{x}(t))^2}_{\geq 0}\, dt \geq \int_0^1 0\, dt = 0 = f(\mathbf{0}).$$

Solutions to the Exercises from Chapter 2

Solution to Exercise 2.3. With $F(\xi, \eta, \tau) := \tau^3 \eta^2$, we have

$$f(\mathbf{x}) := \int_1^2 t^3 (\mathbf{x}'(t))^2 dt = \int_1^2 F(\mathbf{x}(t), \mathbf{x}'(t), t) dt,$$

for $\mathbf{x} \in S := \{\mathbf{x} \in C^1[1, 2] : \mathbf{x}(1) = 5, \ \mathbf{x}(2) = 2\}$. We have

$$\frac{\partial F}{\partial \xi} = 0,$$

$$\frac{\partial F}{\partial \eta} = 2\tau^3 \eta.$$

The Euler-Lagrange equation is

$$0 - \frac{d}{dt}(2t^3 \mathbf{x}'_*(t)) = 0, \quad t \in [1, 2],$$

and so by integrating, $2t^3 \mathbf{x}'_*(t) = C$ for $t \in [1, 2]$. As $t \neq 0$, it follows that

$$\mathbf{x}'_*(t) = \frac{C}{2t^3}, \quad t \in [1, 2],$$

and so $\mathbf{x}_*(t) = -\dfrac{C}{t^2} + D$, $t \in [1, 2]$. As $\mathbf{x}_*(1) = 5$ and $\mathbf{x}_*(2) = 2$, we obtain

$$-C + D = 5,$$
$$-C/4 + D = 2.$$

Hence $-3C/4 = 3$, so that $C = -4$, and then $D = 5 + C = 5 - 4 = 1$. So the only solution of the Euler-Lagrange equation satisfying the endpoint conditions is

$$\mathbf{x}_*(t) = \frac{4}{t^2} + 1, \quad t \in [1, 2].$$

Also, f is convex, because if $\mathbf{x}_1, \mathbf{x}_2 \in S$ and $\alpha \in (0, 1)$, then using the convexity of $\eta \mapsto \eta^2 : \mathbb{R} \to \mathbb{R}$, we obtain

$$
\begin{aligned}
f((1 - \alpha)\mathbf{x}_1 + \alpha \mathbf{x}_2) &= \int_1^2 t^3 ((1 - \alpha)\mathbf{x}'_1(t) + \alpha \mathbf{x}'_2(t))^2 dt \\
&\leq \int_1^2 t^3 ((1 - \alpha)(\mathbf{x}_1(t))^2 + \alpha(\mathbf{x}_2(t))^2) dt \\
&= (1 - \alpha) \int_1^2 t^3 (\mathbf{x}_1(t))^2 dt + \alpha \int_1^2 t^3 (\mathbf{x}_2(t))^2 dt \\
&= (1 - \alpha) f(\mathbf{x}_1) + \alpha f(\mathbf{x}_2).
\end{aligned}
$$

So it follows that $\mathbf{x}_* \in S$ is indeed a minimizer of $f : S \to \mathbb{R}$, and the above shows that in fact it is the only one.

Solution to Exercise 2.4. Let $S := \{\mathbf{x} \in C^1[0,1] : \mathbf{x}(0) = 0, \, \mathbf{x}(1) = 1\}$. The length of a curve $\mathbf{x} \in S$ is given by

$$f(\mathbf{x}) := \int_0^1 \sqrt{1 + (\mathbf{x}'(t))^2}\, dt = \int_0^1 F(\mathbf{x}(t), \mathbf{x}'(t), t)\, dt,$$

where $F(\xi, \eta, \tau) = \sqrt{1 + \eta^2}$. Then

$$\frac{\partial F}{\partial \xi} = 0,$$

$$\frac{\partial F}{\partial \eta} = \frac{\eta}{\sqrt{1 + \eta^2}}.$$

The Euler-Lagrange equation is

$$0 - \frac{d}{dt}\left(\frac{\mathbf{x}'_*(t)}{\sqrt{1 + (\mathbf{x}'_*(t))^2}}\right) = 0, \quad t \in [0,1],$$

and so upon integrating, we obtain $\dfrac{\mathbf{x}'_*(t)}{\sqrt{1 + (\mathbf{x}'_*(t))^2}} = C$ on $[0,1]$ for some constant C. Thus

$$(\mathbf{x}'_*(t))^2 = \frac{C^2}{1 - C^2} =: A, \quad t \in [0,1].$$

So $A \geq 0$. But if $A = 0$, then $\mathbf{x}_*$ is constant, which is impossible, since we must have $\mathbf{x}_*(0) = 0$ and $\mathbf{x}_*(1) = 1$. Thus $A > 0$, which implies that $\mathbf{x}'_*(t) = \pm\sqrt{A}$ for each $t \in [0,1]$. As $\mathbf{x}'_*$ is continuous, we can conclude that $\mathbf{x}'_*$ must be either everywhere equal to $\sqrt{A}$, or everywhere equal to $-\sqrt{A}$. In either case, $\mathbf{x}'_*$ is constant, and so $\mathbf{x}_*$ is given by

$$\mathbf{x}_*(t) = \alpha t + \beta, \quad t \in [0,1].$$

As $\mathbf{x}_*(0) = 0$ and $\mathbf{x}_*(1) = 1$, it follows that $\alpha \cdot 0 + \beta = 0$ and $\alpha \cdot 1 + \beta = 1$, so that $\alpha = 1$ and $\beta = 0$. Consequently, $\mathbf{x}_*(t) = t$, $t \in [0,1]$.

That this is indeed a minimizer can be concluded either by noticing that F (see Exercise 1.52) and hence also f is convex, or by using the inequality that for $a, b \in \mathbb{R}$,

$$\sqrt{a^2 + b^2} \geq \frac{a+b}{\sqrt{2}}$$

(which is just a rearrangement of the fact that $(a - b)^2 \geq 0$). Indeed, we have for $\mathbf{x} \in S$ that

$$f(\mathbf{x}) = \int_0^1 \sqrt{1 + (\mathbf{x}'(t))^2}\, dt \;\geq\; \int_0^1 \frac{1 + \mathbf{x}'(t)}{\sqrt{2}}\, dt = \frac{1 + \mathbf{x}(1) - \mathbf{x}(0)}{\sqrt{2}} = \frac{1 + 1 - 0}{\sqrt{2}} = \sqrt{2}$$

$$= \int_0^1 \sqrt{1 + (1)^2}\, dt = \int_0^1 \sqrt{1 + (\mathbf{x}'_*(t))^2}\, dt = f(\mathbf{x}_*).$$

Hence $\mathbf{x}_*$ is the minimizer. (This is of course expected geometrically, since the straight line is the curve of shortest length between two points in the plane.)

Solution to Exercise 2.5. With $F(\xi, \eta, \tau) := \xi^3$, we have

$$f(\mathbf{x}) = \int_0^1 (\mathbf{x}(t))^3 dt = \int_0^1 F(\mathbf{x}(t), \mathbf{x}'(t), t) dt, \quad \mathbf{x} \in S.$$

We have

$$\frac{\partial F}{\partial \xi} = 3\xi^2,$$

$$\frac{\partial F}{\partial \eta} = 0.$$

The Euler-Lagrange equation is

$$3(\mathbf{x}_*(t))^2 - \frac{d}{dt}0 = 0, \quad t \in [0, 1],$$

and so $\mathbf{x}_* = \mathbf{0}$. So the only solution of the Euler-Lagrange equation is $\mathbf{0} \in S$. However, this $\mathbf{x}_*$ is not a minimizer since with $\mathbf{x} := t(t - 1) \in S$ we have

$$f(\mathbf{x}) = -\int_0^1 t^3 (1 - t)^3 dt < 0 = f(\mathbf{0}).$$

Solution to Exercise 2.6.

(1) With $F(\xi, \eta, \tau) = \eta$, we have that

$$f(\mathbf{x}) = \int_0^1 \mathbf{x}'(t) dt = \int_0^1 F(\mathbf{x}(t), \mathbf{x}'(t), t) dt.$$

We have

$$\frac{\partial F}{\partial \xi} = 0,$$

$$\frac{\partial F}{\partial \eta} = 1.$$

The Euler-Lagrange equation is

$$0 - \frac{d}{dt}1 = 0, \quad t \in [0, 1],$$

which is true for all $\mathbf{x} \in S$. So all $\mathbf{x} \in S$ satisfy the Euler-Lagrange equation. This is to be expected because

$$f(\mathbf{x}) = \int_0^1 \mathbf{x}'(t) dt = \mathbf{x}(1) - \mathbf{x}(0) = 1 - 0 = 1, \quad \mathbf{x} \in S.$$

So $f : S \to \mathbb{R}$ is constant on S, and hence every $\mathbf{x} \in S$ is a minimizer.

(2) With $F(\xi, \eta, \tau) = \xi\eta$, we have that

$$f(\mathbf{x}) = \int_0^1 \mathbf{x}(t)\mathbf{x}'(t) dt = \int_0^1 F(\mathbf{x}(t), \mathbf{x}'(t), t) dt.$$

We have

$$\frac{\partial F}{\partial \xi} = \eta,$$

$$\frac{\partial F}{\partial \eta} = \xi.$$

The Euler-Lagrange equation is

$$\mathbf{x}'(t) - \frac{d}{dt}\mathbf{x}(t) = 0, \quad t \in [0, 1],$$

which is true for all $\mathbf{x} \in S$. So once again all $\mathbf{x} \in S$ satisfy the Euler-Lagrange equation. Again this is not surprising since

$$f(\mathbf{x}) = \int_0^1 \mathbf{x}(t)\mathbf{x}'(t)dt = \int_0^1 \frac{d}{dt}\frac{(\mathbf{x}(t))^2}{2}dt = \frac{(\mathbf{x}(1))^2 - (\mathbf{x}(0))^2}{2} = \frac{1}{2}, \quad \mathbf{x} \in S.$$

So $f : S \to \mathbb{R}$ is constant on S, and hence every $\mathbf{x} \in S$ is a minimizer.

(3) With $F(\xi, \eta, \tau) = \xi + \tau\eta$, we have that

$$f(\mathbf{x}) = \int_0^1 (\mathbf{x}(t) + t\mathbf{x}'(t))dt = \int_0^1 F(\mathbf{x}(t), \mathbf{x}'(t), t)dt.$$

We have

$$\frac{\partial F}{\partial \xi} = 1,$$

$$\frac{\partial F}{\partial \eta} = \tau.$$

The Euler-Lagrange equation is

$$1 - \frac{d}{dt}t = 0, \quad t \in [0, 1],$$

which is true for all $\mathbf{x} \in S$. So all $\mathbf{x} \in S$ satisfy the Euler-Lagrange equation. This is to be expected because

$$f(\mathbf{x}) = \int_0^1 (\mathbf{x}(t) + t\mathbf{x}'(t))dt = \int_0^1 \frac{d}{dt}(t\mathbf{x}(t))dt = 1 \cdot \mathbf{x}(1) - 0 \cdot \mathbf{x}(0) = 1, \quad \mathbf{x} \in S.$$

So $f : S \to \mathbb{R}$ is constant on S, and hence every $\mathbf{x} \in S$ is a minimizer.

Solution to Exercise 2.7.

(1) If F doesn't depend on ξ, then $\dfrac{\partial F}{\partial \xi} = 0$, and so the Euler-Lagrange equation is

$$0 - \frac{d}{dt}\left(\frac{\partial F}{\partial \eta}(\mathbf{x}_*(t), \mathbf{x}'_*(t), t)\right) = 0, \quad t \in [a, b].$$

Using the Fundamental Theorem of Calculus, we obtain

$$\frac{\partial F}{\partial \eta}(\mathbf{x}_*(t), \mathbf{x}'_*(t), t) - \frac{\partial F}{\partial \eta}(\mathbf{x}_*(a), \mathbf{x}'_*(a), a) = \int_a^t \frac{d}{dt}\left(\frac{\partial F}{\partial \eta}(\mathbf{x}_*(t), \mathbf{x}'_*(t), t)\right)dt = \int_a^t 0\,dt = 0.$$

Thus, if we define $C := \dfrac{\partial F}{\partial \eta}(\mathbf{x}_*(a), \mathbf{x}'_*(a), a)$, then $\dfrac{\partial F}{\partial \eta}(\mathbf{x}_*(t), \mathbf{x}'_*(t), t) = C$, $t \in [a, b]$.

(2) If F doesn't depend on η, then $\dfrac{\partial F}{\partial \eta} = 0$, and so the Euler-Lagrange equation is

$$\frac{\partial F}{\partial \xi}(\mathbf{x}_*(t), \mathbf{x}'_*(t), t) - \frac{d}{dt}0 = 0, \quad t \in [a, b],$$

that is, $\dfrac{\partial F}{\partial \xi}(\mathbf{x}_*(t), \mathbf{x}'_*(t), t) = 0$, $t \in [a, b]$.

(3) If F doesn't depend on τ, then $\dfrac{\partial F}{\partial \tau} = 0$. We have

$$\frac{d}{dt}\left(F(\mathbf{x}_*(t), \mathbf{x}_*'(t), t) - \mathbf{x}_*'(t)\frac{\partial F}{\partial \eta}(\mathbf{x}_*(t), \mathbf{x}_*'(t), t)\right)$$

$$= \frac{\partial F}{\partial \xi}(\mathbf{x}_*(t), \mathbf{x}_*'(t), t)\cdot \mathbf{x}_*'(t) + \frac{\partial F}{\partial \eta}(\mathbf{x}_*(t), \mathbf{x}_*'(t), t)\cdot \mathbf{x}_*''(t) + \frac{\partial F}{\partial \tau}(\mathbf{x}_*(t), \mathbf{x}_*'(t), t)\cdot 1$$

$$-\mathbf{x}_*''(t)\cdot \frac{\partial F}{\partial \eta}(\mathbf{x}_*(t), \mathbf{x}_*'(t), t) - \mathbf{x}_*'(t)\cdot \frac{d}{dt}\left(\frac{\partial F}{\partial \eta}F(\mathbf{x}_*(t), \mathbf{x}_*'(t), t)\right)$$

$$= \frac{\partial F}{\partial \xi}(\mathbf{x}_*(t), \mathbf{x}_*'(t), t)\cdot \mathbf{x}_*'(t) - \mathbf{x}_*'(t)\cdot \frac{d}{dt}\left(\frac{\partial F}{\partial \eta}F(\mathbf{x}_*(t), \mathbf{x}_*'(t), t)\right) = 0.$$

Thus upon integrating, we obtain

$$F(\mathbf{x}_*(t), \mathbf{x}_*'(t), t) - \mathbf{x}_*'(t)\frac{\partial F}{\partial \eta}(\mathbf{x}_*(t), \mathbf{x}_*'(t), t) = C, \quad t \in [a, b].$$

Solution to Exercise 2.8. With $F : \mathbb{R}^3 \to \mathbb{R}$ defined by $F(\alpha, \beta, \gamma) = \gamma\beta^2$, $(\alpha, \beta, \gamma) \in \mathbb{R}^3$,

$$f(\mathbf{x}) = \int_a^1 F(\mathbf{x}(t), \mathbf{x}'(t), t)dt, \quad \mathbf{x} \in S.$$

We have

$$\frac{\partial F}{\partial \alpha}(\alpha, \beta, \gamma) = 0, \quad \frac{\partial F}{\partial \beta}(\alpha, \beta, \gamma) = 2\gamma\beta.$$

The Euler-Lagrange equation is

$$\frac{\partial F}{\partial \alpha}(\mathbf{x}(t), \mathbf{x}'(t), t) - \frac{d}{dt}\left(\frac{\partial F}{\partial \beta}(\mathbf{x}(t), \mathbf{x}'(t), t)\right) = 0 \quad (t \in [a, 1]),$$

that is, $0 - \dfrac{d}{dt}(2t\mathbf{x}'(t)) = 0$, $t \in [a, 1]$.

(1) By integrating, we obtain that there is a constant A such that $t\mathbf{x}'(t) = A$ for all $t \in [a, 1]$. As $a > 0$, it follows that

$$\mathbf{x}'(t) = \frac{A}{t}, \quad t \in [a, 1].$$

Thus, by integrating again, there is a constant B such that

$$\mathbf{x}(t) = A\log t + B, \quad t \in [a, 1].$$

As $\mathbf{x} \in S$, we must have $\mathbf{x}(a) = 0$, $\mathbf{x}(1) = 1$, which give

$$\begin{aligned} A\log a + B &= 0, \\ A\log 1 + B &= 1. \end{aligned}$$

Hence $B = 1$, and $A = -\dfrac{1}{\log a}$.

Consequently, the unique solution of the Euler-Lagrange equation is given by

$$\mathbf{x}(t) = 1 - \frac{\log t}{\log a}, \quad t \in [0, 1].$$

(2) Proceeding as in the previous part, we obtain by integrating the Euler-Lagrange equation that there is a constant A such that $t\mathbf{x}'(t) = A$ for all $t \in [0,1]$. But by setting $t = 0$, we see that $A = 0$, and so $\mathbf{x}'(t) = 0$ for $t \in (0,1]$. As $\mathbf{x}' \in C[0,1]$, it follows that $\mathbf{x}'(t) = 0$ for all $t \in [0,1]$. Hence, by integrating from $t = 0$ to $t \in [0,1]$, we see that $\mathbf{x}(t) = \mathbf{x}(0)$ for all $t \in [0,1]$. But this is impossible for an $\mathbf{x} \in S$ since $\mathbf{x}(0) = 0$ while $\mathbf{x}(1) = 1$.

Consequently, there is no solution to the optimization problem in this case, since the solvability of the Euler-Lagrange equation is necessary for any minimizer.

Solution to Exercise 2.9. Discretization yields the auxiliary finite dimensional problem of determining the points $y_1, \cdots, y_{n-1}$ that minimizes

$$f_n(y_1, \cdots, y_{n-1}) = \frac{1}{n} \sum_{k=0}^{n-1} \sqrt{1 + \left(\frac{y_{k+1} - y_k}{1/n}\right)^2},$$

where $y_0 := 0$ and $y_n := 1$. We know that the function $\eta \mapsto \sqrt{1 + \eta^2}$ is convex. Using Exercise 1.49, we have for $a_1, \cdots, a_n \in \mathbb{R}$ that

$$\frac{\sqrt{1 + a_1^2} + \cdots + \sqrt{1 + a_n^2}}{n} \geq \sqrt{1 + \left(\frac{a_1 + \cdots + a_n}{n}\right)^2}$$

and we note that if $a_1 = \cdots = a_n$, then equality holds. With $a_k := \dfrac{y_{k+1} - y_k}{1/n}$, we obtain

$$f_n(y_1, \cdots, y_{n-1}) \geq \sqrt{1 + \left(\frac{1}{n} \sum_{k=0}^{n-1} \frac{y_{k+1} - y_k}{1/n}\right)^2}$$

$$= \sqrt{1 + ((\cancel{y_1} - y_0) + (\cancel{y_2} - \cancel{y_1}) + \cdots + (\cancel{y_{n-1}} - \cancel{y_{n-2}}) + (y_n - \cancel{y_{n-1}}))^2}$$

$$= \sqrt{1 + (y_n - y_0)^2} = \sqrt{1 + (1 - 0)^2} = \sqrt{2},$$

and that there is equality if

$$y_k = \frac{k}{n}, \quad k = 1, \cdots, n-1.$$

The corresponding points clearly lie on the straight line $\mathbf{x}(t) = t$, $t \in [0,1]$ (which we have already seen, in Exercise 2.4, is the curve that minimizes f).

Solution to Exercise 2.12. We have

$$\int_0^1 \left(\frac{(\mathbf{x}'(t))^2}{2} + \mathbf{x}(t)\mathbf{x}'(t) + \mathbf{x}'(t) + \mathbf{x}(t)\right) dt = \int_0^1 F(\mathbf{x}(t), \mathbf{x}'(t), t) dt,$$

where $F(\xi, \eta, \tau)$ is given by

$$F(\xi, \eta, \tau) = \frac{\eta^2}{2} + \xi\eta + \eta + \xi.$$

We have

$$\frac{\partial F}{\partial \xi} = \eta + 1,$$

$$\frac{\partial F}{\partial \eta} = \eta + \xi + 1.$$

The Euler-Lagrange equation is

$$\mathbf{x}'_*(t) + 1 - \frac{d}{dt}(\mathbf{x}'_*(t) + \mathbf{x}_*(t) + 1) = 0, \quad t \in [0, 1],$$

that is, $\mathbf{x}''_*(t) = 1$, $t \in [0, 1]$. Hence we obtain by integrating twice that

$$\mathbf{x}_*(t) = \frac{t^2}{2} + At + B, \quad t \in [0, 1],$$

for some constants A, B. The endpoint conditions at $t = 0$ and at $t = 1$ are:

$$\frac{\partial F}{\partial \eta}(\mathbf{x}_*(0), \mathbf{x}'_*(0), 0) = A + B + 1 = 0,$$

$$\frac{\partial F}{\partial \eta}(\mathbf{x}_*(1), \mathbf{x}'_*(1), 1) = (1 + A) + \left(\frac{1}{2} + A + B\right) + 1 = 0,$$

and so $A = -3/2$, $B = 1/2$. Given that there is a minimizer $\mathbf{x}_*$, it follows that $\mathbf{x}_*$ must be given by

$$\mathbf{x}_*(t) = \frac{t^2}{2} - \frac{3}{2}t + \frac{1}{2} = \frac{t^2 - 3t + 1}{2}, \quad t \in [0, 1].$$

Solution to Exercise 2.14. We have

$$\int_0^1 \cos(\mathbf{x}'(t))dt = \int_0^1 F(\mathbf{x}(t), \mathbf{x}'(t), t)dt,$$

where $F(\xi, \eta, \tau)$ is given by $F(\xi, \eta, \tau) = \cos\eta$. We have

$$\frac{\partial F}{\partial \xi} = 0,$$

$$\frac{\partial F}{\partial \eta} = -\sin\eta.$$

The Euler-Lagrange equation is

$$0 - \frac{d}{dt}(-\sin(\mathbf{x}'_*(t))) = 0, \quad t \in [0, 1].$$

Integrating, we obtain $\sin(\mathbf{x}'_*(t)) = C$, $t \in [0, 1]$, for some constant C. By the endpoint condition at $t = 1$, we have

$$0 = \frac{\partial F}{\partial \eta}(\mathbf{x}_*(1), \mathbf{x}'_*(1), 1) = -\sin(\mathbf{x}'_*(1)),$$

and so it follows that $C = 0$. Thus $\sin(\mathbf{x}'_*(t)) = 0$, $t \in [0, 1]$. Hence at each fixed $t \in [0, 1]$, we have $\mathbf{x}'_*(t) \in \pi\mathbb{Z}$. But $\mathbf{x}'_*$ is continuous on $[0, 1]$, and so it follows from here (using the Intermediate Value Theorem) that there is a fixed $n \in \mathbb{Z}$ such that for all $t \in [0, 1]$, $\mathbf{x}'_*(t) = n\pi$. Since $\mathbf{x}_*(0) = 0$, we obtain by integrating that $\mathbf{x}_*(t) = n\pi t$, $t \in [0, 1]$. So there are countably many solutions of the Euler-Lagrange equation in S, and they are given by

$$\mathbf{x}_{*,n}(t) = n\pi t, \quad t \in [0, 1], \quad n \in \mathbb{Z}.$$

As $-1 \leq \cos\theta \leq 1$ for each $\theta \in \mathbb{R}$, it follows that

$$-1 = \int_0^1 -1dt \leq f(\mathbf{x}) = \int_0^1 \cos(\mathbf{x}'(t))dt \leq \int_0^1 1dt = 1.$$

But

$$f(\mathbf{x}_{*,2n}) = \int_0^1 \cos(2n\pi)dt = \int_0^1 1dt = 1, \text{ and}$$

$$f(\mathbf{x}_{*,2n+1}) = \int_0^1 \cos((2n+1)\pi)dt = \int_0^1 -1dt = -1.$$

So it follows that the elements $\mathbf{x}_{*,2n} \in S$, $n \in \mathbb{Z}$, are all maximizers of f, while the elements $\mathbf{x}_{*,2n+1} \in S$, $n \in \mathbb{Z}$, are all minimizers of f. See Figure 3.

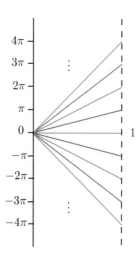

Figure 3. Minimizers and maximizers.

Solution to Exercise 2.15. We have

$$f(\mathbf{x}) = \int_0^1 \left(\frac{(\mathbf{x}'(t))^2}{2} + \mathbf{x}(t) \right)dt = \int_0^1 F(\mathbf{x}(t), \mathbf{x}'(t), t)dt,$$

where $F(\xi, \eta, \tau) = \frac{\eta^2}{2} + \xi$. Since $\eta \mapsto \eta^2$ is convex, it follows that F is convex, which in turn implies that f is convex. So in order to find minimizers we just have to solve the Euler-Lagrange equation. We have

$$\frac{\partial F}{\partial \xi} = 1,$$

$$\frac{\partial F}{\partial \eta} = \eta.$$

The Euler-Lagrange equation is

$$1 - \frac{d}{dt}\mathbf{x}'(t) = 0, \quad t \in [0,1].$$

Upon integrating, we obtain that $\mathbf{x}'(t) = t + A$, $t \in [0,1]$, for some constant A, and integrating a second time, we have

$$\mathbf{x}(t) = \frac{t^2}{2} + At + B, \quad t \in [0,1].$$

(1) Using $\mathbf{x}_*(0) = 1$ and $\mathbf{x}_*(1) = 0$, we obtain

$$A \cdot 0 + B = 1,$$
$$A \cdot 1 + B = -\frac{1}{2},$$

and so $A = -\frac{3}{2}$, $B = 1$. Thus the minimizer $\mathbf{x}_* \in S$ of $f : S \to \mathbb{R}$ is given by

$$\mathbf{x}_*(t) = \frac{t^2}{2} - \frac{3}{2}t + 1, \quad t \in [0,1].$$

The minimum cost is

$$
\begin{aligned}
f(\mathbf{x}_*) &= \int_0^1 \left(\frac{1}{2}\left(t - \frac{3}{2}\right)^2 + \frac{t^2}{2} - \frac{3}{2}t + 1 \right) dt \\
&= \int_0^1 \left(t^2 - 3t + \frac{17}{8} \right) dt = \frac{1}{3} - \frac{3}{2} + \frac{17}{8} = \frac{23}{24}.
\end{aligned}
$$

(2) If $\mathbf{x}(1)$ is free, we use the endpoint condition at $t = 1$:

$$0 = \frac{\partial F}{\partial \eta}(\tilde{\mathbf{x}}_*(1), \tilde{\mathbf{x}}_*'(1), 1) = 1 + A,$$

and so $A = -1$. Since $\tilde{\mathbf{x}}_*(0) = 1$, we have $0 + A \cdot 0 + B = 1$, and so $B = 1$. Consequently,

$$\tilde{\mathbf{x}}_*(t) = \frac{t^2}{2} - t + 1, \quad t \in [0,1].$$

The minimum cost is

$$
\begin{aligned}
f(\tilde{\mathbf{x}}_*) &= \int_0^1 \left(\frac{1}{2}(t - 1)^2 + \frac{t^2}{2} - t + 1 \right) dt \\
&= \int_0^1 \left(t^2 - 2t + \frac{3}{2} \right) dt = \frac{1}{3} - 1 + \frac{3}{2} = \frac{5}{6} = \frac{20}{24}.
\end{aligned}
$$

(3) Clearly,

$$f(\mathbf{x}_*) = \frac{23}{24} > \frac{20}{24} = f(\tilde{\mathbf{x}}_*),$$

and this is expected because

$$S = \{\mathbf{x} \in C^1[0,1] : \mathbf{x}(0) = 1 \text{ and } \mathbf{x}(1) = 0\} \subset \tilde{S} = \{\mathbf{x} \in C^1[0,1] : \mathbf{x}(0) = 1\},$$

which has the consequence that

$$f(\mathbf{x}_*) = \min_{\mathbf{x} \in S} f(\mathbf{x}) \geq \min_{\mathbf{x} \in \tilde{S}} f(\mathbf{x}) = f(\tilde{\mathbf{x}}_*).$$

Solution to Exercise 2.16. Proceeding as in the solution to Exercise 2.8, the Euler-Lagrange equation is

$$\frac{\partial F}{\partial \alpha}(\mathbf{x}(t), \mathbf{x}'(t), t) - \frac{d}{dt}\left(\frac{\partial F}{\partial \beta}(\mathbf{x}(t), \mathbf{x}'(t), t)\right) = 0 \quad (t \in [a, 1]),$$

that is,

$$0 - \frac{d}{dt}(2t\mathbf{x}'(t)) = 0 \quad (t \in [a, 1]).$$

By integrating, we obtain that there is a constant A such that $t\mathbf{x}'(t) = A$ for all $t \in [a, 1]$. As $a > 0$, it follows that

$$\mathbf{x}'(t) = \frac{A}{t}, \quad t \in [a, 1].$$

Hence by integrating again, there is a constant B such that

$$\mathbf{x}(t) = A \log t + B, \quad t \in [a, 1].$$

As $\mathbf{x} \in \widetilde{S}$, we must have $\mathbf{x}(a) = 0$. Thus

$$A \log a + B = 0.$$

As the endpoint $\mathbf{x}(1)$ is free, we have an end condition at $t = 1$, namely

$$\frac{\partial F}{\partial \beta}(\mathbf{x}(1), \mathbf{x}'(1), 1) = 0,$$

that is,

$$2t\frac{A}{t}\Big|_{t=1} = 2A = 0,$$

and so $A = 0$. From the above equation $A \log a + B = 0$, it now follows that also $B = 0$. Hence, the only solution $\mathbf{x} \in \widetilde{S}$ to the Euler-Lagrange equation is the zero function $\mathbf{0}$. Clearly this is a minimizer, since for any $\mathbf{x} \in \widetilde{S}$,

$$\widetilde{f}(\mathbf{x}) = \int_a^1 t(\mathbf{x}(t))^2 dt \geq \int_a^1 0 dt = 0 = \widetilde{f}(\mathbf{0}).$$

Thus there is a unique solution to the optimization problem, namely the zero function $\mathbf{0} \in \widetilde{S}$.

Solution to Exercise 2.22. Suppose that $\mathbf{x}_*$ is a solution. We have

$$F(\xi, \eta, \tau) = \tau\xi,$$
$$G(\xi, \eta, \tau) = \xi^2.$$

If for all $t \in [0, 1]$ there holds that

$$\frac{\partial G}{\partial \xi}(\mathbf{x}_*(t), \mathbf{x}_*'(t), t) - \frac{d}{dt}\left(\frac{\partial G}{\partial \eta}(\mathbf{x}_*(t), \mathbf{x}_*'(t), t)\right) = 2\mathbf{x}_*(t) - \frac{d}{dt}0 = 0,$$

then it follows that $\int_0^1 (\mathbf{x}_*(t))^2 dt = 0 \neq \frac{1}{12}$.

For $\lambda \in \mathbb{R}$, we have $F + \lambda G = \tau\xi + \lambda\xi^2$, and so

$$\frac{\partial(F + \lambda G)}{\partial \xi}(\xi, \eta, \tau) = \tau + 2\lambda\xi,$$
$$\frac{\partial(F + \lambda G)}{\partial \eta}(\xi, \eta, \tau) = 0.$$

So the Euler-Lagrange equation for the solution $\mathbf{x}_*$ is given by

$$t + 2\lambda \mathbf{x}_*(t) - \frac{d}{dt}0 = 0, \quad t \in [0,1].$$

Clearly we can't have $\lambda = 0$. So

$$\mathbf{x}_*(t) = \frac{-t}{2\lambda}, \quad t \in [0,1].$$

Using the fact that $\int_0^1 (\mathbf{x}_*(t))^2 dt = \dfrac{1}{12}$, we obtain that

$$\int_0^1 \frac{t^2}{4\lambda^2} dt = \frac{1}{12\lambda^2} = \frac{1}{12},$$

and so $\lambda = \pm 1$. So if at all there is a minimizer, it has to be $\pm \dfrac{t}{2}$. But since

$$\int_0^1 t \cdot \frac{-t}{2} dt = -\frac{1}{6} < \frac{1}{6} = \int_0^1 t\frac{t}{2} dt,$$

it follows that if a minimizer exists, then it must be $\dfrac{-t}{2}$.

We can now check that $\mathbf{x}_* := \dfrac{-t}{2}$ is indeed a minimizer. Indeed, if

$$\mathbf{x} \in S := \left\{ \mathbf{x} \in C^1[0,1] : \int_0^1 (\mathbf{x}(t))^2 dt = \frac{1}{12} \right\},$$

then we have

$$
\begin{aligned}
0 &\le \int_0^1 \left(\mathbf{x}(t) + \frac{t}{2} \right)^2 dt = \int_0^1 \left(\frac{t^2}{4} + t\mathbf{x}(t) + (\mathbf{x}(t))^2 \right) dt \\
&= \frac{1}{12} + \int_0^1 t\mathbf{x}(t) dt + \frac{1}{12} = \frac{1}{6} + \int_0^1 t\mathbf{x}(t) dt,
\end{aligned}
$$

and so $\int_0^1 t\mathbf{x}(t) dt \ge -\dfrac{1}{6} = \int_0^1 t \cdot \dfrac{-t}{2} dt.$

Solutions to the Exercises from Chapter 3

Solution to Exercise 3.1.

(1) With $\mathbf{x}_1 := \mathbf{x}$ and $\mathbf{x}_2 := \mathbf{x}'$, we have

$$\mathbf{x}_1' = \mathbf{x}' = \mathbf{x}_2 \quad \text{and} \quad \mathbf{x}_2' = \mathbf{x}'' = -\omega^2 \mathbf{x} = -\omega^2 \mathbf{x}_1,$$

and so we have

$$\begin{cases} \mathbf{x}_1'(t) = \mathbf{x}_2(t), \\ \mathbf{x}_2'(t) = -\omega^2 \mathbf{x}_1(t). \end{cases}$$

(2) With $\mathbf{x}_1 := \mathbf{y}$, $\mathbf{x}_2 := \mathbf{y}'$, $\mathbf{x}_3 := \mathbf{z}$, $\mathbf{x}_4 := \mathbf{z}'$, we have

$$\begin{aligned}
\mathbf{x}_1' &= \mathbf{y}' = \mathbf{x}_2, \\
\mathbf{x}_2' &= \mathbf{y}'' = -\mathbf{y} = -\mathbf{x}_1, \\
\mathbf{x}_3' &= \mathbf{z}' = \mathbf{x}_4, \\
\mathbf{x}_4' &= \mathbf{z}'' = -\mathbf{z}' - \mathbf{z} = -\mathbf{x}_4 - \mathbf{x}_3,
\end{aligned}$$

and so

$$\begin{cases} \mathbf{x}_1'(t) = \mathbf{x}_2(t), \\ \mathbf{x}_2'(t) = -\mathbf{x}_1(t), \\ \mathbf{x}_3'(t) = \mathbf{x}_4(t), \\ \mathbf{x}_4'(t) = -\mathbf{x}_3(t) - \mathbf{x}_4(t). \end{cases}$$

(3) With $\mathbf{x}_1 := \mathbf{x}$ and $\mathbf{x}_2 := \mathbf{x}'$, we have $\mathbf{x}_1' = \mathbf{x}_2$, $\mathbf{x}_2' = \mathbf{x}'' = -t\sin\mathbf{x} = -t\sin\mathbf{x}_1$, and so

$$\begin{cases} \mathbf{x}_1'(t) = \mathbf{x}_2(t), \\ \mathbf{x}_2'(t) = -t\sin(\mathbf{x}_1(t)). \end{cases}$$

Solution to Exercise 3.10.

(1) Autonomous.

(2) Nonautonomous.

(3) Autonomous.

Solution to Exercise 3.11.

(1) We have

$$\frac{d}{dt}\mathbf{x}(t) = \frac{d}{dt}(e^{ta}x_i) = ae^{ta}x_i = a(e^{ta}x_i) = a\mathbf{x}(t), \quad t \geq 0,$$

and $\mathbf{x}(0) = e^{0 \cdot a}x_i = e^0 x_i = 1 \cdot x_i = x_i$.

(2) We have

$$\begin{aligned}
\mathbf{x}_1'(t) &= \frac{d}{dt}\sin(2t) = 2\cos(2t) = 2\mathbf{x}_2(t), \\
\mathbf{x}_2'(t) &= \frac{d}{dt}\cos(2t) = -2\sin(2t) = -2\mathbf{x}_1(t),
\end{aligned}$$

for $t \geq 0$, and moreover,

$$\begin{aligned}
\mathbf{x}_1(0) &= \sin(2 \cdot 0) = 0, \\
\mathbf{x}_2(0) &= \cos(2 \cdot 0) = 1.
\end{aligned}$$

(3) We have

$$\frac{d}{dt}\mathbf{x}(t) = \frac{d}{dt}\frac{1}{1-t^2} = -\frac{1}{(1-t^2)^2} \cdot (-2t) = 2t\left(\frac{1}{1-t^2}\right)^2 = 2t(\mathbf{x}(t))^2,$$

for $t \in [0,1)$. Moreover, $\mathbf{x}(0) = \dfrac{1}{1-0^2} = 1$.

Solution to Exercise 3.13. We have

$$\begin{aligned}
\frac{d}{dt}((\boldsymbol{\lambda}(t))^\top \mathbf{x}(t)) &= (\boldsymbol{\lambda}'(t))^\top \mathbf{x}(t) + (\boldsymbol{\lambda}(t))^\top \mathbf{x}'(t) \\
&= -((A(t))^\top \boldsymbol{\lambda}(t))^\top \mathbf{x}(t) + (\boldsymbol{\lambda}(t))^\top (A(t)\mathbf{x}(t)) \\
&= -(\boldsymbol{\lambda}(t))^\top A(t)\mathbf{x}(t) + (\boldsymbol{\lambda}(t))^\top A(t)\mathbf{x}(t) \\
&= 0.
\end{aligned}$$

Thus $(\boldsymbol{\lambda}(t))^\top \mathbf{x}(t)$ is a constant on $[t_i, t_f]$, and so $(\boldsymbol{\lambda}(t))^\top \mathbf{x}(t) = (\boldsymbol{\lambda}(t_f))^\top \mathbf{x}(t_f)$, that is,

$$(\boldsymbol{\lambda}(t))^\top \Phi(t, t_i) x_i = (\lambda_f)^\top \Phi(t_f, t_i) x_i.$$

But the choice of x_i was arbitrary, and the same computation can be carried out with any other $x_i \in \mathbb{R}^n$ as well. Hence we can conclude that $(\boldsymbol{\lambda}(t))^\top \Phi(t, t_i) = (\lambda_f)^\top \Phi(t_f, t_i)$. We had also seen that $\Phi(t, t_i)$ is invertible, which yields that $(\boldsymbol{\lambda}(t))^\top = (\lambda_f)^\top \Phi(t_f, t)$. Taking transposes, we obtain $\boldsymbol{\lambda}(t) = (\Phi(t_f, t))^\top \lambda_f$, $t \in [t_i, t_f]$.

Solution to Exercise 3.14. From the general expression for the solution given in Section 3.3.1, we have

$$\mathbf{x}(t) = \exp\left(\int_{t_i}^t a(\tau)d\tau\right)x_i = \exp\left(\int_{t_i}^t a d\tau\right)x_i = \exp(a(t - t_i))x_i = e^{(t-t_i)a}x_i, \quad t \geq t_i.$$

Solution to Exercise 3.16. Since D is diagonal with diagonal entries $\lambda_1, \cdots, \lambda_n$, it is easy to see that D^k is also diagonal, with diagonal entries $\lambda_1^k, \cdots, \lambda_n^k$, and so

$$\begin{aligned}
e^D &= I + \frac{1}{1!}D + \frac{1}{2!}D^2 + \frac{1}{3!}D^3 + \cdots \\
&= \begin{bmatrix} 1 & & \\ & \ddots & \\ & & 1 \end{bmatrix} + \frac{1}{1!}\begin{bmatrix} \lambda_1 & & \\ & \ddots & \\ & & \lambda_n \end{bmatrix} + \frac{1}{2!}\begin{bmatrix} \lambda_1^2 & & \\ & \ddots & \\ & & \lambda_n^2 \end{bmatrix} + \frac{1}{3!}\begin{bmatrix} \lambda_1^3 & & \\ & \ddots & \\ & & \lambda_n^3 \end{bmatrix} + \cdots \\
&= \begin{bmatrix} e^{\lambda_1} & & \\ & \ddots & \\ & & e^{\lambda_n} \end{bmatrix}.
\end{aligned}$$

Since $A = PDP^{-1}$, it follows that for all $k \in \mathbb{N}$, $(PDP^{-1})^k = PD^kP^{-1}$. Thus

$$
\begin{aligned}
e^A &= I + A + \frac{1}{2!}A^2 + \frac{1}{3!}A^3 + \cdots = I + PDP^{-1} + \frac{1}{2!}PD^2P^{-1} + \frac{1}{3!}PD^3P^{-1} + \cdots \\
&= PIP^{-1} + PDP^{-1} + \frac{1}{2!}PD^2P^{-1} + \frac{1}{3!}PD^3P^{-1} + \cdots \\
&= P\left(I + D + \frac{1}{2!}D^2 + \frac{1}{3!}D^3 + \cdots\right)P^{-1} \\
&= Pe^DP^{-1}.
\end{aligned}
$$

Solution to Exercise 3.21. From Theorem 3.17 and Corollary 3.19, we conclude that for large enough real s,

$$
(sI - A)^{-1} = \begin{bmatrix} \dfrac{s}{s^2 - 1} & \dfrac{1}{s^2 - 1} \\ \dfrac{1}{s^2 - 1} & \dfrac{s}{s^2 - 1} \end{bmatrix},
$$

and so

$$
sI - A = ((sI - A)^{-1})^{-1} = \begin{bmatrix} \dfrac{s}{s^2 - 1} & \dfrac{1}{s^2 - 1} \\ \dfrac{1}{s^2 - 1} & \dfrac{s}{s^2 - 1} \end{bmatrix}^{-1} = \begin{bmatrix} s & -1 \\ -1 & s \end{bmatrix}.
$$

Consequently, $A = \begin{bmatrix} 0 & 1 \\ 1 & 0 \end{bmatrix}$.

Alternately, in light of Theorem 3.15, one could simply observe that for any vector $x_i \in \mathbb{R}^2$,

$$
Ax_i = \frac{d}{dt}(e^{tA}x_i)\Big|_{t=0} = \frac{d}{dt}\begin{bmatrix} \cosh t & \sinh t \\ \sinh t & \cosh t \end{bmatrix}\Big|_{t=0} x_i = \begin{bmatrix} \sinh t & \cosh t \\ \cosh t & \sinh t \end{bmatrix}\Big|_{t=0} x_i = \begin{bmatrix} 0 & 1 \\ 1 & 0 \end{bmatrix} x_i.
$$

As the choice of $x_i \in \mathbb{R}^2$ was arbitrary, it follows that $A = \begin{bmatrix} 0 & 1 \\ 1 & 0 \end{bmatrix}$.

Solution to Exercise 3.22. We have that

$$
sI - A = \begin{bmatrix} s - \lambda & -1 & 0 \\ 0 & s - \lambda & -1 \\ 0 & 0 & s - \lambda \end{bmatrix},
$$

and so $\det(sI - A) = (s - \lambda)^3$. A short computation gives

$$
\operatorname{adj}(sI - A) = \begin{bmatrix} (s - \lambda)^2 & s - \lambda & 1 \\ 0 & (s - \lambda)^2 & s - \lambda \\ 0 & 0 & (s - \lambda)^2 \end{bmatrix}.
$$

By Cramer's Rule,

$$
(sI - A)^{-1} = \frac{1}{\det(sI - A)}\operatorname{adj}(sI - A) = \begin{bmatrix} \dfrac{1}{s - \lambda} & \dfrac{1}{(s - \lambda)^2} & \dfrac{1}{(s - \lambda)^3} \\ 0 & \dfrac{1}{s - \lambda} & \dfrac{1}{(s - \lambda)^2} \\ 0 & 0 & \dfrac{1}{s - \lambda} \end{bmatrix}.
$$

By using Theorem 3.18 ("taking the inverse Laplace transform"), we obtain

$$e^{tA} = \begin{bmatrix} e^{\lambda t} & te^{\lambda t} & \frac{t^2}{2!}e^{\lambda t} \\ 0 & e^{\lambda t} & te^{\lambda t} \\ 0 & 0 & e^{\lambda t} \end{bmatrix}.$$

Solution to Exercise 3.23. (Since for commuting matrices A and B, we know that $e^{A+B} = e^A e^B$, we must choose noncommuting A and B.) Let

$$A = \begin{bmatrix} 0 & 1 \\ 0 & 0 \end{bmatrix} \quad \text{and} \quad B = \begin{bmatrix} 0 & 0 \\ -1 & 0 \end{bmatrix}.$$

Then

$$AB = \begin{bmatrix} 0 & 1 \\ 0 & 0 \end{bmatrix}\begin{bmatrix} 0 & 0 \\ -1 & 0 \end{bmatrix} = \begin{bmatrix} -1 & 0 \\ 0 & 0 \end{bmatrix} \neq \begin{bmatrix} 0 & 0 \\ 0 & -1 \end{bmatrix} = \begin{bmatrix} 0 & 0 \\ -1 & 0 \end{bmatrix}\begin{bmatrix} 0 & 1 \\ 0 & 0 \end{bmatrix} = BA,$$

and so A and B do not commute. Since $A^2 = B^2 = 0$, we obtain

$$e^A = I + A + \frac{1}{2!}A^2 + \cdots = I + A = \begin{bmatrix} 1 & 1 \\ 0 & 1 \end{bmatrix}, \quad \text{and}$$

$$e^B = I + B + \frac{1}{2!}B^2 + \cdots = I + B = \begin{bmatrix} 1 & 0 \\ -1 & 1 \end{bmatrix},$$

and so

$$e^A e^B = \begin{bmatrix} 1 & 1 \\ 0 & 1 \end{bmatrix}\begin{bmatrix} 1 & 0 \\ -1 & 1 \end{bmatrix} = \begin{bmatrix} 0 & 1 \\ -1 & 1 \end{bmatrix},$$

while

$$e^B e^A = \begin{bmatrix} 1 & 0 \\ -1 & 1 \end{bmatrix}\begin{bmatrix} 1 & 1 \\ 0 & 1 \end{bmatrix} = \begin{bmatrix} 1 & 1 \\ -1 & 0 \end{bmatrix}.$$

Clearly $e^A e^B \neq e^B e^A$, while if it was the case that $e^A e^B = e^{A+B}$ and $e^B e^A = e^{B+A}$, then we would have had $e^A e^B = e^{A+B} = e^{B+A} = e^B e^A$. So either $e^A e^B \neq e^{A+B}$ or $e^B e^A \neq e^{B+A}$. (In the latter case, we switch the roles of A and B.) We can also explicitly find that in fact $e^A e^B \neq e^{A+B}$ by directly computing e^{A+B}:

$$A + B = \begin{bmatrix} 0 & 1 \\ -1 & 0 \end{bmatrix},$$

and so

$$sI - (A + B) = \begin{bmatrix} s & -1 \\ 1 & s \end{bmatrix},$$

$$(sI - (A + B))^{-1} = \frac{1}{s^2 + 1}\begin{bmatrix} s & 1 \\ -1 & s \end{bmatrix},$$

$$e^{t(A+B)} = \begin{bmatrix} \cos t & \sin t \\ -\sin t & \cos t \end{bmatrix}.$$

Hence

$$e^{A+B} = \begin{bmatrix} \cos 1 & \sin 1 \\ -\sin 1 & \cos 1 \end{bmatrix},$$

and so we see that $e^{A+B} \neq e^A e^B$, for example, because otherwise, comparing the entries of the second column, we would have $(\cos 1)^2 + (\sin 1)^2 = 1^2 + 1^2 = 2 \neq 1$ in violation of Pythagoras's Theorem!

Solution to Exercise 3.24. We have

$$\int_0^t A(\tau)d\tau = \int_0^t \begin{bmatrix} 1 & 2\tau \\ 0 & 0 \end{bmatrix} d\tau = \begin{bmatrix} t & t^2 \\ 0 & 0 \end{bmatrix}.$$

Note that for $t \neq 0$,

$$A(t) \int_0^t A(\tau)d\tau = \begin{bmatrix} 1 & 2t \\ 0 & 0 \end{bmatrix} \begin{bmatrix} t & t^2 \\ t & 0 \end{bmatrix} = \begin{bmatrix} t & t^2 \\ 0 & 0 \end{bmatrix}$$

$$\int_0^t A(\tau)d\tau A(t) = \begin{bmatrix} t & t^2 \\ 0 & 0 \end{bmatrix} \begin{bmatrix} 1 & 2t \\ 0 & 0 \end{bmatrix} = \begin{bmatrix} t & 2t^2 \\ 0 & 0 \end{bmatrix}.$$

A computation reveals that if $\alpha \in \mathbb{R}$, then

$$\exp \begin{bmatrix} \alpha & \alpha^2 \\ 0 & 0 \end{bmatrix} = \begin{bmatrix} e^\alpha & \alpha(e^\alpha - 1) \\ 0 & 1 \end{bmatrix}.$$

Thus with $\mathbf{x}$ defined by

$$\mathbf{x}(t) := \exp \left(\int_{t_i}^t A(\tau)d\tau \right) x_i, \quad t \geq t_i,$$

we have using the values of $t_i = 0$, $x_i = e_2$ and with the given $A(\cdot)$ that

$$\mathbf{x}(t) = \exp \left(\begin{bmatrix} t & t^2 \\ 0 & 0 \end{bmatrix} \right) \begin{bmatrix} 0 \\ 1 \end{bmatrix} = \begin{bmatrix} e^t & t(e^t - 1) \\ 0 & 1 \end{bmatrix} \begin{bmatrix} 0 \\ 1 \end{bmatrix} = \begin{bmatrix} t(e^t - 1) \\ 1 \end{bmatrix}.$$

But this $\mathbf{x}$ does not satisfy the differential equation $\mathbf{x}'(t) = A(t)\mathbf{x}(t)$ for any positive time instant $t > 0$ because

$$\mathbf{x}'(t) = \frac{d}{dt} \begin{bmatrix} t(e^t - 1) \\ 1 \end{bmatrix} = \begin{bmatrix} e^t - 1 + te^t \\ 0 \end{bmatrix}$$

$$A(t)\mathbf{x}(t) = \begin{bmatrix} 1 & 2t \\ 0 & 0 \end{bmatrix} \begin{bmatrix} t(e^t - 1) \\ 1 \end{bmatrix} = \begin{bmatrix} t + te^t \\ 0 \end{bmatrix}.$$

(For $t > 0$, we have the inequality above because $e^t = 1 + t + \cdots > 1 + t$, and so it follows that $t + te^t \neq e^t - 1 + te^t$.)

Solution to Exercise 3.29.

(1) By the chain rule,

$$\begin{aligned} \mathbf{q}'(t) &= -\frac{1}{(\mathbf{p}(t) + \alpha)^2} \mathbf{p}'(t) = -\frac{\gamma(\mathbf{p}(t) + \alpha)(\mathbf{p}(t) + \beta)}{(\mathbf{p}(t) + \alpha)^2} = -\frac{\gamma(\mathbf{p}(t) + \beta)}{\mathbf{p}(t) + \alpha} \\ &= -\gamma \frac{\mathbf{p}(t) + \alpha + \beta - \alpha}{\mathbf{p}(t) + \alpha} = -\gamma \left(\frac{\mathbf{p}(t) + \alpha}{\mathbf{p}(t) + \alpha} + (\beta - \alpha)\frac{1}{\mathbf{p}(t) + \alpha} \right) \\ &= -\gamma \left(1 + (\beta - \alpha)\mathbf{q}(t) \right) = \gamma(\alpha - \beta)\mathbf{q}(t) - \gamma, \end{aligned}$$

for all $t \in [0, T]$.

(2) We first note that if at some time instant $t_0 \in [0, 1]$, we have that $\mathbf{p}(t_0) + 1 = 0$, then a solution to the differential equation

$$\mathbf{p}'(t) = (\mathbf{p}(t))^2 - 1, \quad t \in [0, 1],$$

is $\mathbf{p} \equiv -1$. As the solution is unique (why?), it follows that this is *the* solution. But then $\mathbf{p}(1) = 0$ can't be satisfied. So we conclude that at each $t \in [0, 1]$, $\mathbf{p}(t) + 1 \neq 0$.

Define

$$\mathbf{q}(t) := \frac{1}{\mathbf{p}(t) + 1}, \quad t \in [0, 1].$$

Then by the previous exercise, we have that $\mathbf{q}'(t) = 1(1 - (-1))\mathbf{q}(t) - 1 = 2\mathbf{q}(t) - 1$, so that

$$
\begin{aligned}
\mathbf{q}(t) &= e^{t \cdot 2}\mathbf{q}(0) + e^{2t} \int_0^t e^{-\tau \cdot 2}(-1)1 d\tau = e^{2t}\mathbf{q}(0) - e^{2t} \int_0^t e^{-2\tau} d\tau \\
&= e^{2t}\mathbf{q}(0) + \frac{e^{2t}}{2}(e^{-2t} - 1) = e^{2t}\mathbf{q}(0) + \frac{1 - e^{2t}}{2}.
\end{aligned}
$$

Using $\mathbf{p}(1) = 0$ and the fact that $\mathbf{q}(1) = 1/(\mathbf{p}(1) + 1) = 1$, we obtain from the above that $1 = e^2\mathbf{q}(0) + (1 = e^2)/2$. Hence $\mathbf{q}(0) = (e^{-2} + 1)/2$. Consequently,

$$\frac{1}{\mathbf{p}(t) + 1} = e^{2t}\left(\frac{e^{-2} + 1}{2}\right) + \frac{1 - e^{2t}}{2},$$

and so

$$\mathbf{p}(t) = \frac{1 - e^{2(t-1)}}{1 + e^{2(t-1)}}, \quad t \in [0, 1].$$

Solutions to the Exercises from Chapter 4

Solution to Exercise 4.6. The Hamiltonian is given by

$$H(t, x, u, \lambda) = \frac{1}{2}(3x^2 + u^2) + \lambda(x + u).$$

By the Pontryagin Minimum Principle, $\mathbf{u}_T$ minimizes $u \mapsto H(t, \mathbf{x}_T(t), u, \boldsymbol{\lambda}(t)) : \mathbb{R} \to \mathbb{R}$, and so

$$\mathbf{u}_T(t) = \arg\min_{u \in \mathbb{R}} \left(\frac{1}{2}(3(\mathbf{x}_T(t))^2 + u^2) + \boldsymbol{\lambda}(t)(\mathbf{x}_T(t) + u) \right) = -\boldsymbol{\lambda}(t),$$

since the quadratic expression in u on the right-hand side is minimized when $u + \boldsymbol{\lambda}(t) = 0$. So we obtain

$$\mathbf{x}_T'(t) = \mathbf{x}_T(t) + \mathbf{u}_T(t) = \mathbf{x}_T(t) - \boldsymbol{\lambda}(t),$$
$$\boldsymbol{\lambda}'(t) = -\frac{\partial H}{\partial x}(t, \mathbf{x}_T(t), \mathbf{u}_T(t), \boldsymbol{\lambda}(t)) = -3\mathbf{x}_T(t) - \boldsymbol{\lambda}(t),$$

with $\mathbf{x}_T(0) = x_i$ and $\boldsymbol{\lambda}(T) = \varphi'(\mathbf{x}_T(T)) = 0$ (since $\varphi \equiv 0$). Rewriting, we have

$$\begin{bmatrix} \mathbf{x}_T' \\ \boldsymbol{\lambda}' \end{bmatrix} = \begin{bmatrix} 1 & -1 \\ -3 & -1 \end{bmatrix} \begin{bmatrix} \mathbf{x}_T \\ \boldsymbol{\lambda} \end{bmatrix}, \quad t \in [0, T].$$

Hence uisng the diagonalization method of finding the exponential of a matrix (see Exercise 3.16),

$$\begin{bmatrix} \mathbf{x}_T(t) \\ \boldsymbol{\lambda}(t) \end{bmatrix} = \left(\exp \begin{bmatrix} 1 & -1 \\ -3 & -1 \end{bmatrix} \right) \begin{bmatrix} x_i \\ \boldsymbol{\lambda}(0) \end{bmatrix}$$

$$= \begin{bmatrix} 1 & 1 \\ -1 & 3 \end{bmatrix} \begin{bmatrix} e^{2t} & 0 \\ 0 & e^{-2t} \end{bmatrix} \begin{bmatrix} 1 & 1 \\ -1 & 3 \end{bmatrix}^{-1} \begin{bmatrix} x_i \\ \boldsymbol{\lambda}(0) \end{bmatrix}$$

$$= \begin{bmatrix} 1 & 1 \\ -1 & 3 \end{bmatrix} \begin{bmatrix} e^{2t} & 0 \\ 0 & e^{-2t} \end{bmatrix} \begin{bmatrix} \alpha \\ \beta \end{bmatrix},$$

where $\begin{bmatrix} \alpha \\ \beta \end{bmatrix} = \begin{bmatrix} 1 & 1 \\ -1 & 3 \end{bmatrix}^{-1} \begin{bmatrix} x_i \\ \boldsymbol{\lambda}(0) \end{bmatrix}$. As $\mathbf{x}_T(0) = x_i$ and $\boldsymbol{\lambda}(T) = 0$, we obtain

$$\alpha + \beta = x_i,$$
$$-\alpha e^{2T} + 3\beta e^{-2T} = 0,$$

which gives $\alpha = \dfrac{3x_i e^{-4T}}{3e^{-4T} + 1}$ and $\beta = \dfrac{x_i}{3e^{-4T} + 1}$. Consequently,

$$\mathbf{x}_T(t) = \frac{x_i}{3e^{-4T} + 1}(3e^{-4T}e^{2t} + e^{-2t}), \quad t \in [0, T],$$

$$\mathbf{u}_T(t) = -\boldsymbol{\lambda}(t) = \frac{3x_i}{3e^{-4T} + 1}(3e^{-4T}e^{2t} - e^{-2t}), \quad t \in [0, T].$$

For a fixed $t \in \mathbb{R}$, we have

$$\lim_{T \to \infty} \mathbf{x}_T(t) = \lim_{T \to \infty} \frac{x_i}{3e^{-4T} + 1}(3e^{-4T}e^{2t} + e^{-2t}) = \frac{x_i}{3 \cdot 0 + 1}(3 \cdot 0 \cdot e^{2t} + e^{-2t}) = x_i e^{-2t},$$

$$\lim_{T \to \infty} \mathbf{u}_T(t) = \lim_{T \to \infty} \frac{3x_i}{3e^{-4T} + 1}(3e^{-4T}e^{2t} - e^{-2t}) = \frac{3x_i}{3 \cdot 0 + 1}(3 \cdot 0 \cdot e^{2t} - e^{-2t}) = -3x_i e^{-2t},$$

and so $\lim_{T \to \infty} \mathbf{u}_T(t) = -3(x_i e^{-2t}) = -3 \cdot \lim_{T \to \infty} \mathbf{x}_T(t)$, so that $k = -3$.

Solution to Exercise 4.7. We note that the problem is a *maximization* problem. The Hamiltonian is given by $H(t, x, u, \lambda) = u^2 + tu - 1 + \lambda u$. By the Pontryagin Minimum Principle, $\mathbf{u}_*$ minimizes $u \mapsto H(t, \mathbf{x}_*(t), u, \boldsymbol{\lambda}(t)) : \mathbb{R} \to \mathbb{R}$, and so

$$\mathbf{u}_*(t) = \arg\min_{u \in \mathbb{R}} \left(u^2 + tu - 1 + \boldsymbol{\lambda}(t)u \right) = -\frac{t + \boldsymbol{\lambda}(t)}{2}, \quad t \in [0, 1],$$

since the quadratic expression in u on the right-hand side is minimized when

$$2u + t + \boldsymbol{\lambda}(t) = 0.$$

We have

$$\frac{\partial H}{\partial x} = 0,$$

and so the co-state equation is $\boldsymbol{\lambda}'(t) = 0$ for all $t \in [0, 1]$, that is, $\boldsymbol{\lambda}(t) = C$ on $[0, 1]$ for some constant C. Finally,

$$\boldsymbol{\lambda}(1) = \varphi'(\mathbf{x}_*(1)) = -2,$$

since $\varphi(x) = -2x - 3$. Hence $C = \boldsymbol{\lambda}(1) = -2$. Consequently,

$$\mathbf{u}_*(t) = -\frac{t + \boldsymbol{\lambda}(t)}{2} = -\frac{t - 2}{2} = 1 - \frac{t}{2}, \quad t \in [0, 1].$$

Solution to Exercise 4.8. We note that the problem is a *maximization* problem. The Hamiltonian is given by

$$H(t, x, u, \lambda) = \frac{1}{2}u^2 + \lambda(x + u).$$

By the Pontryagin Minimum Principle, $\mathbf{u}_*$ minimizes $u \mapsto H(t, \mathbf{x}_*(t), u, \boldsymbol{\lambda}(t)) : \mathbb{R} \to \mathbb{R}$, and so

$$\mathbf{u}_*(t) = \arg\min_{u \in \mathbb{R}} \left(\frac{1}{2}u^2 + \boldsymbol{\lambda}(t)(\mathbf{x}_*(t) + u) \right) = -\boldsymbol{\lambda}(t), \quad t \in [0, t],$$

since the quadratic expression in u on the right-hand side is minimized when $u + \boldsymbol{\lambda}(t) = 0$. We have

$$\frac{\partial H}{\partial x} = \lambda,$$

and so the co-state equation is given by $\boldsymbol{\lambda}'(t) = -\boldsymbol{\lambda}(t)$, $t \in [0, 1]$. Thus $\boldsymbol{\lambda}(t) = e^{-t}C$, $t \in [0, 1]$, for some constant C. We have $\varphi(x) = -\sqrt{x}$, and so

$$\frac{\partial \varphi}{\partial x} = -\frac{1}{2\sqrt{x}}.$$

Hence

$$\frac{C}{e} = \boldsymbol{\lambda}(1) = \frac{\partial \varphi}{\partial x}(\mathbf{x}_*(1)) = -\frac{1}{2\sqrt{\mathbf{x}_*(1)}}.$$

On the other hand, we also have $\mathbf{x}'_*(t) = \mathbf{x}_*(t) + \mathbf{u}_*(t) = \mathbf{x}_*(t) + \boldsymbol{\lambda}(t) = \mathbf{x}_*(t) + Ce^{-t}$, and so

$$\begin{aligned}
\mathbf{x}_*(t) &= e^t \mathbf{x}_*(0) + \int_0^t e^{(t-\tau)\cdot 1} \cdot 1 \cdot Ce^{-\tau} d\tau \\
&= e^t \cdot 0 + \int_0^t e^t \cdot C \cdot e^{-2\tau} d\tau = e^t \cdot C \cdot \frac{(e^{-2t} - 1)}{-2} = -\frac{C}{2}(e^{-t} - e^t).
\end{aligned}$$

Thus $\mathbf{x}_*(1) = -\dfrac{C}{2}(e^{-1} - e^1).$

Using $\dfrac{C}{e} = -\dfrac{1}{2\sqrt{\mathbf{x}_*(1)}}$, we obtain $C = -\dfrac{e}{\sqrt[3]{2(e^2-1)}}$, and so

$$\mathbf{u}_*(t) = -\boldsymbol{\lambda}(t) = -e^{-t}C = \frac{e}{\sqrt[3]{2(e^2-1)}}e^{-t} = \frac{e^{1-t}}{\sqrt[3]{2(e^2-1)}}, \quad t \in [0,1].$$

Solution to Exercise 4.9. We note that the problem is a *maxi*mization problem. The Hamiltonian is given by $H(t,x,u,\lambda) = (u^2-1)x^2 + \lambda ux$, and so

$$\frac{\partial H}{\partial x} = 2(u^2-1)x + \lambda u,$$

$$\frac{\partial H}{\partial u} = 2ux^2 + \lambda x.$$

Hence for $t \in [0,1]$, we have

$$\mathbf{x}_*'(t) = \mathbf{u}_*(t)\mathbf{x}_*(t)$$
$$\boldsymbol{\lambda}'(t) = 2(1-(\mathbf{u}_*(t))^2)\mathbf{x}_*(t) - \boldsymbol{\lambda}(t)\mathbf{u}_*(t)$$
$$0 = 2\mathbf{u}_*(t)(\mathbf{x}_*(t))^2 + \boldsymbol{\lambda}(t)\mathbf{x}_*(t),$$

with $\mathbf{x}_*(0) = 0$ and $\boldsymbol{\lambda}(1) = 0$.

We have for all $t \in [0,1]$ that

$$0 = \frac{d}{dt}\Big(2\mathbf{u}_*(t)(\mathbf{x}_*(t))^2 + \boldsymbol{\lambda}(t)\mathbf{x}_*(t)\Big),$$

and so

$$0 = 2\mathbf{u}_*'(t)(\mathbf{x}_*(t))^2 + 2\mathbf{u}_*(t)2\mathbf{x}_*(t)\mathbf{x}_*'(t) + \boldsymbol{\lambda}'(t)\mathbf{x}_*(t) + \boldsymbol{\lambda}(t)\mathbf{x}_*'(t)$$

$$= 2\mathbf{u}_*'(t)(\mathbf{x}_*(t))^2 + 4\mathbf{u}_*(t)\mathbf{x}_*(t)\mathbf{u}_*(t)\mathbf{x}_*(t)$$

$$+ \Big(2(1-(\mathbf{u}_*(t))^2)\mathbf{x}_*(t) - \boldsymbol{\lambda}(t)\mathbf{u}_*(t)\Big)\mathbf{x}_*(t) + \boldsymbol{\lambda}(t)\mathbf{u}_*(t)\mathbf{x}_*(t)$$

$$= 2\mathbf{u}_*'(t)(\mathbf{x}_*(t))^2 + 4(\mathbf{u}_*(t))^2(\mathbf{x}_*(t))^2 + 2(1-(\mathbf{u}_*(t))^2)(\mathbf{x}_*(t))^2.$$

Thus

$$2\mathbf{u}_*'(t)(\mathbf{x}_*(t))^2 = \Big(-4(\mathbf{u}_*(t))^2 - 2 + 2(\mathbf{u}_*(t))^2\Big)(\mathbf{x}_*(t))^2 = -2\Big((\mathbf{u}_*(t))^2 + 1\Big)(\mathbf{x}_*(t))^2.$$

If $\mathbf{x}_*(t) \neq 0$ for all $t \in [0,1]$, then we obtain

$$\mathbf{u}_*'(t) = -1 - (\mathbf{u}_*(t))^2, \quad t \in [0,1].$$

As $\boldsymbol{\lambda}(t)\mathbf{x}_*(t) = -2\mathbf{u}_*(t)(\mathbf{x}_*(t))^2$ for all $t \in [0,1]$, we have in particular when $t = 1$ that

$$2\mathbf{u}_*(1)(\mathbf{x}_*(1))^2 = -\boldsymbol{\lambda}(1)\mathbf{x}_*(1) = -0 \cdot \mathbf{x}_*(1) = 0,$$

and as $\mathbf{x}_*(1) \neq 0$, it follows that $\mathbf{u}_*(1) = 0$.

So we need to solve

$$\begin{cases} \mathbf{u}_*'(t) = -1 - (\mathbf{u}_*(t))^2, & t \in [0,1], \\ \mathbf{u}_*(1) = 0. \end{cases}$$

If $\arctan : \mathbb{R} \to (-\pi/2, \pi/2)$ denotes the inverse of $\tan : (-\pi/2, \pi/2) \to \mathbb{R}$, then we have

$$\frac{d}{dt}\arctan \mathbf{u}_*(t) = \frac{1}{1+(\mathbf{u}_*(t))^2}\mathbf{u}_*'(t) = -1, \quad t \in [0,1].$$

Thus integrating from t to 1, we get using the Fundamental Theorem of Calculus that

$$\arctan \mathbf{u}_*(1) - \arctan \mathbf{u}_*(t) = \int_t^1 -1 dt = t - 1.$$

But $\arctan \mathbf{u}_*(1) = \arctan 0 = 0$, and so we obtain $\arctan \mathbf{u}_*(t) = 1 - t$, that is,

$$\mathbf{u}_*(t) = \tan(1 - t), \quad t \in [0, 1].$$

Also, using the state equation we have $\mathbf{x}'_*(t) = \mathbf{u}_*(t)\mathbf{x}_*(t) = \tan(1 - t)\mathbf{x}_*(t)$, for $t \in [0, 1]$, that is, upon rearranging,

$$\cos(1 - t)\mathbf{x}'_*(t) - \sin(1 - t)\mathbf{x}_*(t) = 0, \quad t \in [0, 1].$$

So

$$\frac{d}{dt}\frac{\mathbf{x}_*(t)}{\cos(1 - t)} = \frac{(\cos(1 - t))\mathbf{x}'_*(t) - (-\sin(1 - t))(-1)\mathbf{x}_*(t)}{(\cos(1 - t))^2} = 0, \quad t \in [0, 1],$$

and by integrating from 0 to t, we have

$$\frac{\mathbf{x}_*(t)}{\cos(1 - t)} = \frac{\mathbf{x}_*(0)}{\cos(1 - 0)} = \frac{1}{\cos 1},$$

and so $\mathbf{x}_*(t) = \dfrac{\cos(1 - t)}{\cos 1}$, $t \in [0, 1]$.

Solution to Exercise 4.14. The Hamiltonian H is given by (taking $\lambda_0 = 1$)

$$H(x, u, \lambda) = x^2 - 2x + \lambda u.$$

Then

$$\begin{aligned}
\mathbf{u}_*(t) &= \arg \min_{u \in [-1,1]} H(\mathbf{x}_*(t), u, \boldsymbol{\lambda}(t)) = \arg \min_{u \in [-1,1]} \left((\mathbf{x}_*(t))^2 - 2\mathbf{x}_*(t) + \boldsymbol{\lambda}(t)u \right) \\
&= \begin{cases} -1 & \text{if } \boldsymbol{\lambda}(t) > 0, \\ 1 & \text{if } \boldsymbol{\lambda}(t) < 0. \end{cases}
\end{aligned}$$

We now investigate when $\boldsymbol{\lambda}(t)$ is 0. The evolution of $\boldsymbol{\lambda}$ is governed by the co-state equation

$$\boldsymbol{\lambda}'(t) = -\frac{\partial H}{\partial x}(\mathbf{x}_*(t), \mathbf{u}_*(t), \boldsymbol{\lambda}(t)) = -(2\mathbf{x}_*(t) - 2) = 2(1 - \mathbf{x}_*(t)), \quad t \in [0, 1].$$

Also, we have from the state equation that

$$\mathbf{x}_*(t) - \underbrace{\mathbf{x}_*(0)}_{=0} = \int_0^t \mathbf{u}_*(t) dt \le \int_0^1 1 dt = t.$$

Thus $\mathbf{x}(t) \le t < 1$ for $t \in [0, 1)$. This gives $\boldsymbol{\lambda}'(t) = 2(1 - \mathbf{x}_*(t)) > 0$ for $t \in [0, 1)$. So $\boldsymbol{\lambda}$ is strictly increasing, and hence it can have at most one zero crossing. In other words, we have the following three cases.

$1°$ Suppose that $\boldsymbol{\lambda}(t) > 0$ in $[0, 1)$. Then $\mathbf{x}_*(1) = \mathbf{x}_*(0) + \int_0^1 -1 dt = 0 - 1 = -1 \ne 0$, and so this case is not possible.

$2°$ Suppose that $\boldsymbol{\lambda}(t) < 0$ in $[0, 1)$. Then $\mathbf{x}_*(1) = \mathbf{x}_*(0) + \int_0^1 1 dt = 0 + 1 = 1 \ne 0$, and so this case is not possible either.

$\underline{3}°$ There is a $t_s \in (0,1)$ such that $\lambda(t) < 0$ on $[0, t_s)$ and $\lambda(t) > 0$ on $(t_s, 1]$. Then we have

$$0 = \mathbf{x}_*(1) = \mathbf{x}_*(0) + \int_0^{t_s} 1 dt + \int_{t_s}^1 -1 dt = 0 + t_s + (-1)(1 - t_s) = 2t_s - 1,$$

and so $t_s = \dfrac{1}{2}$. Consequently, the optimal control $\mathbf{u}_*$ is given by

$$\mathbf{u}_*(t) = \begin{cases} 1 & \text{for } t \in [0, \frac{1}{2}), \\ -1 & \text{for } t \in (\frac{1}{2}, 1]. \end{cases}$$

The optimal state $\mathbf{x}_*$ is given by $\mathbf{x}_*(t) = \displaystyle\int_0^t \mathbf{u}_*(\tau) d\tau = \begin{cases} t & \text{for } t \in [0, \frac{1}{2}], \\ 1 - t & \text{for } t \in [\frac{1}{2}, 1]. \end{cases}$

Solution to Exercise 4.21. The Hamiltonian is $H(x, u, \lambda) = 2u - x^2 + \lambda u$. By (P2) we have that there is a constant C such that for all $t \in [0, 1]$,

$$H(\mathbf{x}_*(t), \mathbf{u}_*(t), \boldsymbol{\lambda}(t)) = \min_{u \in [0,1]} \left(2u - (\mathbf{x}_*(t))^2 + \boldsymbol{\lambda}(t) \cdot u \right) = C.$$

So

$$\mathbf{u}_*(t) = \begin{cases} 0 & \text{if } \boldsymbol{\lambda}(t) > -2, \\ 1 & \text{if } \boldsymbol{\lambda}(t) < -2. \end{cases}$$

Can $\boldsymbol{\lambda}(t) = -2$ in an interval? Let us look at the co-state equation. We have

$$\boldsymbol{\lambda}'(t) = -\frac{\partial H}{\partial x}(\mathbf{x}_*(t), \mathbf{u}_*(t), \boldsymbol{\lambda}(t)) = -(-2\mathbf{x}_*(t)) = 2\mathbf{x}_*(t), \quad t \in [0, 2].$$

But $\mathbf{x}'_*(t) = \mathbf{u}_*(t) \in [0, 1]$, and so

$$\mathbf{x}_*(t) - 1 = \int_0^t \underbrace{\mathbf{u}_*(\tau)}_{\geq 0} d\tau \geq 0.$$

Hence $\mathbf{x}_*(t) \geq 1$ for all $t \in [0, 2]$. So $\boldsymbol{\lambda}'(t) = 2\mathbf{x}_*(t) \geq 2$, and so $\boldsymbol{\lambda}$ is strictly increasing. Since $\mathbf{x}_*(2)$ is free, we obtain from (P1b) that $\boldsymbol{\lambda}(2) = 0$. Thus

$$\underbrace{\boldsymbol{\lambda}(2)}_{=0} - \boldsymbol{\lambda}(t) = \int_t^2 \boldsymbol{\lambda}'(\tau) d\tau \geq \int_t^2 2 d\tau = 2(2 - t),$$

and so $\boldsymbol{\lambda}(t) \leq 2(t - 2)$, $t \in [0, 2]$. In particular, $\boldsymbol{\lambda}(0) \leq 2(0 - 2) = -4 < -2$. On the other hand, $\boldsymbol{\lambda}(2) = 0$. So by the Intermediate Value Theorem, there is a time instant $t_s \in (0, 2)$ such that $\boldsymbol{\lambda}(t_s) = -2$. Moreover, since $\boldsymbol{\lambda}$ is strictly increasing, this is the only time instant where $\boldsymbol{\lambda}$ assumes the value -2. Consequently,

$$\mathbf{u}_*(t) = \begin{cases} 0 & \text{for } t \in [0, t_s], \\ 1 & \text{for } t \in (t_s, 2]. \end{cases}$$

Using the state equation $\mathbf{x}'_*(t) = \mathbf{u}_*(t)$, $t \in [0, 2]$, and the initial condition $\mathbf{x}_*(0) = 1$, we obtain

$$\mathbf{x}_*(t) = \begin{cases} t + 1 & \text{for } t \in [0, t_s], \\ t_s + 1 & \text{for } t \in (t_s, 2]. \end{cases}$$

The co-state equation gives for $t \in (t_s, 2]$ that

$$\underbrace{\lambda(2)}_{=0} - \lambda(t) = \int_t^2 \lambda'(\tau)d\tau = \int_t^2 2 \cdot \mathbf{x}'_*(\tau)d\tau = \int_t^2 2 \cdot (t_s + 1)d\tau = 2(t_s + 1)(2 - t),$$

and so $\lambda(t) = 2(t_s + 1)(t - 2)$, $t \in (t_s, 2]$. In particular, $\lambda(t_s) = -2(t_s + 1)(2 - t_s)$. We will use

$$H(\mathbf{x}_*(t), \mathbf{u}_*(t), \lambda(t)) = C, \quad t \in [0, 1],$$

to determine t_s. We have

$$2 \cdot \mathbf{u}_*(t) - (\mathbf{x}_*(t))^2 + \lambda(t) \cdot \mathbf{u}_*(t) = C, \quad t \in [0, 2].$$

In particular,

$$\lim_{t \nearrow t_s} \left(2 \cdot \mathbf{u}_*(t) - (\mathbf{x}_*(t))^2 + \lambda(t) \cdot \mathbf{u}_*(t) \right) = \lim_{t \searrow t_s} \left(2 \cdot \mathbf{u}_*(t) - (\mathbf{x}_*(t))^2 + \lambda(t) \cdot \mathbf{u}_*(t) \right),$$

and so

$$2 \cdot 0 - (t_s + 1)^2 - 2(t_s + 1)(2 - t_s) \cdot 0 = 2 \cdot 1 - (t_s + 1)^2 - 2(t_s + 1)(2 - t_s) \cdot 1.$$

Thus $t_s^2 - t_s - 1 = 0$, giving

$$t_s = \frac{1 + \sqrt{5}}{2} \text{ or } {}_s = \frac{1 - \sqrt{5}}{2}.$$

But $t_s \geq 0$, and so we have $t_s = \dfrac{1 + \sqrt{5}}{2}$. Thus

$$\mathbf{u}_*(t) = \begin{cases} 0 & \text{for } t \in \left[0, \dfrac{1 + \sqrt{5}}{2}\right], \\[2mm] 1 & \text{for } t \in \left(\dfrac{1 + \sqrt{5}}{2}, 2\right]. \end{cases}$$

Solution to Exercise 4.22. The Hamiltonian is $H(x, u, \lambda) = u_1^2 + u_2^2 + \lambda_1 \cdot u_1 + \lambda_2 \cdot u_2$. Let $\mathbf{u}_* = (\mathbf{u}_{1,*}, \mathbf{u}_{2,*})$ be optimal. Then there is a constant C such that

$$H(\mathbf{x}_*(t), \mathbf{u}_*(t), \lambda(t)) = \min_{u_1, u_2 \in \mathbb{R}} \left(u_1^2 + u_2^2 + \lambda_1(t) \cdot u_1 + \lambda_2(t) \cdot u_2 \right) = C, \quad t \in [0, 2].$$

Hence

$$\mathbf{u}_{1,*}(t) = -\frac{\lambda_1(t)}{2},$$

$$\mathbf{u}_{2,*}(t) = -\frac{\lambda_2(t)}{2},$$

for all $t \in [0, 2]$. As H is independent of x, the co-state equations are given by

$$\lambda'_1(t) = 0,$$
$$\lambda'_2(t) = 0.$$

Hence $\lambda_1 \equiv \alpha$ and $\lambda_2 \equiv \beta$ for some constants α, β. Consequently,

$$\mathbf{u}_{1,*}(t) = -\frac{\alpha}{2},$$

$$\mathbf{u}_{2,*}(t) = -\frac{\beta}{2},$$

$t \in [0, 2]$. Using the state equations and the initial condition, we obtain

$$\mathbf{x}_{1,*}(t) = -\frac{\alpha}{2}t,$$

$$\mathbf{x}_{2,*}(t) = -\frac{\beta}{2}t,$$

$t \in [0, 2]$. Since the final state belongs to $\mathbb{X}_f$, we have $(-\beta)^2 - \left(-\frac{\alpha}{2} \cdot 2\right) + 1 = 0$, that is,

$$\beta^2 + \alpha + 1 = 0. \tag{5.4}$$

We now use (P1b). We have that $\mathbb{X}_f = \{x \in \mathbb{R}^2 : g(x) = 0\}$, where $g(x) := x_2^2 - x_1 + 1$. Thus

$$\nabla g(\mathbf{x}_*(2)) = \begin{bmatrix} -1 & 2 \cdot \left(-\frac{\beta}{2} \cdot 2\right) \end{bmatrix} = \begin{bmatrix} -1 & -2\beta \end{bmatrix}.$$

Hence

$$\ker(\nabla g(\mathbf{x}_*(2))) = \text{span} \begin{bmatrix} 2\beta \\ -1 \end{bmatrix}.$$

So the condition that $\boldsymbol{\lambda}(2)^\top v = 0$ for all $v \in \ker(\nabla g(\mathbf{x}_*(2)))$ gives

$$\begin{bmatrix} \alpha & \beta \end{bmatrix} \begin{bmatrix} 2\beta \\ -1 \end{bmatrix} = 0,$$

that is, $2\alpha\beta = \beta$. Hence $\alpha = 1/2$ or $\beta = 0$. But if $\alpha = 1/2$, then the equation (5.4) gives $\beta^2 = -3/2$, which is impossible. Thus $\beta = 0$, and again (5.4) gives $\alpha = -1$. So we obtain

$$\mathbf{u}_{1,*}(t) = \frac{1}{2},$$

$$\mathbf{u}_{2,*}(t) = 0.$$

Solution to Exercise 4.26. The Hamiltonian is given by

$$H(x, u, \lambda_0, \lambda) = \lambda_0(ax + bu^2) + \lambda u.$$

Thus the co-state equation is given by

$$\boldsymbol{\lambda}'(t) = -\frac{\partial H}{\partial x}(\mathbf{x}_*(t), \mathbf{u}_*(t), \lambda_0, \boldsymbol{\lambda}(t)) = -\lambda_0 \cdot a,$$

and so $\boldsymbol{\lambda}(t) = -\lambda_0 \cdot a \cdot t + C$ for $t \in [0, T_*]$ for some constant C.

Let us first consider the case when $\lambda_0 = 0$. In this case, $\boldsymbol{\lambda} \equiv C$, and $C \neq 0$ by (P3). By (P2), we have

$$\mathbf{u}_*(t) = \arg\min_{u \geq 0} \left(0 + \lambda(t) \cdot u\right) = \arg\min_{u \geq 0} C \cdot u.$$

From here we see that C can't be negative. Thus $C > 0$ and so $\mathbf{u}_* = \mathbf{0}$ on $[0, T_*]$. But then from the state equation $\mathbf{x}'_* = \mathbf{u}*$ and the initial condition $\mathbf{x}_*(0) = 0$, it follows that $\mathbf{x}_* = \mathbf{0}$ too and so the control task $\mathbf{x}_*(T_*) = B$ is violated. So λ_0 must be nonzero.

Without loss of generality, we take $\lambda_0 = 1$, and obtain $\boldsymbol{\lambda}(t) = -a \cdot t + C$ for $t \in [0, T_*]$. So we see that $\boldsymbol{\lambda}_*$ can't be zero in an interval. Again from (P2), we obtain

$$\mathbf{u}_*(t) = \arg\min_{u \geq 0} \left(a \cdot \mathbf{x}_*(t) + b \cdot u^2 + \lambda(t) \cdot u\right),$$

and so

$$\mathbf{u}_*(t) = \begin{cases} -\dfrac{\boldsymbol{\lambda}(t)}{2b} & \text{if } \boldsymbol{\lambda}(t) < 0, \\ 0 & \text{if } \boldsymbol{\lambda}(t) \geq 0. \end{cases}$$

But it cannot be the case that for all $t \in [0, T_*]$, $\boldsymbol{\lambda}(t) \geq 0$. This is because we would then have $\mathbf{u}_* = \mathbf{0}$ on $[0, T_*]$; but then the control task that $\mathbf{x}_*(T_*) = B$ cannot be achieved. So $\boldsymbol{\lambda} \neq \mathbf{0}$. As $a > 0$, it follows that $\boldsymbol{\lambda}$ is decreasing. So let us now suppose that $t_s \geq 0$ is such that $\boldsymbol{\lambda}(t_s) = 0$ (so that $C = a \cdot t_s$). Then

$$\mathbf{u}_*(t) = \begin{cases} 0 & \text{for } t \in [0, t_s], \\ \dfrac{a(t - t_s)}{2b} & \text{for } t \in [t_s, T_*]. \end{cases}$$

Using the state equation, and the initial condition $\mathbf{x}_*(0) = 0$, we obtain for $t \geq t_s$ that

$$\mathbf{x}_*(t) = \int_{t_s}^t \mathbf{u}_*(\tau) d\tau = \frac{a(t - t_s)^2}{4b}.$$

As $\mathbf{x}_*(T_*) = B$, and $T_* \geq t_s$, we obtain $T_* = t_s + \sqrt{\dfrac{4bB}{a}}$.

So the class of optimal solutions consists of pairs $(\mathbf{u}_*, T_*)$ given by

$$T_* = t_s + \sqrt{\frac{4bB}{a}},$$

where t_s is any nonnegative number, and

$$\mathbf{u}_*(t) = \begin{cases} 0 & \text{for } t \in [0, t_s], \\ \dfrac{a(t - t_s)}{2b} & \text{for } t \in [t_s, T_*]. \end{cases}$$

In particular, when $t_s = 0$, we get the pair $(\mathbf{u}_{*,0}, T_{*,0})$ given by

$$T_{*,0} = \sqrt{\frac{4bB}{a}}, \quad \mathbf{u}_{*,0}(t) = \frac{at}{2b}, \ 0 \leq t \leq T_{*,0} = \sqrt{\frac{4bB}{a}}.$$

The other solutions (when we have $t_s > 0$) correspond to the situation where the firm "does nothing" up to a certain (arbitrary) time instant t_s. Each of these solutions give the same cost:

$$\int_0^{T_*} \left(a \cdot \mathbf{x}_*(t) + b \cdot (\mathbf{u}_*(t))^2 \right) dt = \int_{t_s}^{T_*} \left(a \cdot \frac{a(t - t_s)^2}{4b} + b \cdot \left(\frac{a(t - t_s)}{2b} \right)^2 \right) dt$$

$$= \frac{a^2}{2b} \frac{(T_* - t_s)^3}{3} = \frac{4}{3} \sqrt{abB^3}.$$

Solution to Exercise 4.27. The Hamiltonian is given by

$$H(x, u, \lambda_0, \lambda) = \lambda_0 \left(9 + \frac{u^2}{4} \right) + \lambda u.$$

Suppose that $\lambda_0 = 0$. Then (P2) gives

$$\mathbf{u}_*(t) = \arg\min_{u \in \mathbb{R}} \left(0 + \boldsymbol{\lambda}(t) \cdot u \right),$$

and so it must be the case that for all t, $\boldsymbol{\lambda}(t) = 0$. But then for all t, $(\lambda_0, \boldsymbol{\lambda}(t)) = 0$, violating (P3).

Hence $\lambda_0 \neq 0$, and without loss of generality, we may take it equal to 1. Thus (P2) gives

$$\mathbf{u}_*(t) = \arg\min_{u \in \mathbb{R}} \left(9 + \frac{u^2}{4} + \lambda(t) \cdot u \right) = -2\lambda(t).$$

The co-state equation is

$$\lambda'(t) = -\frac{\partial H}{\partial x}(\mathbf{x}_*(t), \mathbf{u}_*(t), \lambda_0, \lambda(t)) = 0.$$

Hence there is a constant C such that $\lambda \equiv C$. From the state equation, we obtain

$$\mathbf{x}'_*(t) = \mathbf{u}_*(t) = -2\lambda(t) = -2C,$$

and so there is a constant D such that $\mathbf{x}_*(t) = -2Ct + D$. Using the fact that $\mathbf{x}_*(0) = 0$, we obtain $D = 0$. Thus $\mathbf{x}_*(t) = -2Ct$ on $[0, T_*]$. As the final state $\mathbf{x}_*(T) = 16$, we must have

$$-2CT_* = 16. \tag{5.5}$$

Since the final time T is variable, we have $H(\mathbf{x}_*(t), \mathbf{u}_*(t), \lambda_0, \lambda(t)) = 0$, for all t, that is,

$$9 + \frac{(-2C)^2}{4} + C(-2C) = 0,$$

and so $C^2 = 9$. Hence $C \in \{3, -3\}$. Since $T_* \geq 0$, it follows from (5.5) that $C = -3$, and that

$$T_* = \frac{8}{3}.$$

Thus $\mathbf{u}_*(t) = -2\lambda(t) = -2C = -2(-3) = 6$ for $t \in [0, T_*] = \left[0, \frac{8}{3}\right]$.

Solution to Exercise 4.30. The Hamiltonian is given by

$$H(x, u, \lambda_0, \lambda) = \lambda_0 \cdot \frac{u^2}{2} + \lambda(x + u).$$

From (P2), we have that there exists a C such that for all $t \in [0, 1]$,

$$C = \lambda_0 \cdot \frac{(\mathbf{u}_*(t))^2}{2} + \lambda(t)(\mathbf{x}_*(t) + \mathbf{u}_*(t)) = \min_{u \in \mathbb{R}} \left(\lambda_0 \cdot \frac{u^2}{2} + \lambda(t)(\mathbf{x}_*(t) + u) \right).$$

From here and (P3) it follows that $\lambda_0 \neq 0$, and so we take $\lambda_0 = 1$. Also, $\mathbf{u}_*(t) = -\lambda(t)$, for $t \in [0, 1]$. The co-state equation is given by

$$\lambda'(t) = -\lambda(t), \quad t \in [0, 1],$$

and so $\lambda(t) = e^{-t}D$ for $t \in [0, 1]$. From (P1b) we obtain

$$\lambda(1) = \varphi'(\mathbf{x}_*(1)) = -\frac{1}{2\sqrt{\mathbf{x}_*(1)}}.$$

So we need to find $\mathbf{x}_*(1)$. We have $\mathbf{x}'_*(t) = \mathbf{x}_*(t) + \mathbf{u}_*(t) = \mathbf{x}_*(t) - \lambda(t) = \mathbf{x}_*(t) - De^{-t}$, and so

$$\mathbf{x}_*(t) = e^t \mathbf{x}_*(0) + \int_0^t e^{t-\tau}(-D)e^{-\tau}d\tau = 0 - De^t \int_0^t e^{-2\tau}d\tau = -D \cdot \frac{e^t - e^{-t}}{2}.$$

Hence we obtain

$$De^{-1} = \lambda(1) = -\frac{1}{2\sqrt{\mathbf{x}_*(1)}} = -\frac{1}{2\sqrt{-D \cdot \dfrac{e^1 - e^{-1}}{2}}},$$

and so $D < 0$. Thus $(-D)^{3/2} = \dfrac{e^{3/2}}{2^{1/2}(e^2 - 1)^{1/2}}$, and so

$$D = -\frac{e}{\sqrt[3]{2(e^2 - 1)}}.$$

Consequently, $\mathbf{u}_*(t) = \dfrac{e^{1-t}}{\sqrt[3]{2(e^2 - 1)}}, \; t \in [0, 1]$.

Solution to Exercise 4.33. The Hamiltonian is given by

$$H(t, x, u, \lambda_0, \lambda) = -\lambda_0 e^{-t}\sqrt{u} + \lambda(x - u) = \lambda x - \lambda_0 e^{-t}\sqrt{u} - \lambda u.$$

By (P2a), we have

$$-\lambda_0 e^{-t}\sqrt{\mathbf{u}_*(t)} + \boldsymbol{\lambda}(t)(\mathbf{x}_*(t) - \mathbf{u}_*(t)) = \min_{u \geq 0}\left(\boldsymbol{\lambda}(t)\mathbf{x}_*(t) - \lambda_0 e^{-t}\sqrt{u} - \boldsymbol{\lambda}(t)u\right). \quad (5.6)$$

The co-state equation is given by

$$\boldsymbol{\lambda}'(t) = -\boldsymbol{\lambda}(t), \quad t \in [0, 1].$$

Thus $\boldsymbol{\lambda}(t) = e^{-t}C, \; t \in [0, 1]$, for some constant C. So (5.6) becomes

$$\mathbf{u}_*(t) = \arg\min_{u \geq 0}\left(-\lambda_0\sqrt{u} - Cu\right). \quad (5.7)$$

If λ_0 and $C = 0$, then

$$H_*(t) := H(t, \mathbf{x}_*(t), \mathbf{u}_*(t), \lambda_0, \boldsymbol{\lambda}(t)) = 0,$$

and so $(\lambda_0, \boldsymbol{\lambda}(t), H_*(t_f) - H_*(t)) = 0$, violating (P3). Hence not both λ_0 and C can be zero.

Now if $C = 0$, then $\lambda_0 = 1$, and (5.7) is contradicted. So we know that $C \neq 0$. Furthermore, if $C > 0$, again (5.7) is contradicted. Thus $C < 0$. From (5.7), we obtain

$$\mathbf{u}_*(t) = \frac{1}{4C^2} =: \alpha.$$

We now use the state equation to determine the value of α. We have

$$\mathbf{x}_*(t) = e^t\mathbf{x}_*(0) + \int_0^t e^{t-\tau} \cdot (-\alpha)d\tau.$$

Thus $0 = \mathbf{x}_*(1) = e^1 + \displaystyle\int_0^1 e^{1-\tau} \cdot (-\alpha)d\tau$, and so $\alpha = \dfrac{e}{e - 1}$. Consequently,

$$\mathbf{u}_*(t) = \frac{e}{e - 1},$$

for all $t \in [0, 1]$.

Solutions to the Exercises from Chapter 5

Solution to Exercise 5.5.

(1) Bellman's equation is given by

$$\frac{\partial V}{\partial t}(x,t) + \min_{u \in \mathbb{R}}\left(\frac{\partial V}{\partial x}(x,t) \cdot u + x^4 + x^2 u^2\right) = 0, \quad x \in \mathbb{R}, \ t \in [0,1],$$

with the boundary condition $V(x,1) = 0$ for all $x \in \mathbb{R}$.

(2) With $V(x,t) := x^4 \cdot \mathbf{p}(t)$, we have

$$\frac{\partial V}{\partial x}(x,t) = 4x^3 \cdot \mathbf{p}(t),$$
$$\frac{\partial V}{\partial t}(x,t) = x^4 \cdot \mathbf{p}'(t).$$

Thus

$$\frac{\partial V}{\partial t}(x,t) + \min_{u \in \mathbb{R}}\left(\frac{\partial V}{\partial x}(x,t) \cdot u + x^4 + x^2 u^2\right)$$
$$= x^4 \cdot \mathbf{p}'(t) + \min_{u \in \mathbb{R}}\left(4x^3 \cdot \mathbf{p}(t) \cdot u + x^4 + x^2 u^2\right).$$

So if $x = 0$, then

$$\frac{\partial V}{\partial t}(x,t) + \min_{u \in \mathbb{R}}\left(\frac{\partial V}{\partial x}(x,t) \cdot u + x^4 + x^2 u^2\right) = 0 \cdot \mathbf{p}'(t) + \min_{u \in \mathbb{R}} 0 = 0.$$

On the other hand, if $x \neq 0$, then

$$U(x,t) = \arg\min_{u \in \mathbb{R}}\left(4x^3 \cdot \mathbf{p}(t) \cdot u + x^4 + x^2 u^2\right) = -\frac{1}{2} \cdot \frac{4x^3 \cdot \mathbf{p}(t)}{x^2} = -2x \cdot \mathbf{p}(t),$$

and so

$$\frac{\partial V}{\partial t}(x,t) + \min_{u \in \mathbb{R}}\left(\frac{\partial V}{\partial x}(x,t) \cdot u + x^4 + x^2 u^2\right)$$
$$= x^4 \cdot \mathbf{p}'(t) + 4x^3 \cdot \mathbf{p}(t) \cdot (-2x \cdot \mathbf{p}(t)) + x^4 + x^2 \cdot (-2x \cdot \mathbf{p}(t))^2$$
$$= x^4(\mathbf{p}'(t) - 4(\mathbf{p}(t))^2 + 1) = 0.$$

Moreover, $V(x,1) = x^4 \cdot \mathbf{p}(1) = x^4 \cdot 0 = 0$ for all $x \in \mathbb{R}$.

(3) Let $\mathbf{x}_*$ be the solution to

$$\mathbf{x}_*'(t) = f(t, \mathbf{x}_*(t), U(\mathbf{x}_*(t), t)) = U(\mathbf{x}_*(t), t) = -2\mathbf{x}_*(t) \cdot \mathbf{p}(t), \ t \in [0,1],$$
$$\mathbf{x}_*(0) = 1.$$

The optimal control $\mathbf{u}_*$ is given by

$$\mathbf{u}_*(t) = U(\mathbf{x}_*(t), t) = -2\mathbf{x}_*(t) \cdot \mathbf{p}(t), \quad t \in [0,1].$$

The corresponding cost is given by

$$V(1,0) = 1^4 \cdot \mathbf{p}(0) = \mathbf{p}(0) = \frac{e^0 - e^{-4}}{2(e^{-4} + e^0)} = \frac{1 - e^{-4}}{2(1 + e^{-4})}.$$

Solution to Exercise 5.6.

(1) We have for all $x \in \mathbb{R}$ and $t \in [0, 1]$

$$
\begin{aligned}
\frac{\partial V}{\partial t}(x, t) &= -\frac{1}{2} \cdot 2(t-1) - x = 1 - t - x, \\
\frac{\partial V}{\partial x}(x, t) &= -(t-1) = 1 - t.
\end{aligned}
$$

Thus for $t \in [0, 1)$ and $x \in \mathbb{R}$,

$$
\begin{aligned}
\frac{\partial V}{\partial t}(x, t) + \min_{u \in [-1, 1]} \left(\frac{\partial V}{\partial x}(x, t) \cdot u + x \right) &= 1 - t - x + \min_{u \in [-1, 1]} \left(\underbrace{(1-t)}_{> 0} \cdot u + x \right) \\
&= 1 - t - x + (1 - t) \cdot (-1) + x = 0.
\end{aligned}
$$

Also if $t = 1$, then for all $x \in \mathbb{R}$,

$$
\begin{aligned}
\frac{\partial V}{\partial t}(x, t) + \min_{u \in [-1, 1]} \left(\frac{\partial V}{\partial x}(x, t) \cdot u + x \right) &= 1 - 1 - x + \min_{u \in [-1, 1]} (0 \cdot u + x) \\
&= 1 - 1 - x + x = 0.
\end{aligned}
$$

Finally, $V(x, 1) = -\frac{1}{2}(1 - 1)^2 - x \cdot (1 - 1) = 0$ for all $x \in \mathbb{R}$.

(2) We have for $t \in [0, 1]$

$$
U(x, t) := \arg \min_{u \in [-1, 1]} \left(\frac{\partial V}{\partial x}(x, t) \cdot u + x \right) = \arg \min_{u \in [-1, 1]} \left(\underbrace{(1 - t)}_{\geq 0} \cdot u + x \right) = -1.
$$

Now we consider the differential equation

$$
x'_*(t) = f(x_*(t), U(x_*(t), t)) = U(x_*(t), t) = -1, \quad t \in [0, 1]
$$

with $\mathbf{x}_*(0) = x_0$. Integrating, we obtain $\mathbf{x}_*(t) - x_0 = -t$, and so $\mathbf{x}_*(t) = x_0 - t$, $t \in [0, 1]$. Thus the optimal control is given by

$$
\mathbf{u}_*(t) = U(\mathbf{x}_*(t), t) = -1, \quad t \in [0, 1].
$$

The corresponding state is given by

$$
\mathbf{x}_*(t) = x_0 - t, \quad t \in [0, 1].
$$

Finally, the optimal cost is

$$
\int_0^1 \mathbf{x}_*(t) dt = \int_0^1 (x_0 - t) dt = x_0 - \frac{1}{2}.
$$

Or alternately, the optimal cost can also be found out to be

$$
V(x_0, 0) = -\frac{1}{2}(0 - 1)^2 - x_0(0 - 1) = x_0 - \frac{1}{2}.
$$

(3) We have

$$\int_0^1 \int_0^t \mathbf{u}(\tau)\,d\tau\,dt = \int_0^1 \int_\tau^t \mathbf{u}(\tau)\,dt\,d\tau = \int_0^1 \mathbf{u}(\tau) \int_\tau^t dt\,d\tau$$
$$= \int_0^1 (\mathbf{u}(\tau)) \cdot (1 - \tau)\,d\tau.$$

Consequently, for any $\mathbf{u}$,

$$x_0 + \int_0^1 \int_0^t \mathbf{u}(\tau)\,d\tau\,dt = x_0 + \int_0^1 (\mathbf{u}(\tau)) \cdot \underbrace{(1 - \tau)}_{\geq 0}\,d\tau$$
$$\geq x_0 + \int_0^1 (-1) \cdot (1 - \tau)\,d\tau$$
$$= x_0 + (-1)\left(1 - \frac{1}{2}\right) = x_0 - \frac{1}{2} = \text{cost of } \mathbf{u}_*.$$

This shows the desired optimality of u_*.

Solution to Exercise 5.7. Given an $x_0 \in \mathbb{R}$ and a $t_0 \in [0, 1]$, consider the optimization problem given by

$$\begin{cases} \text{maximize} & \int_{t_0}^1 \left(1 - (\mathbf{u}(t))^2\right) \cdot (\mathbf{x}(t))^2\,dt \\ \text{subject to} & \mathbf{x}'(t) = \mathbf{u}(t) \cdot \mathbf{x}(t), \quad t \in [t_0, 1], \\ & \mathbf{x}(t_0) = x_0. \end{cases}$$

Proceeding as in the solution to Exercise 4.9, by using the Pontryagin Minimum Principle, we obtain the candidate optimal control $\mathbf{u}_*$ and corresponding optimal state $\mathbf{x}_*$ as

$$\mathbf{u}_*(t) = \tan(1 - t),$$
$$\mathbf{x}_*(t) = \frac{x_0 \cos(1 - t)}{\cos(1 - t_0)},$$

for all $t \in [t_0, 1]$. Hence the corresponding cost is given by

$$\int_{t_0}^1 \left(1 - (\mathbf{u}_*(t))^2\right) \cdot (\mathbf{x}_*(t))^2\,dt = \int_{t_0}^1 \frac{x_0^2 (\cos(1 - t))^2}{(\cos(1 - t_0))^2} - \frac{x_0^2 (\sin(1 - t))^2}{(\cos(1 - t_0))^2}\,dt$$
$$= \frac{x_0^2}{(\cos(1 - t_0))^2} \int_{t_0}^t \cos(2(1 - t))\,dt$$
$$= \frac{x_0^2}{(\cos(1 - t_0))^2} (\sin(1 - t_0))(\cos(1 - t_0))$$
$$= x_0^2 \tan(1 - t_0).$$

Based on this, we try the following V as a candidate for a solution to Bellman's equation:

$$V(x, t) := -x^2 \tan(1 - t), \quad x \in \mathbb{R}, \ t \in [0, 1].$$

Then we have

$$\frac{\partial V}{\partial t} = -\frac{x^2}{(\cos(1-t))^2}(-1) = \frac{x^2}{(\cos(1-t))^2},$$

$$\frac{\partial V}{\partial x} = -2x\tan(1-t).$$

Hence

$$\frac{\partial V}{\partial t}(x,t) + \min_{u\in\mathbb{R}}\left(\frac{\partial V}{\partial x}(x,t)\cdot xu + (u^2-1)x^2\right)$$

$$= \frac{x^2}{(\cos(1-t))^2} + \min_{u\in\mathbb{R}}\left(-2u\tan(1-t) + (u^2-1)\right)x^2$$

$$= \frac{x^2}{(\cos(1-t))^2} + \min_{u\in\mathbb{R}}\left(-2u\tan(1-t) + (u^2-1)\right)x^2$$

$$= \frac{x^2}{(\cos(1-t))^2} + \left(-2(\tan(1-t))\tan(1-t) + ((\tan(1-t))^2 - 1)\right)x^2$$

$$= \frac{x^2}{(\cos(1-t))^2} - x^2((\tan(1-t))^2 + 1)$$

$$= 0.$$

Moreover, $V(x,1) = -x^2\tan(1-1) = -x^2\tan 0 = -x^2\cdot 0 = 0$ for all $x\in\mathbb{R}$. Hence we have obtained a solution V to Bellman's equation. Also,

$$U(x,t) = \arg\min_{u\in\mathbb{R}}\left(-2u\tan(1-t) + (u^2-1)\right)x^2 = \tan(1-t).$$

Thus we can find the solution to

$$\mathbf{x}'(t) = f(t,\mathbf{x}(t),U(\mathbf{x}(t),t)) = U(\mathbf{x}(t),t)\mathbf{x}(t) = \tan(1-t)\mathbf{x}(t), \quad t\in[0,1],$$

$$\mathbf{x}(0) = 1,$$

giving

$$\mathbf{x}_*(t) = \frac{\cos(1-t)}{\cos 1}, \quad t\in[0,1].$$

The optimal control is

$$\mathbf{u}_*(t) = U(\mathbf{x}_*(t),t) = \tan(1-t), \quad t\in[0,1].$$

Consequently, the solution $\mathbf{u}_*$ we had found in Exercise 4.9 is indeed a maximizer.

Project on Curve Fitting

In this last chapter, we will give an outline of a project on curve fitting, which can be posed as an optimization problem in the function space $C^2[a, b]$, and which can be solved using the tools developed at the very outset of this book, from Chapter 1.

Problem formulation. Often measurements lead to observation data points of the form

$$(t_1, y_1), \cdots, (t_n, y_n) \in [a, b] \times \mathbb{R}.$$

See the figure on the left below.

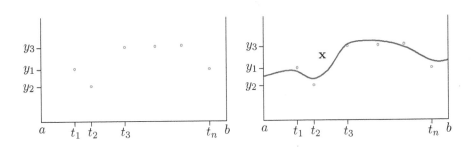

For example, t_i might represent the blood pressure of patients in a clinic who suffer from hypertension, and the y_i might be the corresponding dosage of a drug that was found to be beneficial. Now suppose a patient has a blood pressure that is different from any of $t_1, \cdots, t_n$. The physician would like to fit a reasonably smooth curve $\mathbf{x}$ for this observational data so that he can then read off the appropriate amount of drug from its graph. See the figure on the right above. So we see that a "curve fitting" problem arises quite naturally in applications.

What do we want from our curve $\mathbf{x} : [a, b] \to \mathbb{R}$?

(1) We want it to be "smooth." So it should not be just something obtained by joining the points. (This is reasonable because the observation data itself is not very accurate: there are possibly measurement errors. In this case, a smooth fit might be more desirable as it is more likely to uncover the underlying "law" that is behind the observational data.) One can measure the smoothness of a curve $\mathbf{x} : [a, b] \to \mathbb{R}$ by looking at how fast the derivative $\mathbf{x}'$ changes.

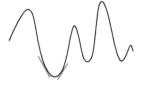

$\mathbf{x}$ not smooth
$(\mathbf{x}''(t))^2 \gg 0$
at several points

smoother $\mathbf{x}$
$(\mathbf{x}''(t))^2 \approx 0$ everywhere

Figure 4. $\displaystyle\int_a^b (\mathbf{x}''(t))^2 dt$ measures the smoothness of $\mathbf{x}$.

Look at Figure 4. In the picture on the left, the magnitude of the derivative changes rapidly, leading to a high value of the second derivative, while in the picture on the left, $\mathbf{x}'$ is slowly changing, giving a small magnitude for the second derivative. So it's natural to use

$$C_1(\mathbf{x}) := \int_a^b (\mathbf{x}''(t))^2 dt$$

as a measure of smoothness of the curve $\mathbf{x}$, with a low value $C_1(\mathbf{x})$ signifying greater smoothness. Hence we want to try to minimize

$$C_1(\mathbf{x}) := \int_a^b (\mathbf{x}''(t))^2 dt.$$

(2) We want it to interpolate as best as possible, that is, the sought-after curve $\mathbf{x} : [a, b] \to \mathbb{R}$ should try to minimize the interpolating error

$$C_2(\mathbf{x}) := \sum_{k=1}^n (\mathbf{x}(t_i) - y_i)^2.$$

Of course the two costs above (very smooth versus good interpolation) are conflicting, as illustrated in Figure 5, where on the left, where we have just fitted a straight line for which the cost $C_1(\mathbf{x}) = 0$, but the interpolation is very poor (that is $C_2(\mathbf{x})$ is high). On the other hand, in the picture on the right, the interpolation is excellent, but the smoothness is very poor!

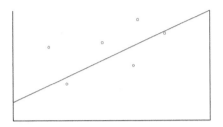 

Figure 5. Greater smoothness versus good interpolation.

Hence the modeler can decide on what he values most, and decide to look at the following optimization problem:

$$\begin{cases} \text{minimize} & \int_a^b (\mathbf{x}''(t))^2 dt + \lambda \sum_{k=1}^n (\mathbf{x}(t_i) - y_i)^2 \\ \text{subject to} & \mathbf{x} \in C^2[a, b]. \end{cases}$$

Here λ is a nonnegative parameter selected by the modeler, which represents the trade-off between smoothness and interpolation. (A high λ corresponds to demanding better interpolation, while the extreme case when $\lambda = 0$ corresponds to just worrying about smoothness.)

The project involves finding theoretically the procedure for constructing a solution to the above problem using the techniques established in Chapter 1, and numerically implementing the resulting algorithm using MATLAB, that is, writing computer code that takes as input

(1) n (the number of data points),

(2) the interval endpoint values a and b,

(3) the data $(t_1, y_1), \cdots, (t_n, y_n)$ and

(4) $\lambda > 0$,

and produces as output a plot of the graph of the minimizing $\mathbf{x}$.

In the following, we give an outline of the key steps as a guide.

Calculating the derivative. We equip $C^2[a, b]$ with the norm

$$\|\mathbf{x}\|_{2,\infty} = \|\mathbf{x}\|_\infty + \|\mathbf{x}'\|_\infty + \|\mathbf{x}''\|_\infty, \quad \mathbf{x} \in C^2[a, b].$$

See Exercise 1.18. Let $f : C^2[a, b] \to \mathbb{R}$ be given by

$$f(\mathbf{x}) = \int_a^b (\mathbf{x}''(t))^2 dt + \lambda \sum_{k=1}^n (\mathbf{x}(t_i) - y_i)^2, \quad \mathbf{x} \in C^2[a, b].$$

If $\mathbf{x}_* \in C^2[a, b]$, then show that $f'(\mathbf{x}_*) : C^2[a, b] \to \mathbb{R}$ is given by

$$(f'(\mathbf{x}_*))\mathbf{h} = 2 \int_a^b \mathbf{x}_*''(t)\mathbf{h}''(t) dt + 2\lambda \sum_{k=0}^n (\mathbf{x}_*(t_i) - y_i)\mathbf{h}(t_i), \quad \mathbf{h} \in C^2[a, b].$$

When is $f'(\mathbf{x}_*) = 0$?

Using the outline provided below, show that if $f'(\mathbf{x}_*) = 0$, then $\mathbf{x}_*$ has the following properties:

(A1) For all $t \in [a, b] \setminus \{t_1, \cdots, t_n\}$, $\mathbf{x}_*''''(t) = 0$.

(A2) $\mathbf{x}_*'''(a+) = 0$,

$\mathbf{x}_*'''(t_i+) - \mathbf{x}_*'''(t_i-) = -\lambda(\mathbf{x}_*(t_i) - y_i)$ for all $i = 1, \cdots, n$,

$\mathbf{x}_*'''(b-) = 0$.

(Here $\mathbf{x}_*'''(\tau+) := \lim_{t \searrow \tau} \mathbf{x}_*'''(t)$ for $\tau \in [a, b)$, and

$\mathbf{x}_*'''(\tau-) := \lim_{t \nearrow \tau} \mathbf{x}_*'''(t)$ for $\tau \in (a, b]$.)

(A3) $\mathbf{x}_*''(a) = 0$,

$\mathbf{x}_*''(b) = 0$.

Note that we are making some smoothness assumptions on $\mathbf{x}_*$ in order to derive the above. In particular, we assume that $\mathbf{x}_*$ is four times continuously differentiable and integrable in the set $[a, b] \setminus \{a, t_1, \cdots, t_n, b\}$, and that the limits $\mathbf{x}_*'''(a+)$, $\mathbf{x}_*'''(t_i+)$, $\mathbf{x}_*'''(t_i-)$, $i = 1, \cdots, n$, and $\mathbf{x}_*'''(b-) = 0$ all exist.
We set $t_0 := a$ and $t_{n+1} := b$.

We also remark that (A1) is telling us that the optimal $\mathbf{x}_*$ is going to be a piecewise cubic function, that is, in each interval (t_i, t_{i+1}),

$$\mathbf{x}_*(t) = c_{0,i} + c_{1,i}t + c_{2,i}t^2 + c_{3,i}t^3.$$

Such curves are called "splines."

(A1) For showing (A1), proceed as follows.

Fix an $i \in \{0, 1, \cdots, n\}$, and let $\mathbf{h} \in C^2[a, b]$ be such that

$$\mathbf{h}(t) = 0 \text{ for all } t \notin [x_i, x_{i+1}], \tag{5.8}$$
$$\mathbf{h}(t_i) = 0, \tag{5.9}$$
$$\mathbf{h}(t_{i+1}) = 0, \tag{5.10}$$
$$\mathbf{h}'(t_i) = 0, \tag{5.11}$$
$$\mathbf{h}'(t_{i+1}) = 0. \tag{5.12}$$

See Figure 6.

Figure 6. The test function $\mathbf{h}$ used for showing (A1).

Then $(f'(\mathbf{x}_*))(\mathbf{h}) = 0$ gives

$$\int_{t_i}^{t_{i+1}} \mathbf{x}_*''(t)\mathbf{h}''(t)dt = 0.$$

Show using integration by parts that

$$\int_{t_i}^{t_{i+1}} \mathbf{x}_*''''(t)\mathbf{h}(t)dt = 0.$$

Note that the above holds for all $\mathbf{h}$ satisfying (5.8)-(5.12). Prove that this gives $\mathbf{x}_*''''(t) = 0$ for $t \in (t_i, t_{i+1})$.

(A2) The property (A2) can be demonstrated as follows.

First fix $i \in \{1, \cdots, n\}$. Take $\mathbf{h} \in C^2[a, b]$ such that $\mathbf{h}(t) = 1$ for all t in a small neighborhood $(t_i - \epsilon, t_i + \epsilon)$ of t_i (with an ϵ small enough), and such that $\mathbf{h}(t) = 0$ outside a slightly bigger neighborhood. See Figure 7.

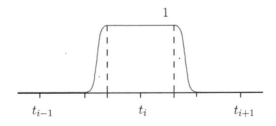

Figure 7. The test function $\mathbf{h}$ used for showing (A2).

Again using integration by parts, show that

$$\mathbf{x}_*'''(t_i+) - \mathbf{x}_*'''(t_i-) = -\lambda(\mathbf{x}_*(t_i) - y_i), \quad i = 1, \cdots, n.$$

The endpoint conditions $\mathbf{x}_*'''(a+) = 0$ and $\mathbf{x}_*'''(b-) = 0$ can be shown in an analogous manner by choosing $\mathbf{h} \in C^2[a, b]$ as shown in Figure 8.

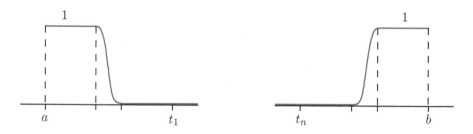

Figure 8. The test functions $\mathbf{h}$ used for showing the endpoint conditions in (A2).

(A3) For showing (A3), we proceed as follows.

Take $\mathbf{h} \in C^2[a, b]$ such that with a small enough ϵ,

$$
\begin{aligned}
\mathbf{h}'(t) &= 1 \text{ for all } t \in [a, a + \epsilon], \\
\mathbf{h}(a) &= 0, \\
\mathbf{h}(t) &= 0 \text{ for all } t \in [a + 2\epsilon, b].
\end{aligned}
$$

See the picture on the left in Figure 9. Show, using integration by parts, that $\mathbf{x}_*''(a) = 0$.

Figure 9. The test functions $\mathbf{h}$ used for showing (A3).

For showing that $\mathbf{x}_*''(b) = 0$, we use the $\mathbf{h}$ shown on the right in Figure 9.

So we have shown that the conditions (A1),(A2),(A3) have to be satisfied by $\mathbf{x}_*$ whenever $f'(\mathbf{x}_*) = \mathbf{0}$ (at least for smooth enough $\mathbf{x}_*$).

Now show that, vice versa, if $\mathbf{x}_* \in C^2[a, b]$ is such that (A1),(A2),(A3) hold, then $f'(\mathbf{x}_*) = \mathbf{0}$.

Convexity of f and uniqueness of the solution. Show that each of the real-valued maps on $C^2[a, b]$ given below

$$\mathbf{x} \;\mapsto\; \int_a^b (\mathbf{x}''(t))^2 dt,$$

$$\mathbf{x} \;\mapsto\; \lambda(\mathbf{x}(t_i) - y_i)^2, \quad i = 1, \cdots, n,$$

are convex. So the pointwise sum f of these convex maps is convex too.

Argue that any $\mathbf{x}_* \in C^2[a, b]$ satisfying (A1),(A2),(A3) is a solution to our optimization problem. So in order to find our curve, we simply have to use (A1),(A2),(A3) to construct it. We will outline the procedure in the next section. But first let us show that if $n \geq 2$ and $\lambda > 0$ and a solution exists, then it must be unique.

Suppose on the contrary that there are two solutions $\mathbf{x}_*$ and $\tilde{\mathbf{x}}_*$ in $C^2[a, b]$. Define $\varphi : \mathbb{R} \to \mathbb{R}$ by $\varphi(t) = f((1-t)\mathbf{x}_* + t\tilde{\mathbf{x}}_*), t \in \mathbb{R}$. Show that $\varphi(t) = At^2 + Bt + C$, $t \in \mathbb{R}$, for constants A, B, C, where

$$A = \int_a^b (\mathbf{x}''(t) - \tilde{\mathbf{x}}_*''(t))^2 dt + \lambda \sum_{k=1}^n (\mathbf{x}(t_i) - \tilde{\mathbf{x}}_*(t_i))^2.$$

We have two cases: $A > 0$ or $A = 0$, which we analyze separately.

If A is positive, then show that we arrive at a contradiction, since a convex combination of $\mathbf{x}_*$ and $\tilde{\mathbf{x}}_*$ gives a cost that is strictly smaller than either of their costs, contradicting the fact that they are minimizers of f.

Now suppose that $A = 0$, and show that this gives the existence of constants α, β such that

$$\tilde{\mathbf{x}}_*(t) = \mathbf{x}_*(t) + \alpha t + \beta, \quad t \in [a, b]$$

(using the fact that $\tilde{\mathbf{x}}_*'' \equiv \mathbf{x}''$ on $[a, b]$), where the constants satisfy

$$\begin{bmatrix} t_1 & 1 \\ \vdots & \vdots \\ t_n & 1 \end{bmatrix} \begin{bmatrix} \alpha \\ \beta \end{bmatrix} = 0$$

(using

$$\lambda \sum_{k=1}^n (\mathbf{x}(t_i) - \tilde{\mathbf{x}}_*(t_i))^2 = 0$$

and $\lambda > 0$). Since $n \geq 2$, it follows that $\alpha = \beta = 0$ and so $\tilde{\mathbf{x}}_* \equiv \mathbf{x}$, a contradiction.

So if there exists a solution and $n \geq 2$ and $\lambda > 0$, then it must be unique.

What happens when either of the conditions $n \geq 2$ or $\lambda > 0$ is violated? Show that if $\lambda = 0$, then "any straight line" is a minimizer of f. On the other hand, if $n = 1$, then show that any straight line passing through the point (t_1, y_1) is a minimizer of f.

Construction of $\mathbf{x}_*$. By (A1), we know that the optimal $\mathbf{x}_*$ we seek is a piecewise cubic function, and so we write $\mathbf{x}_*(t) = \mathbf{p}_k(t)$, $t \in [t_k, t_{k+1}]$, $k = 0, 1, \cdots, n$, where

$$\mathbf{p}_k(t) = A_k(t - t_k)^3 + B_k(t - t_k)^2 + C_k(t - t_k) + D_k, \quad t \in [t_k, t_{k+1}], \ k = 0, 1, \cdots, n.$$

(We will see that the equations determining A_k, B_k, C_k, D_k are nicer to write down owing to the "shift of the origin to $t = t_k$" in the $\mathbf{p}_k$s.) See Figure 10.

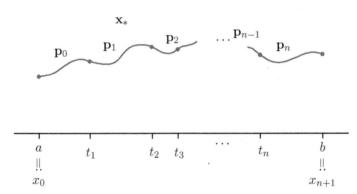

Figure 10. The piecewise cubic $\mathbf{x}_*$.

Note that

$$\begin{aligned}
\mathbf{p}_k(t) &= A_k(t - t_k)^3 + B_k(t - t_k)^2 + C_k(t - t_k) + D_k, \\
\mathbf{p}_k'(t) &= 3A_k(t - t_k)^2 + 2B_k(t - t_k) + C_k, \\
\mathbf{p}_k''(t) &= 6A_k(t - t_k) + 2B_k, \\
\mathbf{p}_k'''(t) &= 6A_k.
\end{aligned}$$

Using the fact that $\mathbf{x}_* \in C^2[a, b]$ and (A2), we have

$$\begin{aligned}
\mathbf{p}_k(t_{k+1}) &= \mathbf{p}_{k+1}(t_{k+1}), \\
\mathbf{p}_k'(t_{k+1}) &= \mathbf{p}_{k+1}'(t_{k+1}), \\
\mathbf{p}_k''(t_{k+1}) &= \mathbf{p}_{k+1}''(t_{k+1}), \\
\mathbf{p}_k'''(t_{k+1}) &= \mathbf{p}_{k+1}'''(t_{k+1}) + \lambda(\mathbf{p}_{k+1}(t_{k+1}) - y_k),
\end{aligned}$$

for $k = 0, 1, \cdots, n - 1$, and so we obtain

$$\begin{aligned}
A_k(t_{k+1} - t_k)^3 + B_k(t_{k+1} - t_k)^2 + C_k(t_{k+1} - t_k) + D_k &= D_{k+1}, \\
3A_k(t_{k+1} - t_k)^2 + 2B_k(t_{k+1} - t_k) + C_k(t_{k+1} - t_k) &= C_{k+1}, \\
6A_k(t_{k+1} - t_k) + 2B_k &= 2B_{k+1}, \\
6A_k &= 6A_{k+1} + \lambda(D_{k+1} - y_{k+1}),
\end{aligned}$$

for $k = 0, 1, \cdots, n-1$. Also the endpoint conditions in (A2) and (A3) give

$$
\begin{aligned}
A_0 &= 0, \\
B_0 &= 0, \\
A_n &= 0, \\
B_n &= 0.
\end{aligned}
$$

Hence with

$$
\Theta_k := \begin{bmatrix}
(t_{k+1}-t_k)^3 & (t_{k+1}-t_k)^2 & t_{k+1}-t_k & 1 \\
3(t_{k+1}-t_k)^2 & 2(t_{k+1}-t_k) & 1 & 0 \\
3(t_{k+1}-t_k) & 1 & 0 & 0 \\
6 & 0 & 0 & 0
\end{bmatrix}, \quad k = 0, 1, \cdots, n-1,
$$

and with

$$
M := \begin{bmatrix}
0 & 0 & 0 & -1 \\
0 & 0 & -1 & 0 \\
0 & -1 & 0 & 0 \\
-6 & 0 & 0 & -\lambda
\end{bmatrix}
$$

we obtain

$$
\Theta_k \underbrace{\begin{bmatrix} A_k \\ B_k \\ C_k \\ D_k \end{bmatrix}}_{=:v_k} + M \underbrace{\begin{bmatrix} A_{k+1} \\ B_{k+1} \\ C_{k+1} \\ D_{k+1} \end{bmatrix}}_{w_{k+1}} = \underbrace{\begin{bmatrix} 0 \\ 0 \\ 0 \\ -\lambda y_{k+1} \end{bmatrix}}
$$

for all $k = 1, \cdots, n-2$.

For $k = 0$, we have

$$
\underbrace{\begin{bmatrix} t_1 - a & 1 \\ 1 & 0 \\ 0 & 0 \\ 0 & 0 \end{bmatrix}}_{=:\Theta_0} \underbrace{\begin{bmatrix} C_0 \\ D_0 \end{bmatrix}}_{=:v_0} + \underbrace{\begin{bmatrix} 0 & 0 & 0 & -1 \\ 0 & 0 & -1 & 0 \\ 0 & -1 & 0 & 0 \\ -6 & 0 & 0 & -\lambda \end{bmatrix}}_{M} \underbrace{\begin{bmatrix} A_1 \\ B_1 \\ C_1 \\ D_1 \end{bmatrix}}_{=:w_1} = \begin{bmatrix} 0 \\ 0 \\ 0 \\ -\lambda y_1 \end{bmatrix}.
$$

Finally, for $k = n-1$, we have

$$
\Theta_{n-1} \begin{bmatrix} A_{n-1} \\ B_{n-1} \\ C_{n-1} \\ D_{n-1} \end{bmatrix} + \underbrace{\begin{bmatrix} 0 & -1 \\ -1 & 0 \\ 0 & 0 \\ 0 & -\lambda \end{bmatrix}}_{N} \underbrace{\begin{bmatrix} C_n \\ D_n \end{bmatrix}}_{=:v_n} = \underbrace{\begin{bmatrix} 0 \\ 0 \\ 0 \\ -\lambda y_n \end{bmatrix}}_{=:w_n}.
$$

Putting it all together, we obtain

$$
\Lambda X = Y,
$$

where

$$X := \begin{bmatrix} v_0 \\ v_1 \\ \vdots \\ v_{n-1} \\ v_n \end{bmatrix}, \quad Y := \begin{bmatrix} w_1 \\ \vdots \\ w_n \end{bmatrix},$$

and

$$\Lambda := \begin{bmatrix} \Theta_0 & M & 0 & 0 & 0 & \cdots & \cdots & 0 \\ 0 & \Theta_1 & M & 0 & 0 & \cdots & \cdots & 0 \\ 0 & 0 & \Theta_2 & M & 0 & \cdots & \cdots & 0 \\ \vdots & & & & & & & \vdots \\ 0 & \cdots & \cdots & 0 & \Theta_{n-3} & M & 0 & 0 \\ 0 & \cdots & \cdots & 0 & 0 & \Theta_{n-2} & M & 0 \\ 0 & \cdots & \cdots & 0 & 0 & 0 & \Theta_{n-1} & N \end{bmatrix}.$$

So if Λ is invertible, we can solve for X:

$$X = \Lambda^{-1}Y,$$

and determine the unknown A_k, B_k, C_k, D_k for all k.

Write a program in MATLAB that numerically implements the above algorithm and plots the resulting minimizer $\mathbf{x}_*$, for example, with $n = 5$ data points. As a test case, suppose the data points are collinear. What is the minimizing optimal $\mathbf{x}_*$ you expect? Run your program to see if the plot matches your expectation. Now suppose you use some generic data points. Study the effect of changing λ on the minimizing optimal $\mathbf{x}_*$. Does the interpolating property get better with increasing λ?

Bibliography

[A] Tom Apostol. *Mathematical analysis*, 2nd Edition. Addison-Wesley, 1974.

[Ar] Michael Artin. *Algebra*. Prentice-Hall, 1991.

[B] David Burghes and Alexander Graham. *Introduction to control theory, including optimal control*. John Wiley, 1980.

[J] Ulf Jönsson, Claes Trygger and Petter Ögren. *Optimal control lecture notes*. Royal Institute of Technology, Stockholm, Sweden, 2010.

[K] Hassan Khalil. *Nonlinear systems*. Macmillan Publishing Company, New York, 1992.

[L] David Luenberger. *Optimization by vector space methods*. John Wiley, New York-London-Sydney, 1969.

[M] Charles MacCluer. *Calculus of variations. Mechanics, control, and other applications*. Prentice Hall, New Jersey, 2005.

[P] L.S. Pontryagin, V.G. Boltyanskii, R.V. Gamkrelidze and E.F. Mishchenko. *The mathematical theory of optimal processes*. Pergamon Press, New York, 1964.

[R] Walter Rudin. *Principles of mathematical analysis*. Third Edition. McGraw-Hill, Singapore, 1976.

[S] Amol Sasane. *The how and why of one variable calculus*. Wiley, 2015.

[Sy] Knut Sydsæter, Peter Hammond, Atle Seierstad and Arne Strøm. *Further mathematics for economic analysis*. Second Edition. Prentice Hall, Harlow, 2008.

[T] John Troutman. *Variational calculus and optimal control.* Second Edition. Springer, New York, 1996.

Index

A CATALOG OF SELECTED
DOVER BOOKS
IN SCIENCE AND MATHEMATICS

Mathematics–Probability and Statistics

BASIC PROBABILITY THEORY, Robert B. Ash. This text emphasizes the probabilistic way of thinking, rather than measure-theoretic concepts. Geared toward advanced undergraduates and graduate students, it features solutions to some of the problems. 1970 edition. 352pp. 5 3/8 x 8 1/2.   0-486-46628-0

PRINCIPLES OF STATISTICS, M. G. Bulmer. Concise description of classical statistics, from basic dice probabilities to modern regression analysis. Equal stress on theory and applications. Moderate difficulty; only basic calculus required. Includes problems with answers. 252pp. 5 5/8 x 8 1/4.   0-486-63760-3

OUTLINE OF BASIC STATISTICS: Dictionary and Formulas, John E. Freund and Frank J. Williams. Handy guide includes a 70-page outline of essential statistical formulas covering grouped and ungrouped data, finite populations, probability, and more, plus over 1,000 clear, concise definitions of statistical terms. 1966 edition. 208pp. 5 3/8 x 8 1/2.   0-486-47769-X

GOOD THINKING: The Foundations of Probability and Its Applications, Irving J. Good. This in-depth treatment of probability theory by a famous British statistician explores Keynesian principles and surveys such topics as Bayesian rationality, corroboration, hypothesis testing, and mathematical tools for induction and simplicity. 1983 edition. 352pp. 5 3/8 x 8 1/2.   0-486-47438-0

INTRODUCTION TO PROBABILITY THEORY WITH CONTEMPORARY APPLICATIONS, Lester L. Helms. Extensive discussions and clear examples, written in plain language, expose students to the rules and methods of probability. Exercises foster problem-solving skills, and all problems feature step-by-step solutions. 1997 edition. 368pp. 6 1/2 x 9 1/4.   0-486-47418-6

CHANCE, LUCK, AND STATISTICS, Horace C. Levinson. In simple, non-technical language, this volume explores the fundamentals governing chance and applies them to sports, government, and business. "Clear and lively ... remarkably accurate." – Scientific Monthly. 384pp. 5 3/8 x 8 1/2.   0-486-41997-5

FIFTY CHALLENGING PROBLEMS IN PROBABILITY WITH SOLUTIONS, Frederick Mosteller. Remarkable puzzlers, graded in difficulty, illustrate elementary and advanced aspects of probability. These problems were selected for originality, general interest, or because they demonstrate valuable techniques. Also includes detailed solutions. 88pp. 5 3/8 x 8 1/2.   0-486-65355-2

EXPERIMENTAL STATISTICS, Mary Gibbons Natrella. A handbook for those seeking engineering information and quantitative data for designing, developing, constructing, and testing equipment. Covers the planning of experiments, the analyzing of extreme-value data; and more. 1966 edition. Index. Includes 52 figures and 76 tables. 560pp. 8 3/8 x 11.   0-486-43937-2

STOCHASTIC MODELING: Analysis and Simulation, Barry L. Nelson. Coherent introduction to techniques also offers a guide to the mathematical, numerical, and simulation tools of systems analysis. Includes formulation of models, analysis, and interpretation of results. 1995 edition. 336pp. 6 1/8 x 9 1/4.   0-486-47770-3

INTRODUCTION TO BIOSTATISTICS: Second Edition, Robert R. Sokal and F. James Rohlf. Suitable for undergraduates with a minimal background in mathematics, this introduction ranges from descriptive statistics to fundamental distributions and the testing of hypotheses. Includes numerous worked-out problems and examples. 1987 edition. 384pp. 6 1/8 x 9 1/4.   0-486-46961-1

Browse over 9,000 books at www.doverpublications.com

Mathematics–Geometry and Topology

PROBLEMS AND SOLUTIONS IN EUCLIDEAN GEOMETRY, M. N. Aref and William Wernick. Based on classical principles, this book is intended for a second course in Euclidean geometry and can be used as a refresher. More than 200 problems include hints and solutions. 1968 edition. 272pp. 5 3/8 x 8 1/2. 0-486-47720-7

TOPOLOGY OF 3-MANIFOLDS AND RELATED TOPICS, Edited by M. K. Fort, Jr. With a New Introduction by Daniel Silver. Summaries and full reports from a 1961 conference discuss decompositions and subsets of 3-space; n-manifolds; knot theory; the Poincaré conjecture; and periodic maps and isotopies. Familiarity with algebraic topology required. 1962 edition. 272pp. 6 1/8 x 9 1/4. 0-486-47753-3

POINT SET TOPOLOGY, Steven A. Gaal. Suitable for a complete course in topology, this text also functions as a self-contained treatment for independent study. Additional enrichment materials make it equally valuable as a reference. 1964 edition. 336pp. 5 3/8 x 8 1/2. 0-486-47222-1

INVITATION TO GEOMETRY, Z. A. Melzak. Intended for students of many different backgrounds with only a modest knowledge of mathematics, this text features self-contained chapters that can be adapted to several types of geometry courses. 1983 edition. 240pp. 5 3/8 x 8 1/2. 0-486-46626-4

TOPOLOGY AND GEOMETRY FOR PHYSICISTS, Charles Nash and Siddhartha Sen. Written by physicists for physics students, this text assumes no detailed background in topology or geometry. Topics include differential forms, homotopy, homology, cohomology, fiber bundles, connection and covariant derivatives, and Morse theory. 1983 edition. 320pp. 5 3/8 x 8 1/2. 0-486-47852-1

BEYOND GEOMETRY: Classic Papers from Riemann to Einstein, Edited with an Introduction and Notes by Peter Pesic. This is the only English-language collection of these 8 accessible essays. They trace seminal ideas about the foundations of geometry that led to Einstein's general theory of relativity. 224pp. 6 1/8 x 9 1/4. 0-486-45350-2

GEOMETRY FROM EUCLID TO KNOTS, Saul Stahl. This text provides a historical perspective on plane geometry and covers non-neutral Euclidean geometry, circles and regular polygons, projective geometry, symmetries, inversions, informal topology, and more. Includes 1,000 practice problems. Solutions available. 2003 edition. 480pp. 6 1/8 x 9 1/4. 0-486-47459-3

TOPOLOGICAL VECTOR SPACES, DISTRIBUTIONS AND KERNELS, François Trèves. Extending beyond the boundaries of Hilbert and Banach space theory, this text focuses on key aspects of functional analysis, particularly in regard to solving partial differential equations. 1967 edition. 592pp. 5 3/8 x 8 1/2.
0-486-45352-9

INTRODUCTION TO PROJECTIVE GEOMETRY, C. R. Wylie, Jr. This introductory volume offers strong reinforcement for its teachings, with detailed examples and numerous theorems, proofs, and exercises, plus complete answers to all odd-numbered end-of-chapter problems. 1970 edition. 576pp. 6 1/8 x 9 1/4. 0-486-46895-X

FOUNDATIONS OF GEOMETRY, C. R. Wylie, Jr. Geared toward students preparing to teach high school mathematics, this text explores the principles of Euclidean and non-Euclidean geometry and covers both generalities and specifics of the axiomatic method. 1964 edition. 352pp. 6 x 9. 0-486-47214-0

Browse over 9,000 books at www.doverpublications.com

Astronomy

CHARIOTS FOR APOLLO: The NASA History of Manned Lunar Spacecraft to 1969, Courtney G. Brooks, James M. Grimwood, and Loyd S. Swenson, Jr. This illustrated history by a trio of experts is the definitive reference on the Apollo spacecraft and lunar modules. It traces the vehicles' design, development, and operation in space. More than 100 photographs and illustrations. 576pp. 6 3/4 x 9 1/4. 0-486-46756-2

EXPLORING THE MOON THROUGH BINOCULARS AND SMALL TELESCOPES, Ernest H. Cherrington, Jr. Informative, profusely illustrated guide to locating and identifying craters, rills, seas, mountains, other lunar features. Newly revised and updated with special section of new photos. Over 100 photos and diagrams. 240pp. 8 1/4 x 11. 0-486-24491-1

WHERE NO MAN HAS GONE BEFORE: A History of NASA's Apollo Lunar Expeditions, William David Compton. Introduction by Paul Dickson. This official NASA history traces behind-the-scenes conflicts and cooperation between scientists and engineers. The first half concerns preparations for the Moon landings, and the second half documents the flights that followed Apollo 11. 1989 edition. 432pp. 7 x 10.
0-486-47888-2

APOLLO EXPEDITIONS TO THE MOON: The NASA History, Edited by Edgar M. Cortright. Official NASA publication marks the 40th anniversary of the first lunar landing and features essays by project participants recalling engineering and administrative challenges. Accessible, jargon-free accounts, highlighted by numerous illustrations. 336pp. 8 3/8 x 10 7/8. 0-486-47175-6

ON MARS: Exploration of the Red Planet, 1958-1978--The NASA History, Edward Clinton Ezell and Linda Neuman Ezell. NASA's official history chronicles the start of our explorations of our planetary neighbor. It recounts cooperation among government, industry, and academia, and it features dozens of photos from Viking cameras. 560pp. 6 3/4 x 9 1/4. 0-486-46757-0

ARISTARCHUS OF SAMOS: The Ancient Copernicus, Sir Thomas Heath. Heath's history of astronomy ranges from Homer and Hesiod to Aristarchus and includes quotes from numerous thinkers, compilers, and scholasticists from Thales and Anaximander through Pythagoras, Plato, Aristotle, and Heraclides. 34 figures. 448pp. 5 3/8 x 8 1/2.
0-486-43886-4

AN INTRODUCTION TO CELESTIAL MECHANICS, Forest Ray Moulton. Classic text still unsurpassed in presentation of fundamental principles. Covers rectilinear motion, central forces, problems of two and three bodies, much more. Includes over 200 problems, some with answers. 437pp. 5 3/8 x 8 1/2. 0-486-64687-4

BEYOND THE ATMOSPHERE: Early Years of Space Science, Homer E. Newell. This exciting survey is the work of a top NASA administrator who chronicles technological advances, the relationship of space science to general science, and the space program's social, political, and economic contexts. 528pp. 6 3/4 x 9 1/4.
0-486-47464-X

STAR LORE: Myths, Legends, and Facts, William Tyler Olcott. Captivating retellings of the origins and histories of ancient star groups include Pegasus, Ursa Major, Pleiades, signs of the zodiac, and other constellations. "Classic." – *Sky & Telescope.* 58 illustrations. 544pp. 5 3/8 x 8 1/2. 0-486-43581-4

A COMPLETE MANUAL OF AMATEUR ASTRONOMY: Tools and Techniques for Astronomical Observations, P. Clay Sherrod with Thomas L. Koed. Concise, highly readable book discusses the selection, set-up, and maintenance of a telescope; amateur studies of the sun; lunar topography and occultations; and more. 124 figures. 26 halftones. 37 tables. 335pp. 6 1/2 x 9 1/4. 0-486-42820-6

Browse over 9,000 books at www.doverpublications.com

Chemistry

MOLECULAR COLLISION THEORY, M. S. Child. This high-level monograph offers an analytical treatment of classical scattering by a central force, quantum scattering by a central force, elastic scattering phase shifts, and semi-classical elastic scattering. 1974 edition. 310pp. 5 3/8 x 8 1/2. 0-486-69437-2

HANDBOOK OF COMPUTATIONAL QUANTUM CHEMISTRY, David B. Cook. This comprehensive text provides upper-level undergraduates and graduate students with an accessible introduction to the implementation of quantum ideas in molecular modeling, exploring practical applications alongside theoretical explanations. 1998 edition. 832pp. 5 3/8 x 8 1/2. 0-486-44307-8

RADIOACTIVE SUBSTANCES, Marie Curie. The celebrated scientist's thesis, which directly preceded her 1903 Nobel Prize, discusses establishing atomic character of radioactivity; extraction from pitchblende of polonium and radium; isolation of pure radium chloride; more. 96pp. 5 3/8 x 8 1/2. 0-486-42550-9

CHEMICAL MAGIC, Leonard A. Ford. Classic guide provides intriguing entertainment while elucidating sound scientific principles, with more than 100 unusual stunts: cold fire, dust explosions, a nylon rope trick, a disappearing beaker, much more. 128pp. 5 3/8 x 8 1/2. 0-486-67628-5

ALCHEMY, E. J. Holmyard. Classic study by noted authority covers 2,000 years of alchemical history: religious, mystical overtones; apparatus; signs, symbols, and secret terms; advent of scientific method, much more. Illustrated. 320pp. 5 3/8 x 8 1/2.
0-486-26298-7

CHEMICAL KINETICS AND REACTION DYNAMICS, Paul L. Houston. This text teaches the principles underlying modern chemical kinetics in a clear, direct fashion, using several examples to enhance basic understanding. Solutions to selected problems. 2001 edition. 352pp. 8 3/8 x 11. 0-486-45334-0

PROBLEMS AND SOLUTIONS IN QUANTUM CHEMISTRY AND PHYSICS, Charles S. Johnson and Lee G. Pedersen. Unusually varied problems, with detailed solutions, cover of quantum mechanics, wave mechanics, angular momentum, molecular spectroscopy, scattering theory, more. 280 problems, plus 139 supplementary exercises. 430pp. 6 1/2 x 9 1/4. 0-486-65236-X

ELEMENTS OF CHEMISTRY, Antoine Lavoisier. Monumental classic by the founder of modern chemistry features first explicit statement of law of conservation of matter in chemical change, and more. Facsimile reprint of original (1790) Kerr translation. 539pp. 5 3/8 x 8 1/2. 0-486-64624-6

MAGNETISM AND TRANSITION METAL COMPLEXES, F. E. Mabbs and D. J. Machin. A detailed view of the calculation methods involved in the magnetic properties of transition metal complexes, this volume offers sufficient background for original work in the field. 1973 edition. 240pp. 5 3/8 x 8 1/2. 0-486-46284-6

GENERAL CHEMISTRY, Linus Pauling. Revised third edition of classic first-year text by Nobel laureate. Atomic and molecular structure, quantum mechanics, statistical mechanics, thermodynamics correlated with descriptive chemistry. Problems. 992pp. 5 3/8 x 8 1/2. 0-486-65622-5

ELECTROLYTE SOLUTIONS: Second Revised Edition, R. A. Robinson and R. H. Stokes. Classic text deals primarily with measurement, interpretation of conductance, chemical potential, and diffusion in electrolyte solutions. Detailed theoretical interpretations, plus extensive tables of thermodynamic and transport properties. 1970 edition. 590pp. 5 3/8 x 8 1/2. 0-486-42225-9

Engineering

FUNDAMENTALS OF ASTRODYNAMICS, Roger R. Bate, Donald D. Mueller, and Jerry E. White. Teaching text developed by U.S. Air Force Academy develops the basic two-body and n-body equations of motion; orbit determination; classical orbital elements, coordinate transformations; differential correction; more. 1971 edition. 455pp. 5 3/8 x 8 1/2.   0-486-60061-0

INTRODUCTION TO CONTINUUM MECHANICS FOR ENGINEERS: Revised Edition, Ray M. Bowen. This self-contained text introduces classical continuum models within a modern framework. Its numerous exercises illustrate the governing principles, linearizations, and other approximations that constitute classical continuum models. 2007 edition. 320pp. 6 1/8 x 9 1/4.   0-486-47460-7

ENGINEERING MECHANICS FOR STRUCTURES, Louis L. Bucciarelli. This text explores the mechanics of solids and statics as well as the strength of materials and elasticity theory. Its many design exercises encourage creative initiative and systems thinking. 2009 edition. 320pp. 6 1/8 x 9 1/4.   0-486-46855-0

FEEDBACK CONTROL THEORY, John C. Doyle, Bruce A. Francis and Allen R. Tannenbaum. This excellent introduction to feedback control system design offers a theoretical approach that captures the essential issues and can be applied to a wide range of practical problems. 1992 edition. 224pp. 6 1/2 x 9 1/4.   0-486-46933-6

THE FORCES OF MATTER, Michael Faraday. These lectures by a famous inventor offer an easy-to-understand introduction to the interactions of the universe's physical forces. Six essays explore gravitation, cohesion, chemical affinity, heat, magnetism, and electricity. 1993 edition. 96pp. 5 3/8 x 8 1/2.   0-486-47482-8

DYNAMICS, Lawrence E. Goodman and William H. Warner. Beginning engineering text introduces calculus of vectors, particle motion, dynamics of particle systems and plane rigid bodies, technical applications in plane motions, and more. Exercises and answers in every chapter. 619pp. 5 3/8 x 8 1/2.   0-486-42006-X

ADAPTIVE FILTERING PREDICTION AND CONTROL, Graham C. Goodwin and Kwai Sang Sin. This unified survey focuses on linear discrete-time systems and explores natural extensions to nonlinear systems. It emphasizes discrete-time systems, summarizing theoretical and practical aspects of a large class of adaptive algorithms. 1984 edition. 560pp. 6 1/2 x 9 1/4.   0-486-46932-8

INDUCTANCE CALCULATIONS, Frederick W. Grover. This authoritative reference enables the design of virtually every type of inductor. It features a single simple formula for each type of inductor, together with tables containing essential numerical factors. 1946 edition. 304pp. 5 3/8 x 8 1/2.   0-486-47440-2

THERMODYNAMICS: Foundations and Applications, Elias P. Gyftopoulos and Gian Paolo Beretta. Designed by two MIT professors, this authoritative text discusses basic concepts and applications in detail, emphasizing generality, definitions, and logical consistency. More than 300 solved problems cover realistic energy systems and processes. 800pp. 6 1/8 x 9 1/4.   0-486-43932-1

THE FINITE ELEMENT METHOD: Linear Static and Dynamic Finite Element Analysis, Thomas J. R. Hughes. Text for students without in-depth mathematical training, this text includes a comprehensive presentation and analysis of algorithms of time-dependent phenomena plus beam, plate, and shell theories. Solution guide available upon request. 672pp. 6 1/2 x 9 1/4.   0-486-41181-8

Browse over 9,000 books at www.doverpublications.com

Mathematics–Bestsellers

HANDBOOK OF MATHEMATICAL FUNCTIONS: with Formulas, Graphs, and Mathematical Tables, Edited by Milton Abramowitz and Irene A. Stegun. A classic resource for working with special functions, standard trig, and exponential logarithmic definitions and extensions, it features 29 sets of tables, some to as high as 20 places. 1046pp. 8 x 10 1/2. 0-486-61272-4

ABSTRACT AND CONCRETE CATEGORIES: The Joy of Cats, Jiri Adamek, Horst Herrlich, and George E. Strecker. This up-to-date introductory treatment employs category theory to explore the theory of structures. Its unique approach stresses concrete categories and presents a systematic view of factorization structures. Numerous examples. 1990 edition, updated 2004. 528pp. 6 1/8 x 9 1/4. 0-486-46934-4

MATHEMATICS: Its Content, Methods and Meaning, A. D. Aleksandrov, A. N. Kolmogorov, and M. A. Lavrent'ev. Major survey offers comprehensive, coherent discussions of analytic geometry, algebra, differential equations, calculus of variations, functions of a complex variable, prime numbers, linear and non-Euclidean geometry, topology, functional analysis, more. 1963 edition. 1120pp. 5 3/8 x 8 1/2. 0-486-40916-3

INTRODUCTION TO VECTORS AND TENSORS: Second Edition–Two Volumes Bound as One, Ray M. Bowen and C.-C. Wang. Convenient single-volume compilation of two texts offers both introduction and in-depth survey. Geared toward engineering and science students rather than mathematicians, it focuses on physics and engineering applications. 1976 edition. 560pp. 6 1/2 x 9 1/4. 0-486-46914-X

AN INTRODUCTION TO ORTHOGONAL POLYNOMIALS, Theodore S. Chihara. Concise introduction covers general elementary theory, including the representation theorem and distribution functions, continued fractions and chain sequences, the recurrence formula, special functions, and some specific systems. 1978 edition. 272pp. 5 3/8 x 8 1/2. 0-486-47929-3

ADVANCED MATHEMATICS FOR ENGINEERS AND SCIENTISTS, Paul DuChateau. This primary text and supplemental reference focuses on linear algebra, calculus, and ordinary differential equations. Additional topics include partial differential equations and approximation methods. Includes solved problems. 1992 edition. 400pp. 7 1/2 x 9 1/4. 0-486-47930-7

PARTIAL DIFFERENTIAL EQUATIONS FOR SCIENTISTS AND ENGINEERS, Stanley J. Farlow. Practical text shows how to formulate and solve partial differential equations. Coverage of diffusion-type problems, hyperbolic-type problems, elliptic-type problems, numerical and approximate methods. Solution guide available upon request. 1982 edition. 414pp. 6 1/8 x 9 1/4. 0-486-67620-X

VARIATIONAL PRINCIPLES AND FREE-BOUNDARY PROBLEMS, Avner Friedman. Advanced graduate-level text examines variational methods in partial differential equations and illustrates their applications to free-boundary problems. Features detailed statements of standard theory of elliptic and parabolic operators. 1982 edition. 720pp. 6 1/8 x 9 1/4. 0-486-47853-X

LINEAR ANALYSIS AND REPRESENTATION THEORY, Steven A. Gaal. Unified treatment covers topics from the theory of operators and operator algebras on Hilbert spaces; integration and representation theory for topological groups; and the theory of Lie algebras, Lie groups, and transform groups. 1973 edition. 704pp. 6 1/8 x 9 1/4. 0-486-47851-3

Browse over 9,000 books at www.doverpublications.com

A SURVEY OF INDUSTRIAL MATHEMATICS, Charles R. MacCluer. Students learn how to solve problems they'll encounter in their professional lives with this concise single-volume treatment. It employs MATLAB and other strategies to explore typical industrial problems. 2000 edition. 384pp. 5 3/8 x 8 1/2. 0-486-47702-9

NUMBER SYSTEMS AND THE FOUNDATIONS OF ANALYSIS, Elliott Mendelson. Geared toward undergraduate and beginning graduate students, this study explores natural numbers, integers, rational numbers, real numbers, and complex numbers. Numerous exercises and appendixes supplement the text. 1973 edition. 368pp. 5 3/8 x 8 1/2. 0-486-45792-3

A FIRST LOOK AT NUMERICAL FUNCTIONAL ANALYSIS, W. W. Sawyer. Text by renowned educator shows how problems in numerical analysis lead to concepts of functional analysis. Topics include Banach and Hilbert spaces, contraction mappings, convergence, differentiation and integration, and Euclidean space. 1978 edition. 208pp. 5 3/8 x 8 1/2. 0-486-47882-3

FRACTALS, CHAOS, POWER LAWS: Minutes from an Infinite Paradise, Manfred Schroeder. A fascinating exploration of the connections between chaos theory, physics, biology, and mathematics, this book abounds in award-winning computer graphics, optical illusions, and games that clarify memorable insights into self-similarity. 1992 edition. 448pp. 6 1/8 x 9 1/4. 0-486-47204-3

SET THEORY AND THE CONTINUUM PROBLEM, Raymond M. Smullyan and Melvin Fitting. A lucid, elegant, and complete survey of set theory, this three-part treatment explores axiomatic set theory, the consistency of the continuum hypothesis, and forcing and independence results. 1996 edition. 336pp. 6 x 9. 0-486-47484-4

DYNAMICAL SYSTEMS, Shlomo Sternberg. A pioneer in the field of dynamical systems discusses one-dimensional dynamics, differential equations, random walks, iterated function systems, symbolic dynamics, and Markov chains. Supplementary materials include PowerPoint slides and MATLAB exercises. 2010 edition. 272pp. 6 1/8 x 9 1/4. 0-486-47705-3

ORDINARY DIFFERENTIAL EQUATIONS, Morris Tenenbaum and Harry Pollard. Skillfully organized introductory text examines origin of differential equations, then defines basic terms and outlines general solution of a differential equation. Explores integrating factors; dilution and accretion problems; Laplace Transforms; Newton's Interpolation Formulas, more. 818pp. 5 3/8 x 8 1/2. 0-486-64940-7

MATROID THEORY, D. J. A. Welsh. Text by a noted expert describes standard examples and investigation results, using elementary proofs to develop basic matroid properties before advancing to a more sophisticated treatment. Includes numerous exercises. 1976 edition. 448pp. 5 3/8 x 8 1/2. 0-486-47439-9

THE CONCEPT OF A RIEMANN SURFACE, Hermann Weyl. This classic on the general history of functions combines function theory and geometry, forming the basis of the modern approach to analysis, geometry, and topology. 1955 edition. 208pp. 5 3/8 x 8 1/2. 0-486-47004-0

THE LAPLACE TRANSFORM, David Vernon Widder. This volume focuses on the Laplace and Stieltjes transforms, offering a highly theoretical treatment. Topics include fundamental formulas, the moment problem, monotonic functions, and Tauberian theorems. 1941 edition. 416pp. 5 3/8 x 8 1/2. 0-486-47755-X

Browse over 9,000 books at www.doverpublications.com

Mathematics–Logic and Problem Solving

PERPLEXING PUZZLES AND TANTALIZING TEASERS, Martin Gardner. Ninety-three riddles, mazes, illusions, tricky questions, word and picture puzzles, and other challenges offer hours of entertainment for youngsters. Filled with rib-tickling drawings. Solutions. 224pp. 5 3/8 x 8 1/2.   0-486-25637-5

MY BEST MATHEMATICAL AND LOGIC PUZZLES, Martin Gardner. The noted expert selects 70 of his favorite "short" puzzles. Includes The Returning Explorer, The Mutilated Chessboard, Scrambled Box Tops, and dozens more. Complete solutions included. 96pp. 5 3/8 x 8 1/2.   0-486-28152-3

THE LADY OR THE TIGER?: and Other Logic Puzzles, Raymond M. Smullyan. Created by a renowned puzzle master, these whimsically themed challenges involve paradoxes about probability, time, and change; metapuzzles; and self-referentiality. Nineteen chapters advance in difficulty from relatively simple to highly complex. 1982 edition. 240pp. 5 3/8 x 8 1/2.   0-486-47027-X

SATAN, CANTOR AND INFINITY: Mind-Boggling Puzzles, Raymond M. Smullyan. A renowned mathematician tells stories of knights and knaves in an entertaining look at the logical precepts behind infinity, probability, time, and change. Requires a strong background in mathematics. Complete solutions. 288pp. 5 3/8 x 8 1/2.

0-486-47036-9

THE RED BOOK OF MATHEMATICAL PROBLEMS, Kenneth S. Williams and Kenneth Hardy. Handy compilation of 100 practice problems, hints and solutions indispensable for students preparing for the William Lowell Putnam and other mathematical competitions. Preface to the First Edition. Sources. 1988 edition. 192pp. 5 3/8 x 8 1/2.   0-486-69415-1

KING ARTHUR IN SEARCH OF HIS DOG AND OTHER CURIOUS PUZZLES, Raymond M. Smullyan. This fanciful, original collection for readers of all ages features arithmetic puzzles, logic problems related to crime detection, and logic and arithmetic puzzles involving King Arthur and his Dogs of the Round Table. 160pp. 5 3/8 x 8 1/2.

0-486-47435-6

UNDECIDABLE THEORIES: Studies in Logic and the Foundation of Mathematics, Alfred Tarski in collaboration with Andrzej Mostowski and Raphael M. Robinson. This well-known book by the famed logician consists of three treatises: "A General Method in Proofs of Undecidability," "Undecidability and Essential Undecidability in Mathematics," and "Undecidability of the Elementary Theory of Groups." 1953 edition. 112pp. 5 3/8 x 8 1/2.   0-486-47703-7

LOGIC FOR MATHEMATICIANS, J. Barkley Rosser. Examination of essential topics and theorems assumes no background in logic. "Undoubtedly a major addition to the literature of mathematical logic." – *Bulletin of the American Mathematical Society.* 1978 edition. 592pp. 6 1/8 x 9 1/4.   0-486-46898-4

INTRODUCTION TO PROOF IN ABSTRACT MATHEMATICS, Andrew Wohlgemuth. This undergraduate text teaches students what constitutes an acceptable proof, and it develops their ability to do proofs of routine problems as well as those requiring creative insights. 1990 edition. 384pp. 6 1/2 x 9 1/4.   0-486-47854-8

FIRST COURSE IN MATHEMATICAL LOGIC, Patrick Suppes and Shirley Hill. Rigorous introduction is simple enough in presentation and context for wide range of students. Symbolizing sentences; logical inference; truth and validity; truth tables; terms, predicates, universal quantifiers; universal specification and laws of identity; more. 288pp. 5 3/8 x 8 1/2.   0-486-42259-3

Browse over 9,000 books at www.doverpublications.com

Mathematics–Algebra and Calculus

VECTOR CALCULUS, Peter Baxandall and Hans Liebeck. This introductory text offers a rigorous, comprehensive treatment. Classical theorems of vector calculus are amply illustrated with figures, worked examples, physical applications, and exercises with hints and answers. 1986 edition. 560pp. 5 3/8 x 8 1/2.   0-486-46620-5

ADVANCED CALCULUS: An Introduction to Classical Analysis, Louis Brand. A course in analysis that focuses on the functions of a real variable, this text introduces the basic concepts in their simplest setting and illustrates its teachings with numerous examples, theorems, and proofs. 1955 edition. 592pp. 5 3/8 x 8 1/2.   0-486-44548-8

ADVANCED CALCULUS, Avner Friedman. Intended for students who have already completed a one-year course in elementary calculus, this two-part treatment advances from functions of one variable to those of several variables. Solutions. 1971 edition. 432pp. 5 3/8 x 8 1/2.   0-486-45795-8

METHODS OF MATHEMATICS APPLIED TO CALCULUS, PROBABILITY, AND STATISTICS, Richard W. Hamming. This 4-part treatment begins with algebra and analytic geometry and proceeds to an exploration of the calculus of algebraic functions and transcendental functions and applications. 1985 edition. Includes 310 figures and 18 tables. 880pp. 6 1/2 x 9 1/4.   0-486-43945-3

BASIC ALGEBRA I: Second Edition, Nathan Jacobson. A classic text and standard reference for a generation, this volume covers all undergraduate algebra topics, including groups, rings, modules, Galois theory, polynomials, linear algebra, and associative algebra. 1985 edition. 528pp. 6 1/8 x 9 1/4.   0-486-47189-6

BASIC ALGEBRA II: Second Edition, Nathan Jacobson. This classic text and standard reference comprises all subjects of a first-year graduate-level course, including in-depth coverage of groups and polynomials and extensive use of categories and functors. 1989 edition. 704pp. 6 1/8 x 9 1/4.   0-486-47187-X

CALCULUS: An Intuitive and Physical Approach (Second Edition), Morris Kline. Application-oriented introduction relates the subject as closely as possible to science with explorations of the derivative; differentiation and integration of the powers of x; theorems on differentiation, antidifferentiation; the chain rule; trigonometric functions; more. Examples. 1967 edition. 960pp. 6 1/2 x 9 1/4.   0-486-40453-6

ABSTRACT ALGEBRA AND SOLUTION BY RADICALS, John E. Maxfield and Margaret W. Maxfield. Accessible advanced undergraduate-level text starts with groups, rings, fields, and polynomials and advances to Galois theory, radicals and roots of unity, and solution by radicals. Numerous examples, illustrations, exercises, appendixes. 1971 edition. 224pp. 6 1/8 x 9 1/4.   0-486-47723-1

AN INTRODUCTION TO THE THEORY OF LINEAR SPACES, Georgi E. Shilov. Translated by Richard A. Silverman. Introductory treatment offers a clear exposition of algebra, geometry, and analysis as parts of an integrated whole rather than separate subjects. Numerous examples illustrate many different fields, and problems include hints or answers. 1961 edition. 320pp. 5 3/8 x 8 1/2.   0-486-63070-6

LINEAR ALGEBRA, Georgi E. Shilov. Covers determinants, linear spaces, systems of linear equations, linear functions of a vector argument, coordinate transformations, the canonical form of the matrix of a linear operator, bilinear and quadratic forms, and more. 387pp. 5 3/8 x 8 1/2.   0-486-63518-X

Browse over 9,000 books at www.doverpublications.com